THE LAST ELEPHANTS

Compiled by Don Pinnock & Colin Bell

Foreword by HRH Prince William, Duke of Cambridge

Smithsonian
Books
Washington, DC

These desert-adapted elephants in north-west Namibia encapsulate part of what this book is about – the often problematic relationship between impoverished rural communities and wildlife. It also examines how rural communities can successfully be incorporated into the wildlife and tourism industry so that elephants and people can live harmoniously, side by side.

© Hannes Lochner, Chobe National Park

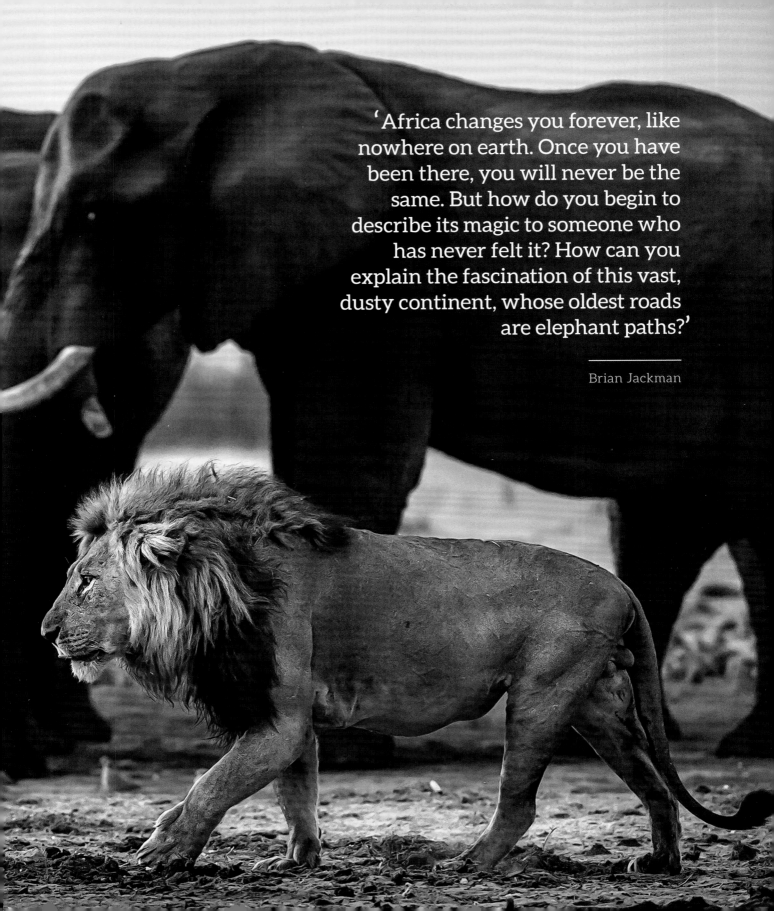

'Africa changes you forever, like nowhere on earth. Once you have been there, you will never be the same. But how do you begin to describe its magic to someone who has never felt it? How can you explain the fascination of this vast, dusty continent, whose oldest roads are elephant paths?'

Brian Jackman

'The generation that destroys the environment is not the generation that pays the price. That is the problem.'

Wangari Maathai

© Martin Harvey. The Amboseli plains and Mount Kilimanjaro

Contents

An elephant is poached somewhere in Africa
every 15 to 20 minutes of every day,
every week and every month.

The Great Elephant Census confirmed what many of us have feared for some time – that one of our planet's most treasured species is on course for extinction at the hands of poachers, criminal syndicates, warlords and traffickers.

When I was born, there were a million elephants roaming Africa. By the time my daughter Charlotte was born in 2015, the numbers of savannah elephants had crashed to just 350 000. At the current pace of illegal poaching, when Charlotte turns 25 the African elephant could be gone from the wild.

We cannot let this happen. I am not prepared to be part of a generation that lets these iconic species disappear and have to explain to our children why we lost this battle when we had the tools to win it.

We have the opportunity to end the mixed messages we have sent for too long about the value and desirability of wildlife products. We need to make it quite clear and broadcast widely that ivory is a symbol of destruction, not of luxury and not something that anyone needs to buy or sell. We must say that rhino horn does not cure anything and does not need a legal market. We must send a message to the world that it is no longer acceptable to buy and sell ivory, rhino horn or other illegal wildlife products.

This crisis is not just about animals - it's also about people. It is some of the world's poorest who will suffer when their natural resources are stripped from them illegally and brutally. It is families in the world's most vulnerable regions who suffer when two rangers a week are killed on the frontline of this fight. It is fragile democratic systems in many nations that are at risk from the scourge of war, violence and corruption that the illegal wildlife trade funds and fuels.

We cannot undo the mistakes of the past. But we can and must take moral responsibility for the decisions we make today. Sometimes we can feel powerless to make a difference for species that we desperately care about. But there are encouraging signs: China and America have banned domestic trade in ivory and the International Union for the Conservation of Nature (IUCN) has called on all countries to do likewise. In a number of countries confiscated ivory is burned or crushed. These are all significant steps, but progress is fragile and we cannot be complacent.

The Last Elephants carries this message from many people in different fields working for the protection of elephants in Africa. Its rich collection of images from top wildlife photographers are windows into the world of these extraordinary creatures. I hope this book is an encouragement to keep the pressure up and to move fast and effectively to conserve our natural heritage.

Preface

Is the name of this book prophetic? We hope not, but the signs are worrying so it's worth pondering. In Africa there's one elephant for every 20 000 people; fewer than 450 000 according to the Great Elephant Census of 2016, and well down from the 3 to 5 million just 100 years ago. And while the human population is rapidly increasing, elephant numbers are plummeting. Human-elephant conflict and poaching seem inevitable and we know who wins those contests.

This book began as an idea over a cup of coffee in Cape Town. The results of the 2016 Elephants Without Borders census of savanna elephants – co-ordinated by Mike Chase and Kelly Landen – had just been made public. It was a shock. On average, three savanna elephants were being killed every hour somewhere in Africa. Add in their forest cousins and that number was an elephant being shot every 15 minutes.

Both of us know Africa well, have many contacts throughout the continent and love elephants. Could we do another kind of census – narrative and photographic – enlisting people on the ground who deal with elephant issues every day? Self-funding the book, we couldn't afford to pay them. Would they be prepared to donate their time, words and images? We composed an email, listed people across the continent we knew to be involved with elephants, pressed SEND and waited.

The response was heart-warming and humbling. Everyone had seen the census results and, like us, was very worried. Chapters poured in from an extraordinary range of people: scientists, poets, game guards, activists, academics, lodge owners and NGO workers. Top wildlife photographers sent portfolios with the message: Take your pick.

The project rapidly expanded. Each chapter was a gem so we couldn't refuse any. The photographs were breathtaking and how could we not display them large – and on good recycled paper, with a top designer and first-class printer? We swallowed hard and went for the best of everything. The elephants deserve it.

This book is a tribute to the many people who work for the welfare of elephants, especially those who risk their lives for wildlife each day: field rangers and the anti-poaching teams, in particular. It is an acknowledgement, also, of the many communities around Africa that have elected to work with elephants and not against them.

The book is also a tribute to such researchers as Iain and Oria Douglas-Hamilton, Joyce Poole, Cynthia Moss, Daphne Sheldrick, Paula Kahumbu, Mike Chase, Kelly Landen, Michelle Henley, Sharon Pincott and many others – the Jane Goodalls and Diane Fosseys of the elephant world who have dedicated their lives to these great, graceful and engaging animals.

We hope this book will help to fulfil two wishes. The first is that the Congress of Parties of the Convention on International Trade in Endangered Species of Wild Fauna and Flora (CITES) uplists all elephants in all countries to Appendix I, forbidding trade of elephants or elephant parts across international borders.

A second wish is that those countries that receive and use legal and poached ivory – primarily China, Vietnam, Laos and Japan – seriously and strenuously ban its trade and use, both within their borders and on their internet platforms.

We're losing elephants fast. They *could* become extinct. In many countries where they once roamed, particularly in West Africa, they already are. They need and deserve our protection. Let us not have to bear witness to the last wild elephants. That would cause a terrible, unforgivable hurt to the Earth's living fabric. And in a deep, ancient way, the loss of such intelligent, thoughtful minds that have been with us throughout the existence of our species would leave us so lonely.

Don Pinnock & Colin Bell

Elephants:
a human-animal crisis

Dr Ian McCallum

Elephants are more than a keystone
species in their ecosystems. They are
an indicator species, large grey mirrors
of the fate of all other wild creatures.
If we can't protect an animal this large,
how can we be expected to protect the
little things?

The future of our wildlife and the habitats in which it thrives cannot be summed up as a purely animal crisis. It is a human crisis, and by this I mean a crisis of human character. Our greatest challenge, I believe, is to combat not only human criminality but the enemies within ourselves. There are four of them: human ignorance, human indifference, human entitlement and, the saddest of all: the cynical language of human defeatism – that there is nothing we can do about it ... that it's too late.

It is time to take a long, hard look at ourselves. It is time to renew our language and with it, our sense of connection and relationship to and with our wild kin. Let's never again refer to them as non-human. They are human-others ... other than human. Large, small, scaled and feathered, they are in our blood and in our psyche. Who and what would we be without them?

© Kelly Landen, Elephants Without Borders, Near Nkasa Rupara National Park

01

Counting elephants

The Great Elephant Census, the results of which were published in 2016, was the first continent-wide survey of African elephants and provides a baseline for assessing change in savanna populations throughout the continent.

Dr Mike Chase & Kelly Landen

We were shocked at the carnage – we counted one dead elephant for nearly every four live elephants that we spotted on our aerial survey over south-east Angola. Just 10 years earlier, after surveying Angola in 2005, we had had so much hope for the region's elephants. Within 18 months after the end of the Angolan Civil War, our counts and satellite-tracking studies indicated that elephants were streaming back into the region, recolonising their ancestral range. But by 2015 they were again being slaughtered for their ivory. This time, though, it wasn't just in Angola. The slaughter stretched across the entire continent.

After flying nearly 300 000 kilometres over 24 months as part of the Great Elephant Census (GEC), we ended up in Angola in 2015 – the last country we surveyed. The world now knew the truth about the tragic plight of African savanna elephants. Conceived and led by one of the authors of this chapter, principal investigator Dr Mike Chase of Elephants Without Borders, and funded by the philanthropist Paul G Allen and his sister Jody, the GEC was the first co-ordinated survey of savanna elephants across Africa. In a project involving dozens of elephant researchers, government wildlife agencies and conservation groups conducting aerial surveys from small

planes and helicopters, elephants were counted across 18 countries (see pages 20 and 23).

The goals of the GEC were to accurately determine the number and distribution of African savanna elephants over the great majority of their range and to provide governments and wildlife conservation organisations a baseline to co-ordinate their conservation efforts across Africa.

Despite their large size, obtaining accurate counts of elephants is not easy, especially over large areas, much less a continent. It would have been impractical and prohibitively expensive to census all savanna elephants, so we focused on the largest and densest populations within each country, with the goal of counting at least 90% of them continent-wide.

Direct counts of elephants are commonly used in open savanna and woodland habitats, and counts of elephant signs, especially dung piles, are typically used in forested areas. Given the spatial scale of the GEC, we used aerial sample counts, the most common survey technique for African elephants and other wildlife in eastern and southern Africa.

GEC surveys were conducted by a partner research organisation or government agency in each country, with all survey teams led by experienced surveyors following consistent standards. Standards required included:

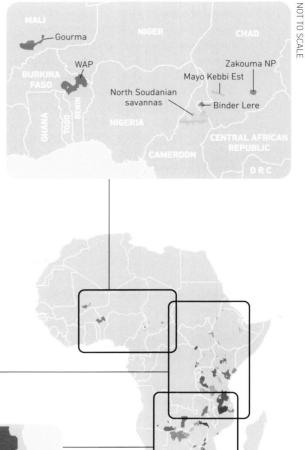

MALI
Gourma
NIGER
CHAD
BURKINA FASO
WAP
Zakouma NP
Mayo Kebbi Est
North Soudanian savannas
Binder Lere
GHANA
TOGO
BENIN
NIGERIA
CAMEROON
CENTRAL AFRICAN REPUBLIC
DRC

SUDAN
ETHIOPIA
SOUTH SUDAN
North-west Ethiopia
Babile Elephant Sanctuary
CAR
Kidepo Valley NP & Karenga CWA
Garamba NP
Murchison Falls Protected Area
KENYA
SOMALIA
Greater Virunga Landscape
UGANDA
Samburu-Laikipia
R
B
Burigi
Maasai Mara
Lamu
DEMOCRATIC REPUBLIC OF THE CONGO
Malagarasi-Muyovosi
Serengeti
Tsavo-Amboseli
Katavi-Rukwa
TANZANIA
Tarangire-Manyara
Ruaha-Rungwa
Selous-Mikumi
ZAMBIA
MALAWI
MOZAMBIQUE

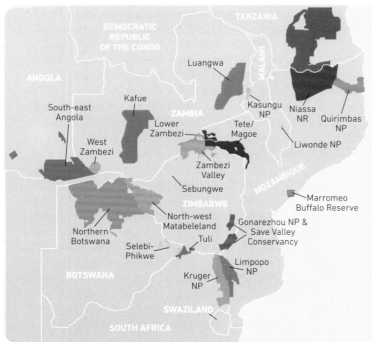

TANZANIA
DEMOCRATIC REPUBLIC OF THE CONGO
Luangwa
MALAWI
ANGOLA
Kafue
South-east Angola
Lower Zambezi
ZAMBIA
Kasungu NP
Niassa NR
Quirimbas NP
West Zambezi
Tete/Magoe
Liwonde NP
Zambezi Valley
Sebungwe
MOZAMBIQUE
Marromeo Buffalo Reserve
ZIMBABWE
North-west Matabeleland
Gonarezhou NP & Save Valley Conservancy
Northern Botswana
Tuli
Selebi-Phikwe
Limpopo NP
Kruger NP
BOTSWANA
SWAZILAND
SOUTH AFRICA

AREAS OF GREATEST PROTECTION

Of the surveyed elephant population, an estimated 84% (295 978) were in protected areas, while 16% (56 262) were in unprotected areas. Most elephants in unprotected areas were in Mali and Angola, while the majority were in the protected areas indicated here. This underscores the critical importance of protected areas for the future of savanna elephants.

- use of the latest technology, such as GPS receivers and digital cameras, to verify herd counts, voice recorders to document observations, and laser altimeters to ensure that flight altitudes were within standards;
- adherence to specified flight parameters (height, speed, search rate) to reduce the likelihood of observers' missing elephants during surveys;
- sound survey design, including appropriate stratification and full coverage of study areas;
- use of experienced, well-trained crews as well as survey schedules that minimised crew fatigue; and
- appropriate analytic methods for estimating elephant populations and carcass ratios.

We used fixed-wing aircraft in all areas except Kruger National Park in South Africa. In keeping with past practices, Kruger used helicopters to conduct a total count of elephants by searching along drainages; helicopters allow for heightened visibility in the rough terrain and dense vegetation of the park.

On GEC sites, each stratum was surveyed with one of two survey methods: total count, a complete census of all elephants present using closely spaced transects; or sample count, counting elephants on a subset of a stratum and then extrapolating to the entire stratum. For GEC surveys, sample counts typically covered 5 to 20% of a stratum. On sample counts, survey intensity (the percentage of the stratum actually sampled) generally increased with the expected number of elephants in a stratum. As an example, for the northern Botswana survey, survey intensity was highest in the two regions where elephant populations were expected to be especially dense – the Okavango Delta in the western part of the ecosystem and the Chobe River region in the north-eastern part of the ecosystem.

On GEC surveys, sample counts that used transects had nominal strip widths of 150 to 200 metres on either side of the plane at the target altitude of 91.4 metres. Transects were generally oriented perpendicularly to rivers or ecological gradients to minimise the variation in elephant density between transects. In mountainous areas, block counts were used instead of transect counts. During survey flights, observers took photographs of larger herds to ensure that herd sizes were estimated accurately.

In addition to counting live elephants, we counted elephant carcasses. Dead elephants remain visible for several years, so the 'carcass ratio' – the number of dead elephants divided by the sum of live plus dead elephants – is correlated with recent mortality rates. Carcass ratios are often used as an index of population growth rates, as ratios <8% are typical of stable or growing populations over the previous 4 years, while higher carcass ratios may indicate mortality in excess of births in the previous 4 years. Survey teams also counted other large and medium-sized mammal species, including livestock.

Most surveys were conducted during the local dry season, when clear weather and lack of leaves on deciduous trees make elephants more visible. Some of the surveyed populations straddled international boundaries. In such cases, seasonal movements of elephants could potentially result in the same animals being counted in two different countries. To avoid double-counting in these situations, in most cases, survey teams along international borders co-ordinated survey timing so that populations in both countries were counted at roughly the same time.

We estimated the total savanna elephant population for the Great Elephant Census at 352 271 elephants (see page 23). Botswana had 37% of this total, with Zimbabwe (23%) and Tanzania (12%) also having large populations. Elephant densities were highest in Botswana and Zimbabwe, and amounted to less than one elephant per square kilometre in all other countries. We also recorded 201 poacher camps and an estimated 3.39 million head of livestock within-GEC surveyed areas.

The all-carcass ratio for the entire GEC was 11.9%. Carcass ratios >8% generally indicate a declining population. Ratios varied greatly by country, with the highest ratios in Cameroon (83%), Mozambique (32%), Angola (30%) and Tanzania (26%), suggesting declining populations in these and other countries over the 4 years prior to the GEC. The highest fresh carcass ratios were found in Angola (10%), Cameroon (10%), the north-western Benin park extension of W-Arly-Pendjari (WAP) Ecosystem (3%) and Mozambique (3%), suggesting high levels of recent elephant mortality in these countries (see table on page 23).

Of the total GEC elephant population, an estimated 84% (295 978) were observed in protected areas, while 16% (56 262) were in unprotected areas. The proportion of elephants in protected areas on GEC sites varied by country, with most elephants in unprotected areas in Mali and Angola and most in protected areas in all other surveyed countries. A large majority of estimated elephant populations were in protected areas – in nine countries, all elephants observed were in protected habitats. This underscores the critical importance of protected areas for the future of savanna elephants and the need to better protect their habitats.

Survey-wide, carcass ratios were 12% in protected areas and 13.2% in unprotected areas, indicating that poaching is serious in both protected and unprotected areas. These high carcass ratios also suggest that deaths likely exceeded births over the 4 years preceding our GEC surveys in both protected and unprotected areas. For protected areas, the clear implication is that many reserves are failing to adequately shield elephants from poaching and human-elephant conflict.

We recorded notably high carcass ratios, potentially indicating high poaching levels, in the northern section of Tsavo East National Park, Kenya (52% carcass ratio); Niassa National Reserve, Mozambique (42%); and Rungwa Game Reserve, Tanzania (36%). Heightened anti-poaching measures are needed in these and other protected areas to ensure that they do not become mere 'paper parks' for elephants. At the same time, we estimated that over 50 000 savanna elephants occur in unprotected areas. So improved protection in these areas with substantial numbers of elephants may also benefit the species.

Looking at historical trends, we see that from 1995 to around 2007, elephant populations were recovering from the poaching outbreak of the 1980s (see graph, top right). After that time trends reversed, with large declines observed in many countries and for the GEC survey area as a whole. If populations continue to decline at the 8% rate we estimated for 2010 to 2014, GEC survey areas will lose half of their savanna elephants every 9 years. Extirpation of some populations is possible, especially in countries such as Mali, Chad and Cameroon, with small and isolated savanna elephant populations.

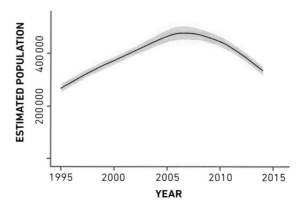

Trends in elephant populations in GEC survey areas from 1995 to 2016.

These dramatic declines are almost certainly due to poaching for ivory. Elephant poaching has increased substantially over the past 5–10 years, especially in eastern and western Africa, with estimates as high as 100 000 elephants killed between 2010 and 2012. Our trend model indicates a comparable decline of 79 413 elephants on GEC sites with historical data over those 3 years. Similarly, genetic analysis of intercepted ivory shipments shows that Mozambique and Tanzania were the primary sources for ivory from savanna elephants. According to our trend model, elephant populations in these two countries were declining at 14% and 17% per year, respectively, as of 2014. The Democratic Republic of the Congo (DRC), which had the second fastest population decline of any country in our dataset, was also a major origination point for ivory.

The Niassa (Mozambique) and Selous (Tanzania) ecosystems were especially frequent sources of poached ivory, and elephant populations have decreased by over 75% in the past 10 years in these two ecosystems. Thus, the illegal ivory trade appears to be the major driver of recent population trends among savanna elephants.

Poaching is not the only anthropogenic factor affecting elephant populations. The large number of livestock observed on our surveys suggests that conflict between elephants and human populations is widespread. By 2050, human populations are projected to double in 12 GEC countries, and many of these countries already have small elephant populations that may be susceptible to poaching and habitat loss. As human populations grow, so

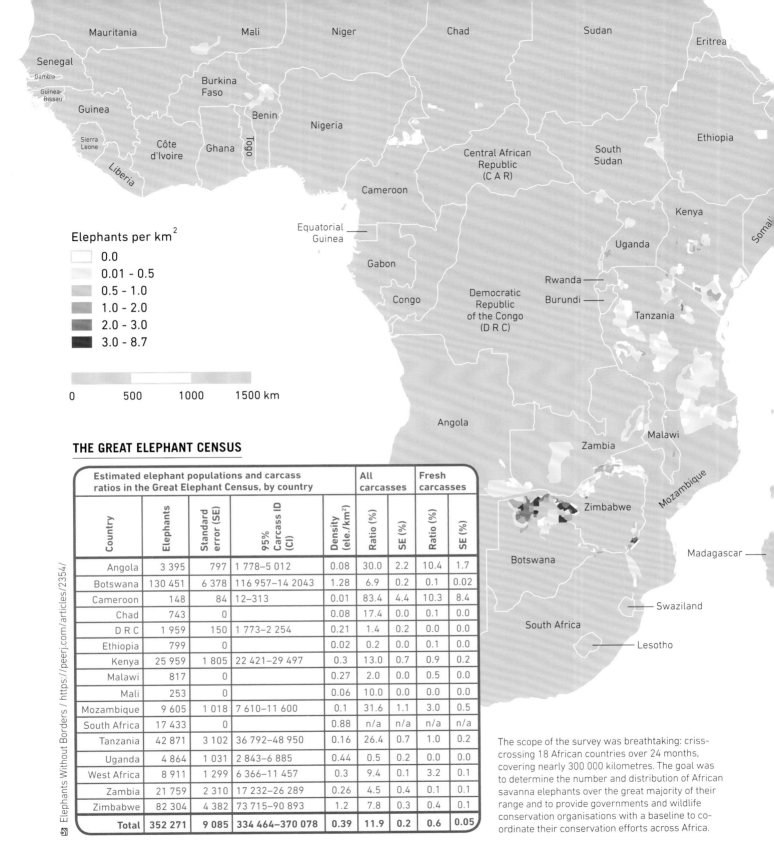

Elephants per km^2

	0.0
	0.01 - 0.5
	0.5 - 1.0
	1.0 - 2.0
	2.0 - 3.0
	3.0 - 8.7

0   500   1000   1500 km

THE GREAT ELEPHANT CENSUS

Country	Elephants	Standard error (SE)	95% Carcass ID (CI)	Density (ele./km²)	All carcasses Ratio (%)	All carcasses SE (%)	Fresh carcasses Ratio (%)	Fresh carcasses SE (%)
Angola	3 395	797	1 778–5 012	0.08	30.0	2.2	10.4	1.7
Botswana	130 451	6 378	116 957–14 2043	1.28	6.9	0.2	0.1	0.02
Cameroon	148	84	12–313	0.01	83.4	4.4	10.3	8.4
Chad	743	0		0.08	17.4	0.0	0.1	0.0
D R C	1 959	150	1 773–2 254	0.21	1.4	0.2	0.0	0.0
Ethiopia	799	0		0.02	0.2	0.0	0.1	0.0
Kenya	25 959	1 805	22 421–29 497	0.3	13.0	0.7	0.9	0.2
Malawi	817	0		0.27	2.0	0.0	0.5	0.0
Mali	253	0		0.06	10.0	0.0	0.0	0.0
Mozambique	9 605	1 018	7 610–11 600	0.1	31.6	1.1	3.0	0.5
South Africa	17 433	0		0.88	n/a	n/a	n/a	n/a
Tanzania	42 871	3 102	36 792–48 950	0.16	26.4	0.7	1.0	0.2
Uganda	4 864	1 031	2 843–6 885	0.44	0.5	0.2	0.0	0.0
West Africa	8 911	1 299	6 366–11 457	0.3	9.4	0.1	3.2	0.1
Zambia	21 759	2 310	17 232–26 289	0.26	4.5	0.4	0.1	0.1
Zimbabwe	82 304	4 382	73 715–90 893	1.2	7.8	0.3	0.4	0.1
Total	**352 271**	**9 085**	**334 464–370 078**	**0.39**	**11.9**	**0.2**	**0.6**	**0.05**

Estimated elephant populations and carcass ratios in the Great Elephant Census, by country

The scope of the survey was breathtaking: criss-crossing 18 African countries over 24 months, covering nearly 300 000 kilometres. The goal was to determine the number and distribution of African savanna elephants over the great majority of their range and to provide governments and wildlife conservation organisations with a baseline to co-ordinate their conservation efforts across Africa.

does the potential for human-elephant conflict, leading to elephant deaths, loss of habitat to agriculture or fires, and potential extirpation of elephant populations.

The GEC revealed large regional differences in the status of savanna elephants. Countries in western and Central Africa, such as Chad, Cameroon, Mali and the DRC, had savanna elephant populations that were small, isolated and declining in the face of poaching and expanding human populations. The WAP population, on the borders of Niger, Burkina Faso and Benin, is the only savanna elephant population in western or Central Africa with over 2 000 elephants. Elephant populations in the WAP ecosystem have grown in recent years, but the high fresh-carcass ratios recorded may be a warning sign of increased poaching that has yet to noticeably affect population estimates.

In eastern Africa, Mozambique and Tanzania have experienced large declines in elephant populations. Though numbers of elephants are still relatively high in these two countries, poaching has had major impacts on their populations. Elephant populations elsewhere in eastern Africa, including Kenya, Uganda and Malawi, show more positive recent trends, indicating that the poaching crisis does not appear to have affected all East African countries equally. In southern Africa, four countries, Botswana, South Africa, Zambia and Zimbabwe, have relatively large elephant populations and show either increasing trends

or mild and non-significant declines recently. Southern Africa has experienced less poaching than any other part of Africa. Angola, however, is an exception, with extremely high carcass ratios and large numbers of fresh carcasses, suggesting high levels of ongoing poaching.

The GEC was the first-ever continent-scale survey of African elephants. These results will serve as a baseline for assessing change in savanna elephant populations throughout Africa. Because elephant populations can change rapidly, surveys on the scale of the GEC should be conducted regularly to measure population trends, gauge the effectiveness of conservation measures and identify populations at risk of extinction. Future surveys may also allow detection of emerging threats such as drought and climate change.

Ideally, results from this survey will encourage people across Africa and around the world to protect and conserve elephant populations. Preliminary results from the GEC have already motivated the governments of Mozambique and Tanzania to implement new measures to stabilise elephant populations. The future of African savanna elephants ultimately depends on the resolve of governments, conservation organisations and people to apply the GEC's findings by fighting poaching, conserving elephant habitats and mitigating human-elephant conflict. Over 350 000 elephants still roam Africa's savannas, but with populations plunging in many areas, action is needed to reverse ongoing declines.

'There is mystery behind that masked gray visage and ancient life force, delicate and mighty, awesome and enchanted, commanding the silence ordinarily reserved for mountain peaks, great fires, and the sea.'

Peter Matthiessen

© Thierry Prieur, Chobe, Botswana

02

Why elephants matter

To wake up to the web of intelligence,
to the wild origins of sentience,
to find your voice and raise it,
that others may raise theirs
for elephants.

Dr Ian McCallum

In 2012, under the auspices of the Wilderness Foundation and its US sister organisation, the WILD Foundation, my close friend Ian Michler and I undertook a 5 000-kilometre expedition across southern Africa. The overall aim of the expedition was to highlight and support corridor conservation initiatives allowing wild animals to link with neighbouring wild habitats across international boundaries.

Calling our expedition the Tracks of Giants and following ancient and current elephant clusters and migration routes, we walked, kayaked and cycled through Namibia, Botswana, southern Zambia, Zimbabwe, Mozambique and South Africa. As important as it was to promote the notion of corridor conservation, of equal importance was to pay attention to and explore the 'corridors of the minds' of those people who would ultimately protect and maintain the actual geographical corridors. It was clear that trust, respect, ownership and intellectual and economic empowerment of the human communities linked to the geographical corridors and reserves were essential to the future of these initiatives.

Of the many insights I gained from this expedition, one of them reminded me of a lesson I learnt a long time ago as a psychiatrist: never underestimate the intelligence of your patients and, in this instance, never underestimate the intelligence of the local people.

It was only after our journey was over that another, more personal reason for our expedition began to emerge. It developed around the answers to a question we put to farmers, wildlife guides, regional chiefs and, on our return, to schoolchildren: can you imagine a world without elephants? A recurring answer was the word 'unthinkable'.

There were questions, too: what would we tell our children? What an indictment of human beings! Every answer resonated with something deep within me ... each one a catalyst for further questions: who and what would we be without elephants in the world? What would their extinction say about the human species? What would be the cost of the loss of this animal to the landscapes, to the ecology of other species such as insects, birds, mammals, plants and trees and more ... and to the human psyche? What role do wild animals and, in this instance, elephants play in the human narrative? If their role is meaningless, why do so many people become activists for their protection? Could it be that they are intrinsic to our sense of identity as human beings?

ESCAPING THE FLOODWATERS

Throughout history, elephants have played an important role in African traditions and folklore. Many believe they are reincarnated from ancient chiefs to live on as chieftains of the savannas and forests. In western Zambia, the elephant is a symbol of power, and the Paramount Chief of the Lozi, the Litunga, has an elephant statue mounted atop his boat. Each year, before the arrival of the Zambezi River's March or April floods, the much-anticipated Kuomboka ('get out of water') ceremony celebrates the move of the Litunga from his low-lying summer home on the open plains to his winter home on higher ground.

Conservation issues are emotional. And so they should be. It's because we are angry, because we are sad, because we are anything but indifferent to the fate of those without a voice, that we become activists. The thought of elephants or rhino or pangolins being slaughtered at their present rate cannot be separated from feelings of outrage and, at times, despair.

Lest we forget, fair play is intrinsic to the integrity not only of human, but also of other primate social evolution. Without the capacity to feel for the fate and situation of others, to 'put yourself into the skin' of the other, our human social systems would disintegrate. Think about it. There would be no morality and personal ethics, no sense of right and wrong, no compassion. The biological evidence is compelling.

The modern mammalian cortex – more specifically, the prefrontal cortex – is home to specialised cells known as mirror neurons. Without these, we cannot mirror or read the body language, intentions and emotions of the other. Individually and socially, the significance of these neurons to our survival should be obvious. To be able to 'read', however vaguely, the intentions and emotions of the 'other', the adversary, the friend, the givers and takers in human relationships, is crucial for decision-making, for reaching out or withdrawing. Our survival depends on it. Is this capacity confined to human beings? The answer is: no.

Thanks to the work of the American neuro-scientist Jaak Panksepp, we can confirm what we have long suspected and sometimes witnessed. We share with all mammals the neuro-circuits responsible for the emotions of rage, fear, panic, nurture and lust, as well as the emotional circuits relating to play. Without mirror neurons, that often-derided word 'anthropomorphism' – the act of projecting human feelings onto animals – would be impossible. We can't help it. We are hard-wired to do so. It's why we become voices for the voiceless.

With rare exceptions, the human being is a moral animal. The outrage of an angered elephant or lion is no less potent than ours. The despair of an animal with a broken spirit is no less saddening than that of a human being with a broken spirit.

Returning to the title of this chapter – why elephants matter – brings me to the heart of an expanding field of environmental thinking called conservation psychology. It's a field with both theoretical and practical applications addressing at least two important questions: what is the place of human beings in nature and what is the place of nature in human beings? Not least is its focus on environmental education and what I believe to be one of its most essential benefits: the cultivation of a greater understanding of the link between the natural world and human identity. Who and what we are as a species is impossible to define outside of our relationships, not only human to human, but also to landscapes, animals, plants, insects and to the biosphere itself. To me, there is no such thing as human nature; there's only nature and the very human expression of it. Every living thing is, in its own way, a manifestation and expression of nature.

And so why *do* elephants matter? They matter because, like anything wild, their lives are biologically and historically linked to ours. Sharing more than 90% of the human genome, they matter because of what they are and for what they represent in the human psyche. They stir our emotions. They come alive in our stories, in folklore, myths and language. They're in our language and metaphors: 'the elephant in the room', 'the white elephant', to have 'a memory like an elephant'.

If humans are storytelling animals, surely the animals that feature in them are part of the human story. They're in our blood and in our psyche. They matter because the elephant crisis – like all environmental crises – is a crisis within each one of us. It's a crisis of character, to act or to turn your head. Who among us is willing to be disturbed, to be a voice for the voiceless?

And lest we forget, the elephant crisis, like that of the rhino, is not an animal tragedy. It is a human-animal tragedy. 'Another day, another dead wildlife ranger' is the headline of a December 2016 article in *The Guardian*. The statistics are chilling. In Africa, 'Two to three rangers die each week in the line of duty … in the last decade, more than 1 000.' Think of the immense loss suffered by families affected by these deaths. Think of their colleagues, their anger, outrage and despair. Think of the rangers who turn to alcohol as medication to numb the despair. Their loss is our loss. Their despair is our despair. We may be the Earth's most effective animal predator, but we have it within us to be the world's most effective protector as well.

You may well ask, Why the emphasis on elephants? Why not rhinos? Why not dung beetles? Why not the world's oceans and forests? I selected elephants not only because of the alarming contemporary slaughter of these iconic species (we are currently losing two to three elephants every hour for their ivory), but for two other reasons. The first is their role as a keystone species in the ecosystems in which they live and function. The second – in keeping with their place in the human psyche – is their symbolic significance: what they represent, what they tell us about ourselves. To me, they're a large, grey mirror of the fate of the wild animal species of the world.

I leave you with a question. If we can't protect an animal this large, how on earth can we be expected to protect the smaller, less charismatic species of our planet? I hope this question will be read not as a demand for an answer but as a personal challenge.

Finally, we can learn a lot about ourselves from the wild … and from elephants. I wrote the poem at the start of this chapter for them and for Michael Chase and Kelly Landen, founders of Elephants Without Borders. For them, elephants matter.

What will we tell our children about elephants if all that remains of their passing is pieces of their teeth?

© Peter Chadwick

'I fear that the African elephant
will have disappeared from
the wild by the time Princess
Charlotte turns 25.'

HRH Prince William

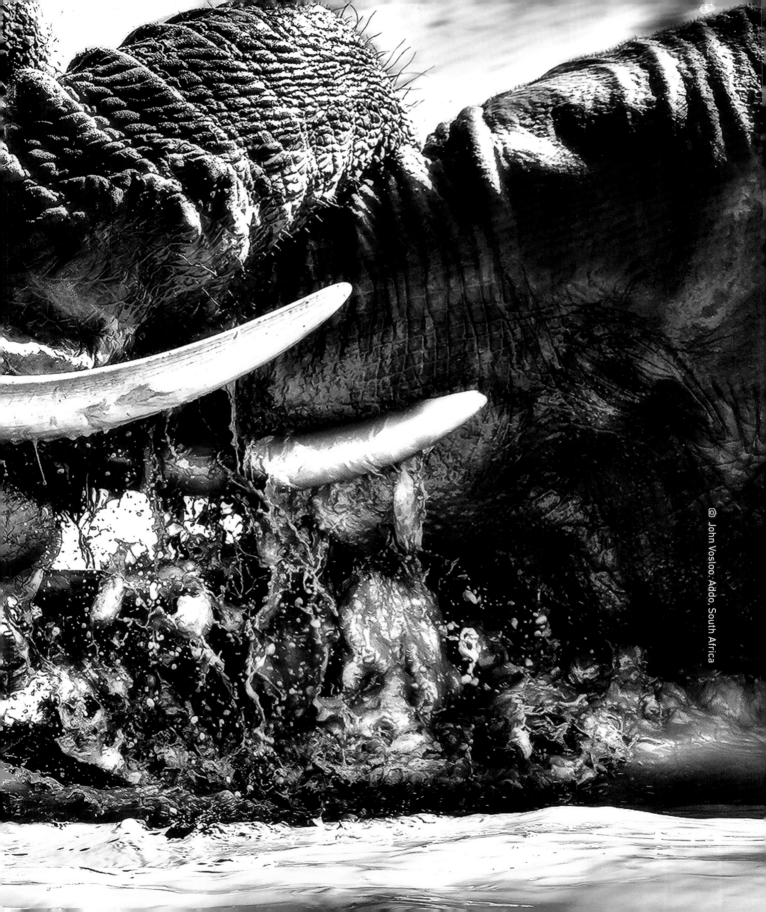

03

Imagining Africa's elephants

We may not know just what ancient African people really thought of elephants, but we do know that they represented them in their earliest forms of expression.

Dan Wylie

About 5 million years ago, when the west coast region of South Africa was lusher than it is today, a flash flood entombed in mud an extraordinary number of animals: the bones of sabre-toothed cats, African bears and short-necked giraffes that together constitute one of the richest fossil deposits of the Pliocene epoch.

Among the specimens is an early four-tusked elephantine species, a gomphothere whose descendants would evolve into the three main genera of modern-day elephants. Four million years ago, *Mammuthus subplanifrons* roamed the region, twice migrating into Europe and Asia, founding the woolly-mammoth genealogies as well as today's Asian elephant. The line of *Loxodonta*, the African elephant, remained in Africa. A tusk shed by one of these great beasts, and fossilised around 130 000 years ago, would eventually be unearthed north of Durban.[1] These recent elephants would have coevolved with humans, and some anthropologists have speculated that humans could not have migrated out of Africa without the elephants preceding them. Human and elephant histories are deeply entangled from the beginnings of memory itself – and so they remain in all our evolving and multifarious art forms.

The fossils of the west coast deposit at Langebaanweg can be viewed today: remnants of once living, eating, fighting, breeding creatures, now carefully preserved beneath a coat of resin. Like fossils, pre-colonial stories tend to be extracted from their original contexts and leached of the vivacity of living performance. Old oral meanings are partly lost, but also partly reworked in literate forms: archived, retold, republished in lavishly illustrated books and in simplified tourist brochures and on internet sites.

We may not know just what ancient African peoples really thought of elephants, but we do know that they represented them in their earliest forms of expression. Rock art, from the Nile to the Limpopo, offered the first concrete manifestation of the mingled reverence and alarm these massive creatures elicited from humans. In southern Libya, which was once a forested and flourishing wetland, basalt slabs bear swirling etchings of elephants, so stylised and yet so accurate that they seem entirely modern, though they date back to between 7 000 and 4 000 years ago.

Skipping over innumerable intermediate examples, the rock paintings of southern Africa also express a high degree of respect for elephants.

'For species such as elephants and rhinos to be fighting for their existence due to human exploitation and interference is unacceptable, and we must do everything within our power to turn this dire situation around. We are responsible for the problem, and we must be held responsible for the solution. It will indeed be a very sad indictment on our species if rhinos and elephants are no more, and that day will come sooner than we think if we do not take action.'

David Attenborough

A life-size painting filling a cave wall in Mtoko, Zimbabwe; etchings from the Karoo semi-desert to the 'Uffizi Gallery of rock art' at Twyfelfontein, Namibia; and exquisite paintings in the Cederberg all attest to observations surely based on lifelong liaisons and, in many cases, totemic identification.

Not that all those liaisons were cordial: elephants were and are dangerous, and it was a mark of extreme courage, as well as materially hugely fruitful, to hunt and kill a grown elephant. Such hunts were probably more opportunistic, occasionally ritualistic, than common. Nevertheless, some rock art depicts elephants being hunted by being driven into pit-traps or being hamstrung and harried to exhaustion and peppered with spears. Other representations appear connected to trance-dance or related ways of embuing humans with animal power. Something of this lies behind a /Xam Bushman folktale or *kukummi*, as recorded by the linguist Wilhelm Bleek in the 1870s in Cape Town:

An Elephant carries away the little Springbok on her back while |kaggen (the Mantis) is inside a hole digging for food. |kaggen calls to the little Springbok from inside the hole but it does not answer him. The Elephant-calf answers him instead, but not with a nice sound. |kaggen thinks that the soil he has thrown out of the hole has stuck in the little Springbok's throat.

|kaggen emerges from the hole to see why the child's throat sounds as it does. He finds the Elephant-calf lying there covered with soil from the hole, strikes the calf and knocks it down, killing it. |kaggen tracks the departed Elephant by its spoor but decides to return home to tell his sister (the Blue Crane) about the theft of the child. The sister scolds him for sleeping in the hole and not hearing what occurred.

|kaggen asks for food so that he may follow the Elephant to its place. He tells his sister she must watch for when the grass blows from another direction, for that is when he will return with the child. He finds the Elephant's house, sees the little Springbok playing with the Elephant children, and calls out to it. The Elephant sees |kaggen coming and swallows the Springbok child. |kaggen demands that the Elephant give the child back and enters her navel to fetch the child, whom he fastens onto his back. The other people try to stab |kaggen to death, so he exits through the Elephant's trunk and flies away to return the child to his sister. [2]

It's difficult to know quite how to interpret such a tale of tangled family relationships, magical occurrences and the role of a slippery trickster-god. But such tales persist right across the continent, passed down from generation to generation, sometimes with astonishing fidelity and stability. Who knows how old might be this fragment of beautifully metaphoric Bushman perception?

Tall-topped acacia, you, full of branches,
Ebony-tree with the big spreading leaves.

This, too, was recorded by Bleek only in the 1870s.[3]

In many cases, folktales – long regarded by invaders, from the Arabs to the British, as fossil-like relics, or as hallmarks of the authentically 'primitive' – have evolved into colourful modern incarnations. Hundreds of children's books, for instance, repackage ancient tales for current ecological purposes, injecting a vein of concern and feelings perhaps entirely alien to the originals.

The drying of the Sahara, combined with imperialist dynamics, contributed to a strong split into an Arab-oriented north and a Europe-oriented centre and south. As early as the Roman Empire, the northern elephant populations were extirpated, partly for war purposes. The Ptolemies of pre-Christian Egypt used and depicted African war elephants, as did Hannibal. In the Christian era, however, contact with and knowledge of elephants virtually vanished from the European consciousness. The fairly accurate representations in Roman mosaics were replaced by fantastic concoctions right through the Middle Ages and the Renaissance until 18th-century imperial ventures began to take captive elephants to Europe as zoo specimens or gifts to royalty. The art then became more realist and empiricist in observation.

Although pre-colonial humans were not incapable of exterminating large mammals, the arrival of imperialists south of the Sahara, as is all too well known, with their muddle of philanthropy, desire for material plunder and general disdain of indigenous societies, spelled doom for elephants. Indigenous hunters probably contributed fatally to the extinction of that other elephantine strain, the mammoths, in the northern stretches of America and Russia, but the arrival of Europeans with firearms was altogether different. Initially, small numbers of hunters encountered such vast numbers of animals that these seemed inexhaustible. Like the bison and passenger pigeon in North America, they were not.

Though the open savannas were especially vulnerable to the new technologies of travel and slaughter, the forest elephants of West and Central Africa were not spared. The portrait of the jungle-crazed, venal ivory-trader Kurtz in the Belgian Congo in Joseph Conrad's novella *Heart of Darkness* (1902), is just one literary testament to the depredation.

Ivory was the spur – often (as in Conrad) in tandem with trade in human slaves. Paradoxically, exquisite artistry has resulted from the destruction of elephants, from Khoisan armbands to European piano keys and cufflinks to Japanese *netsuke* and, today, Indonesian religious shrines. Beautiful artefacts made from one of the most ancient and malleable of materials have doubtless enriched a wide range of human cultural expression, but this has been to elephants' detriment.

Two literary genres were spawned by these imperial ventures: the travelogue and the hunting account. These popular, usually illustrated kinds of publication, designed to titillate armchair adventurers back in Europe, developed quite specific generic commonalities and also laid down attitudes that would continue to inflect actual treatment of elephants into the modern era. The earlier travellers' motives were mixed: sheer curiosity, observing places and people because they could; as scouts for economic entrepreneurs eager for natural resources; and sometimes as self-proclaimed naturalists intent on improving scientific knowledge. That search did not preclude obliterating large numbers of specimens in the service of science. Those early travellers' accounts of elephants were at first coloured by the mediaeval mythologies prevalent in an ignorant Europe, and expressed negligible awareness of either ecological functions or societal dynamics amongst elephants. Elephants were fascinating only in their strange hugeness and as vectors for sundry near-fatal adventures and mishaps. François le Vaillant set the tone in his 1794 *Travels*; subsequent accounts are just a footnote to this attitude:

I was beginning to enjoy this hunting: with practice I came to find it more interesting than dangerous. I could never understand, and have understood even less subsequently, why authors and travellers have stuffed the stories they have told us with so many lies about the powers and tricks of this animal. Why have they excited the reader's imagination about the dangers to which hunters who pursue the elephant expose themselves?

In truth, if anyone were stupid and rash enough to attack an elephant in open countryside, he would be dead if he were to miss his shot. The greatest speed of his horse would never equal the trot of the furious enemy pursuing him. But if the hunter knows how to use his advantage, all the powers of the animal must give way to his ingenuity and his cool-headedness.[4]

The devoted hunting account really took off in the 19th century, especially in eastern and southern Africa (though German, French and Portuguese hunters also roamed their respective imperial territories). In English literature, at least, the names of Gordon Cumming, CH Stigand and Frederick Courtney Selous are just the best-known of the many hunter-writers who made a living from writing and lecturing about their derring-do.

Many of them were of military background, and it shows in the writing:

We had just descended a steep bamboo-covered hillside, crossed a mountain torrent and were slowly climbing the steep opposite side of the valley when we heard a noise from the slope behind us. On looking back we at first only saw the bamboo moving by some unseen agency. Every now and again there would be a trembling in a clump of trees and the top of a stem would bend over and disappear with a cracking sound.

On looking through my glasses I could distinguish here and there a black trunk soaring upwards to reach for a high branch, and occasionally a glimpse of part of a black body between the bamboo clumps. After watching for some time I made out what I took to be three bulls on the right of the herd.

Knowing that I should not, in all probability, get another sight of them, once I left my vantage on the hillside, I took careful stock of their position and of any big trees on the way to serve as landmarks. Then I descended to the bottom of the valley again, crossed the stream and began the steep toil up the slope.

When I finally arrived at the spot at which I had seen them there was nothing but their spoor left; the whole herd had moved on and there was not even the noise of cracking bamboo to be heard. I followed the spoor a little way, and, as I could see or hear nothing of them, I returned to the porters and arranged a site for our camp. [5]

There are acres of this numbingly detailed kind of narrative, followed by accounts of the excision of bullets from corpses and regaling of incidents around the virtual campfire of the publishing industry. The death of the elephant is routinely dismissed in a couple of lines; the run-up predominates; the story is all. Far from being clumsy backwoodsmen unaccustomed to the pen, as they liked to portray themselves, these writers were often well-read, accomplished writers and canny about their audiences. They frequently adopted a tone of light self-deprecation, too, enjoying having survived near-fatal elephant charges and the like. This tone would spill over into a 20th-century successor-genre, the equally popular game-ranger memoir, with its mix of jocund daring and didacticism.

The literatures of the late 19th and early 20th centuries perform an awkward and sometimes ambivalent shift from the hunter's regret at the increasing scarcity of 'game' to the impassioned protectiveness of conservationists battling a variety of hunters, now dubbed 'poachers'. This accompanies the establishment of national parks, or what is now often called 'fortress conservation'. Colonialism's racially, spatially and economically skewed new conservation ethics and strategies are reflected in the literature.

In fiction, the unthinking rapacity of hunter characters in Rider Haggard's imperial adventures, such as Allan Quatermain in *King Solomon's Mines* (1885), is gradually replaced by hunters-turned-conservationists. Stuart Cloete's *The Curve and the*

Tusk (1952), for example, presents characters from all walks of life, including, unusually for the time, passages imagined from the elephant's point of view. In more mystical mode, Romain Gary's *The Roots of Heaven* (1957), set in Central Africa, pits an existentially manic conservationist against human predators:

I defy anyone to look upon elephants without a sense of wonder. Their very enormity, their clumsiness, their giant stature, represent a mass of liberty that sets you dreaming. They're ... yes, they're the last individuals. [...] To tell you the truth, I'd give anything to become an elephant myself. [6]

An increasing number of fictions, especially for younger readers, have to various degrees sympathetically imagined the elephant's perspective, but none go so far as Canadian novelist Barbara Gowdy's *The White Bone* (1998). Gowdy envisions an entire culture, language and historical memory for a group of savanna elephants, zoologically informed but attempting throughout to evoke empathy through the elephants' perspective. Further south, a similar respect underpins Dalene Matthee's novel *Circles in a Forest* (1984), about the famed remnant Knysna elephants, contrasting with the more exploitative thriller matrix of Wilbur Smith's *Elephant Song* (1991), which superficially explores the prickly culling issue.

The burgeoning game-ranger memoirs are also inevitably concerned with so-called culling, as well as a new appreciation of wildlife. Etchings are replaced by increasingly magnificent photographs.

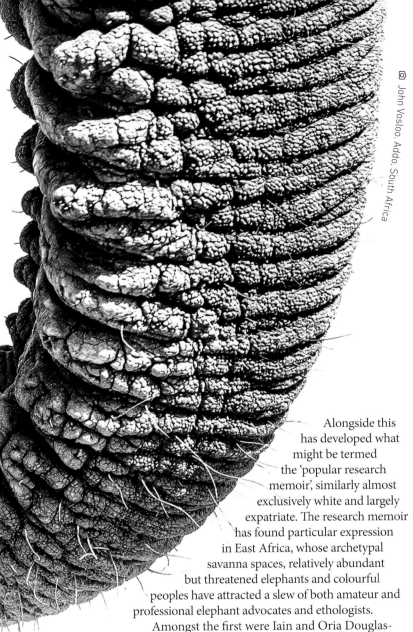

© John Vosloo, Addo, South Africa

elephant orphanage outside Nairobi still operates; and Cynthia Moss, whose book *Elephant Memories* (1988) and film accounts of one particular elephant, Echo, in Amboseli National Park have enjoyed notable success in global media. Similarly, Katy Payne relates her research into elephants' subsonic communications; Joyce Poole records an even more personal story; and Gay Bradshaw incorporates elephant family dynamics into an explicitly feminist agenda in *Elephants on the Edge* (2009). Further south, Sharon Pincott has, through a series of books, campaigned for the so-called Presidential Elephants of Hwange, Zimbabwe; and in *The Elephants' Secret Sense* (2007), American scientist Caitlin O'Connell relates her efforts in the Caprivi to show that elephants also communicate through ground vibrations and the soles of their feet. All this is underpinned by the growing sense that elephants are deeply sentient, intelligent, and emotionally, culturally and beautifully complex.

There is indeed a wealth of literary material. Numerous poems might be represented here by a few lines from Zimbabwean poet NH Brettell's *Elephant*:

> *Slowly the great head turned,*
> *And the late sunlight slept on massive flanks*
> *Like the still slabs of riven krantz,*
> *Immovable, and nonchalantly bearing*
> *The burden of the old enormous lies,*
> *The load of legendary centuries…* [8]

Today's visual culture is also inundated with colourful craftworks, tourist trinkets, coffee-table books, films, documentaries, amateur YouTube clips … But it's important not to forget that in Africa's rural regions, in and out of designated parks, human-elephant conflict exists on many levels. The dangers and setbacks of living alongside elephants are scarcely alluded to by the outpouring of literature, the ubiquitous art from David Shepherd to Paul Bosman, the didactic zoology of endless television documentaries, miniature Chinese ivory carvings and the sentimentalities of Western tourism. In light of such conflict, African agriculturalists may have good reason to hate and fear elephants; and the ivory merchants continue to see elephants as mere repositories of monetary wealth. Until their stories are told, understood and replaced with better stories, elephants will remain imperilled.

Alongside this has developed what might be termed the 'popular research memoir', similarly almost exclusively white and largely expatriate. The research memoir has found particular expression in East Africa, whose archetypal savanna spaces, relatively abundant but threatened elephants and colourful peoples have attracted a slew of both amateur and professional elephant advocates and ethologists.

Amongst the first were Iain and Oria Douglas-Hamilton, whose book *Among the Elephants* (1975) set the standard for highly personalised and vivid accounts of field research into elephant behaviour and ecologies. The Douglas-Hamiltons, while recognising the validity of some economic arguments, conclude that 'elephants fulfil part of man's deep need for the refreshment of his spirit'. This may be true of 'those who are obliged to live in highly industrialised surroundings' – but try telling that to the ivory poacher or the crime-syndicate boss.[7]

Subsequent researchers and campaigners have featured a surprising number of women. Prominent amongst them are Daphne Sheldrick, whose

'Everything is not yet lost, the last hope of freedom has not yet vanished completely from this earth and, who knows, if we stop destroying elephants and save them from extinction, we may yet succeed in protecting our own species.'

Romain Gary

04

Eavesdropping on elephants

I once thought the international ivory trade ban, secured in 1989, would bring lasting peace. I once thought our leaders had learnt something. But I was wrong.

Will Travers OBE

The African elephant's ear bears an uncanny resemblance to the continent of Africa. It's almost as if elephants are eavesdropping on the world.

Imagine ... if these beleaguered giants had been listening in over the past 30 years or so, as their future ebbed and flowed at the hands of a strange, self-important bipedal species, what would they have made of it all?

I have been eavesdropping on the fate of elephants for three decades and longer.

My first dead elephant: a massive bag of putrefying pus and skin draped over a shattered skeleton. The hyenas and vultures kept their distance as we paid our respects. We were upwind, but even there the overpowering smell caught the back of our throats, making us retch. The air was thick with death.

It was in Tsavo East National Park in Kenya in the mid-1970s – a park, a country, a continent in the grip of a poaching epidemic. Hundreds of thousands of elephants wiped out in a handful of years. The figures speak for themselves.

Driven by the ivory markets of Europe, the US and Japan (China had yet to join the bloody triumvirate), aided and abetted by widespread corruption, the massacre was facilitated by the very global wildlife trade control systems that were meant to protect them. Independent expert

analysis at the time (particularly the work of Dr Iain Douglas-Hamilton CBE and his team) revealed the stark reality: between 1979 and 1989, Africa lost 50% of its wild elephants.

The ears of the survivors must have been ringing with gunshots, the rattle of the AK-47, the almost silent flight of the deadly poisoned arrow. But did they also catch a whisper of the change that was afoot, the growing groundswell of public opinion, a mood that caught most politicians and elected representatives unawares?

CITES hears the call

Since its inception, the UN's Convention on International Trade in Endangered Species of Wild Fauna and Flora (CITES) had respectfully – too respectfully – monitored the global trade in ivory, taken money from Hong Kong ivory dealers, and sought to implement an increasingly discredited ivory trade quota system that allowed for the continued legal trade in a certain amount of ivory, seemingly afraid of ruffling too many political feathers. But change was coming and, by the summer of 1989, nobody, including CITES, could ignore the voices calling for an end to the status quo.

'Perhaps the most important lesson I learned is that there are no walls between humans and the elephants except those that we put up ourselves, and that until we allow not only elephants but all living creatures their place in the sun, we can never be whole ourselves.'

Laurence Anthony

I was there. Lausanne, Switzerland, September 1989. One hundred and one countries, the Parties to CITES at the time, gathered to review the evidence and to consider proposals from Tanzania, the US, Somalia and others that would lift Africa's elephants from Appendix II of the CITES convention (permitting regulated commercial trade) to Appendix I, bringing all legal commercial trade in ivory and other elephant products to an end.

It was my first such meeting. Important-looking people, walking and stalking the corridors of power. Government representatives standing in small groups, conversing in hushed tones, casting a suspicious eye over registered observers from non-governmental organisations, like me.

My journey to Lausanne had been an extra-ordinary one by any standards. My organisation, then called Zoo Check (now the Born Free Foundation), had been alerted to the scale of the crisis facing Africa's elephants in the spring of that year. A few months later, in June 1989, an informal group of conservationists and elephant experts, including Daphne Sheldrick, David Shepherd, Ian Redmond and others, called a press conference at the Royal Geographical Society under the banner of Elefriends. A petition – not electronic but ink and paper – was hastily launched, calling on both the UK government (in the twilight years of the Thatcher administration) and CITES as a whole to end the commercial trade in ivory. Our target was 600 000 signatures, one for each living elephant. The responses poured in.

We decided that the petition, the voice of the people, needed to be delivered in person to CITES delegates, but how to get it there? Half a tonne of paper, every sheet carefully bundled, every name scrupulously counted. I acquired a car. Well, in fact, Guy Salmon, the car rental company, lent me a car – a blue Ford Estate. With the back seats down and every box loaded, two days later the petition was delivered to Prince Sadruddin Aga Khan, the internationally recognised humanitarian and wildlife conservationist, who had agreed to present our 20 000-page petition. The sun shone on that auspicious day. In the shadow of an enormous inflatable elephant, the views of 600 000 people could not be ignored.

Nor could the evidence. Report after report (from Save the Elephants, the Environmental Investigation Agency and more) laid out the facts for all to see. They were vigorously resisted, as was to be expected, by the pro-trade lobby, the ivory industry, jobsworth bureaucrats and, more unexpectedly, some of the world's biggest and well-respected wildlife organisations, including WWF.

The day of the vote approached. The United Kingdom (UK) seemed onside, although they sought special dispensation for Hong Kong, at that time a UK-dependent territory. Meetings took place behind locked doors, on the other side of which we chanted for transparency and accountability. The Swiss delegate, as I recall, tried to get the EIA ejected from the meeting for whistling and applauding, until it was pointed out that the person who had broken the stuffy, self-important silence of the committee room was, in fact, a member of the French government delegation.

What would those elephants, clinging on to life itself on the plains and in the forests of Africa, what would they have made of the thunderous roar of approval that went up when delegates voted by more than the required two-thirds to approve a global commercial ivory trade ban? It seemed as if the battle had been won. None of us could have foreseen how short-lived that outbreak of common sense would be.

CITES turns a deaf ear

Roll forward just 5 years to the CITES meeting in Fort Lauderdale, Florida, 1994. Already the siren voices had begun. The price of ivory on the now illegal market had tanked to a few tens of dollars a kilo. Poaching rates were low and elephant populations had, in many cases, stabilised. In some places, they had even started to recover.

Southern Africa had, to a significant degree, been spared from the worst of the poaching epidemic, and now the 'use it or lose it' brigade struck up their mantra. Why should they be 'punished' for successful wildlife management and be denied the opportunity to sell ivory? The groundwork was clearly being laid for the 1997 CITES meeting in Harare, Zimbabwe.

I remember this meeting as if it was yesterday. A full-on pro-trade offensive started on day one. Red

carpets were rolled out. Heads of state and leaders of delegations wined and dined. The state-controlled media machine swung into action with 'expert articles' expounding the 'benefits' of a controlled ivory trade dominating the front pages day after day.

One evening I visited a rally at the offices of the Zimbabwe African National Union – Patriotic Front (ZANU PF), a huge, dimly lit hall, where the crowd was whipped into a frenzy by hoarse voices shouting through a distorted PA system. It felt intimidating. It felt downright dangerous. If the elephants were listening, their great grey ears would have shuddered in fear.

We tried to keep hope alive. At my invitation, Dr Richard Leakey spoke to a packed meeting hosted by the Species Survival Network (SSN), exhorting us to keep the faith and to believe that caring for elephants was the right thing to do. Meanwhile, at the back of that same room, a small but vocal contingent of southern Africans, while helping themselves to our freely offered food and drink, hissed in threatening tones to bewildered SSN members, 'You're evil' and the even more ominous 'You'll regret this'.

In the great Committee Room of the Conference Centre, a roar went up, loud enough to reach the elephants … maybe, maybe not. But if it did, then elephant tears would surely have been shed. Pulverised by a combination of hospitality, rhetoric and dogma, delegates had approved an experimental, one-off sale of ivory, which resulted in tens of thousands of kilos of ivory being sold to Japan. The 'if it pays, it stays' brigade had, despite our best endeavours, seized the day. And, over the next decade, the hard-won international ivory trade ban began to unravel.

The experimental one-off sale, approved in Harare in 1997 and executed (a seemingly appropriate word) in 1999, was followed by another one-off sale (an oxymoron?), agreed when the 21st century was but a couple of years old and delivered in 2008.

The logic supporting this second sale was bizarre and profoundly wrong. Almost every serious conservationist agreed that the meteoric economic growth experienced by China in the '90s, coupled with CITES's decision to permit China to bid alongside Japan for ivory, meant that this second sale was a recipe for devastation and that a surge in the demand for ivory was sure to ensue.

I recall meeting the then UK environment minister in the summer of 2008, just as she, as a member of the CITES Standing Committee, rubber-stamped the impending ivory sale. On a corner seat in a government office overlooking the River Thames, I set out, yet again, my profound fears for Africa's elephants. Her response was as clear as it was chilling. 'By selling this ivory, we hope to satisfy demand and thereby reduce poaching.' It was the kind of soothing (and fantastical) logic that I had heard from ivory traders and dealers time and time again, but I did not expect it from the British government. It was wrong then, and it remains wrong today.

A few months later, over 100 000 kilos of ivory, the last remains of as many as 10 000 elephants, went under the hammer, sold by four southern African countries to China and Japan for about $160 a kilo. The crash of the gavel must have reverberated like a rifle shot across the plains of Africa. It was, in fact, the starting gun in a race for bloody ivory – a race that has continued to this day.

The slaughter resumes

If left in peace, elephants breed at around 4 to 5% net a year. In other words, even taking into account natural mortality, the birth rate will outstrip deaths and numbers will grow. If left in peace. But there has been no peace for Africa's elephants for the last decade or more.

Africa's elephant numbers have not grown. They have been hammered. They fell by at least 144 000 between 2007 and 2014. Tanzania lost, on average, 1 000 elephants a month for 60 months between 2009 and 2014 – just 5 years. This shocking figure accounts for only savanna elephants. We have still to reckon with the devastation wreaked on the smaller, secretive forest elephant, but experts agree that tens of thousands more are gone.

I once thought the killing fields of the 1970s and 1980s were history. I once thought the international ivory trade ban, secured in 1989, would bring lasting peace. I once thought our leaders had learnt something. But I was wrong.

So who is to blame?

Not the majority of African countries with elephants: the 29 African Elephant Range State members of the African Elephant Coalition

have consistently voted as one against further trade. Not the rangers and wardens who put their lives on the line daily (and too often lose them, almost unnoticed) to protect their great grey elephantine heritage. That much is certain.

And, more recently, not the Obama era in the USA, the more enlightened administration in China, nor the leadership in Botswana, which, unlike its immediate neighbours, is committed to the protection of elephants and even to an end to trophy hunting

But what of those wedded to an ideology that seems to place economic values above all others? And those who have resolutely perched on the fence and allowed defeat to be seized from the jaws of victory? They must share responsibility for the current crisis: South Africa, Zimbabwe, Namibia, the European Union and others have much on their collective conscience. Assuming they have one.

Meanwhile, across Africa the slaughter goes on.

In search of compassion

So what have I learnt in 30 years?

To understand the profound stupidity of the human condition; for we are invariably more stupid than we think. To recognise greed and avarice for what they are, base qualities that win more times than they lose. To distrust politics and most politicians; theirs is too often a vision created to serve their own needs and not to reflect yours or mine. To seek those with compassion; this is the only universal value. To hope for the best, for without hope there is nothing. To believe it can be better. To speak truth to power and never stay silent.

The elephants know nothing of all this, despite their great grey continental ears. They can only hope that we humans can do better before it's too late.

05

Ritual elephants

Will only a herd of wooden elephants
and a memory remain?

Patricia Schonstein

**During the Vietnam War, a monk created a
religious sanctuary on an island in the Mekong
River which became known as Peace Island.**

He gathered a community around himself and
set in motion a 24-hour ritual of prayer, chanting,
repeating mantras, bell-ringing and the striking
of gongs.

On Peace Island, he created a symbolic war,
'arming' his followers with fruit and palm-leaf
grenades and cutting spent rocket shells into
flowerpots.

He made a model of the Mekong Delta. Through
this he would walk, showering it with blessing and
the energies of peace, manipulating the symbol in
order to influence the situation itself. There was war
all around Peace Island. Rockets passed over it. In
the face of destruction and chaos, here was order.
Against the thrum of military helicopters was the
sound of mantras and the calm of meditation.

John Steinbeck Jr, son of the Nobel Literature
laureate, wrote of the monk: 'Where there is nothing
but war and turmoil this man, through very curious
if not laughable means, has an island where there is
pure peace. Simple peace.'

I reflected on the monk when I read Iain
Douglas-Hamilton's *Among the Elephants*. He
describes, in 1975, vast numbers of elephants in
Tanzania slaughtered for their ivory. Back then I
began buying small ebony elephants from junk
shops. Most of them were carved in Kenya from
the wood of trees which surely carried a memory
of these majestic beasts. I built up a herd of 25:
matriarchs, lone bulls and youngsters.

Following the same modus as the monk, I gave
sanctuary to this representative wooden herd. In so
doing, I created a narrative in which elephants were
removed from human gunsights. I manipulated the
'small picture' in an attempt to protect the reality
and future dynasty of elephants.

Today, 44 years after forming the carved herd,
the plight of Africa's elephants is worse than ever.
Perhaps the Vietnam War did indeed end because
of the monk's spiritual diligence. My own narrative
herd, however, has not had the alchemical clout
to protect real elephants from being, effectively,
murdered at an unprecedented rate.

If elephants become extinct in our time, on our
watch – as trending statistics warn they might – we'll
have to look upon our species in a new light. Once
the elephants are gone, the other at-risk species will
rapidly follow: rhino, leopard, lion and the smaller,
equally important ones like pangolin and jackal.
When that happens, we'll no longer deserve title to
the *sapiens* part of our name, for we'll have forfeited
wisdom. We'll finally have to recognise ourselves as
warriors against creation, as *Homo guerre*.

The annihilation of elephants will present us with a
new kind of loneliness. Without these majestic beings,
there will be an unimaginable void. Do we dare risk
this, just for the procurement of ivory ornaments
and trinkets, for the illusory wealth these give? Just
to satisfy a narcissistic belief that we've been granted
dominion over the Earth, to do upon it as we wish?

We'd better wise up before it's too late and forge a
new attitude, one of homage and respect for creation.
We are just travellers passing through. We have
no right to murder our fellow species; the killing
of elephants represents impending, irrevocable
specicide. It would be sad beyond measure if all that
was left was herds of little wooden elephants.

'Ivory Coast's [Côte d'Ivoire's] national football team is named the Elephants after these magnificent creatures that are so full of power and grace, yet in my country alone there may be as few as 800 individuals left.'

Yaya Touré

Silent ghosts at a rare waterhole in Chobe, Botswana.

06

A tale of two elephants

Working out elephant management
in a private reserve

Audrey Delsink

One Sunday evening in 2008, an elephant cow named Kwatile passed away of natural causes at Makalali Game Reserve. I sat with her spent body as she breathed her last ragged breath. Aged around 64 (based on molar tooth wear and placental scars), this old girl certainly was a force to be reckoned with; exemplary and instrumental in the wellbeing of her clan.

Kwatile and another elephant, Riff Raff, shaped and touched my life. I came of age with them, and together we loved and lost calves and children that did not survive this Earth. I sought solace in the presence of my grey friends during one of the darkest times of my life. For almost 20 years, I lived among them.

I know that my presence could never enrich their lives the way theirs did mine, but I take comfort in the fact that these individual elephants were part of a grand design that changed the way in which their species is managed. Because of Kwatile and Riff Raff, many more elephants will be spared the bullet.

Kwatile

In 1998, as one of a few female guides in the safari industry and fresh from corporate Johannesburg, I first encountered a breeding herd of elephants in the Manyeleti Game Reserve, and it left me with a lasting impression and a fear of elephants.

The reserve – then a forgotten Eden – was an open system with neighbouring Kruger National Park. Game drive routes were planned to a degree, but you never knew what was waiting behind the next guarri bush. There were lions, buffalo, rhinos and elephant bulls in abundance, though breeding herds were scarce. When such herds did venture across, they were known to be wary, unpredictable and, in 1998, stressed by drought.

During a game drive with a French family, we were charged and chased by two elephant cows – unprovoked and out of nowhere – for several kilometres. My tracker Jules shouting, 'Drive, *mafazi*, drive!', the wide-eyed, horrified gapes of the family and two times four tonnes of fury bearing down on us are details I won't forget.

Later, I moved to the then-12 500 hectare Greater Makalali Private Game Reserve, south-west of Kruger's Orpen Gate, formerly a feldspar and mica mining area and cattle-farming property. It was a completely different game-viewing experience. Between 1994 and 1996, when Makalali Lodge opened, elephant, lion, rhino, hippo and other game were reintroduced. The theme was 'soul safari', and the bush experience was fine dining, eclectic architecture and a safari focused on the senses. Game drives and walks included looking for tiny wonders, and walking where hippo and pachyderms trod.

One of my first elephant encounters at Makalali was with a group of Italian guests under the guidance of Alfred Mathebula. Rifle in hand, he positioned us on a steep road overlooking a drainage line where the elephants were known to cross. My heart thumped loudly as we sat, safe and still, overlooking a small breeding herd silently moving below. As the matriarch, an old one-tusked beauty, appeared, Alfred turned and whispered, 'Kwatile is the only elephant here that you really need to be worried about.' Kwatile (Shangaan for 'angry one') stopped in her path, front leg swinging in a display of displacement behaviour, trunk up, and gazed up at us with wizened old eyes. Little did I know how this aged lady and the clan she ruled would shape my life.

In 1991, private ownership of wildlife and private game ranching in South Africa was legalised[1] – unlike in East Africa, where all the game is owned or controlled by the state in large, unfenced, protected areas. In South Africa, game – including elephants – could occur within enclosed (fenced-in) areas of their former range in private protected areas. Between 1979 and 2001, more than 800 elephants were moved from Kruger, with most translocations occurring between 1990 and 2001.[2] By the early 2000s, many reserves had no record of the numbers and composition of elephants on their properties. Some acquired orphan groups: calves spared from culling in Kruger, which had been grouped together and relocated as 'founder' herds to new reserves.[3]

Makalali was one of the first reserves to acquire relocated, intact family groups through a process pioneered by the late Clem Coetzee of Zimbabwe. This enabled elephants of all ages and sizes (barring very large, mature males) to be relocated, and it began a new era of elephant translocation in South Africa. In 1994 and 1996, four herds were relocated from Kruger to Makalali.

In 1999, in collaboration with Professor Rob Slotow of the University of KwaZulu-Natal's Amarula Elephant Research Programme, Makalali established a research department and practised the concept of adaptive management through learning. Initially, the focus was on monitoring apex predators to understand predator-induced mortalities and home ranges. Following in the footsteps of veteran elephant researchers Cynthia Moss, Joyce Poole and Iain Douglas-Hamilton CBE, I soon shifted my focus to elephants and spent hours documenting individual identifying characteristics.

Slowly, my fear of elephants grew into respect and wonder. In anthropomorphic fashion, I named them according to features or traits I could easily relate to and remember. Kwatile, I soon learnt, was very appropriately named. As a rather grumpy old girl and one who would charge first and ask questions later, her presence immediately made the other elephants a little jittery. I named the other three matriarchs Yvonne (after a generous sponsor), Holey Ear (for a perfectly round hole at the base of her right ear) and Queeny (for her aristocratic bearing). Soon, the rest of the elephants on the reserve had names.

My first elephant identification book comprised hundreds of old-school printed photos that I lovingly pieced together to make up individual templates and photo identikits. This was necessary to gain an understanding of the population demography. Without this basic understanding of herd composition and numbers, we couldn't accurately determine carrying capacity and the faunal biomass of these long-lived megaherbivores.

African elephants are referred to as 'ecosystem engineers', able to modify their landscape, often drastically. They're each capable of consuming 75 to 150 kilograms of browse a day, and can live up to 70 years. The long-term spatial and temporal effects of such feeding need to be monitored and adaptively managed.

With detailed Makalali elephant demographics in hand, it was clear the 12 500-hectare area, with finite resources, was too small to sustain the clan in the long term. Some degree of elephant management was necessary. In South Africa, options were limited to translocation or culling. We opposed culling for fear of its unintended consequences on the remaining elephants, but few places were available to receive relocated elephants. We needed a new alternative.

In late 1999, Makalali's warden, Ross Kettles, and I attended an elephant conference where the head vet of Kruger Park, Dr Douw Grobler, spoke of a pioneering and somewhat controversial new method of elephant population control called immuno-contraception. Field trials funded by the Humane Society of the United States and run in collaboration with Dr Jay Kirkpatrick of the Science and Conservation Centre in Montana and Dr Henk Bertschinger of the University of Pretoria's Onderstepoort had proved successful.[4] Grobler described it as a non-steroidal, non-hormonal, non-invasive method of reversible contraception, administered through a vaccination.[5] It had proven efficacy in 80 species, including elephants.[6] Immuno-contraception had been developed at the University of California, Davis, in 1972 to control the numbers of wild horses and urban white-tailed deer in the USA, and had been used successfully for nearly three decades.[7]

Having demonstrated the vaccine's efficacy in Kruger, the next objective was to determine if it could be used as a remotely deliverable population control mechanism in wild, free-ranging elephants.

It was at this point that my elephant family tree and ID kits became essential. The detailed elephant demography and documented nature and state of each individual made the population at Makalali an ideal test case, and the park's immuno-contraception programme began.[8]

In May 2000, 18 female elephants received the initial and booster shots of the PZP vaccine, applied on the ground by project implementation specialist JJ van Altena, with Makalali head ranger Mark Montgomery as support and me as behavioural monitor. It proved a resounding success. After 17 years of dedication by the sponsors, the research team (Kirkpatrick, Van Altena, Bertschinger and I) and the reserve, the Makalali population demonstrated the vaccine

to be a safe, reversible, remotely deliverable, humane and alternative means of population control which, by 2025, will have prevented this population from trebling.

The vaccine has no social or behavioural consequences with medium-term use (12 years).[9] Long-term monitoring continues. With almost 100% efficacy, the vaccine can be applied on the ground for small populations, or from the air for a mass-darting approach.[10] With increasing numbers of treated elephants, the average cost decreases and is around R1 500 to R2 000 an elephant, fully inclusive of the chopper, vaccine and darter.

The numbers now tell the story. By 2017, with Makalali as the flagship, almost 750 female elephants across 25 reserves and biomes in South Africa were under treatment.[11] Reduction in numbers has been exponential. In the lifetime of a female elephant, assuming she lives to 60 to 65 years, she may produce 8 to 10 calves, and her female calves may produce the same number of offspring. After the first year's treatment, a non-pregnant female is immediately contracepted. Pregnant females carry to term with no effect on the foetus and are contracepted thereafter. Used correctly, immuno-contraception mimics natural episodic events such as droughts, which lengthen the inter-calving interval and result in reduced growth rates while avoiding a zero percent growth rate for indefinite periods, which does not meet the social and behavioural needs of the individual or herd.

© Waldo Swiegers/AP Images

Sedating an animal the size of an elephant is a precise chemical process. Here a wildlife veterinarian prepares drugs to keep Riff Raff asleep during transport to a new location.

In 2008, the *National Norms and Standards for the Management of Elephants in South Africa* recommended immuno-contraception as the choice management option, with culling as the absolute last resort.[12] This, and the reference to elephants as 'sentient beings', certainly marked a paradigm shift in elephant management in South Africa. With 21 years of immuno-contraceptive research (in Kruger and Makalali combined) and countless peer-reviewed scientific publications and book chapters, our research team certainly has done its due diligence. All other management options must be critiqued with the same scrutiny, and lethal control of elephants should no longer be automatically adopted.

Riff Raff

In 1998, I named a beautiful 18- to-20-year-old bull elephant at Makalali, Riff Raff because of his ragged ears and status as one of the top three bulls in the reserve's boys' club. Even when in musth, he remains calm and relaxed, unlike many of his counterparts when they are afflicted by surging testosterone levels, coupled with streaming temporals and 'green-penis syndrome' (there really is such a term).

But Riff Raff's occasional head shakes and mock charges can be intimidating. He's well named: 'riffraff' are troublesome people (or elephants in this case) who are looked down upon. Riff Raff and so many other bulls in his age group are labelled 'problem' or 'damage-causing' elephants.[13]

Elephant biology drives bulls to leave their natal herd and range in search of new areas and females, a strategy that prevents them from mating with their kin. Their need to wander can cause problems for humans. In fenced-in areas (for the purposes of both inclusion and exclusion) with multiple land-use types, ownership and protected-area frameworks, wandering bulls encounter fences, and human-elephant conflict (HEC) can occur. Wanderlust is increased when conditions are not ideal, such as in a drought or with increased population pressure. Irrigated gardens, crops and pools are artificial attractants, and the benefits and rewards they provide are worth the risk of breaking a fence. When fences are poorly maintained or fail elephant-proof specifications,

bulls quickly learn to breach them. This invokes the labels of 'problem' or 'damage-causing' animals. The standard solution is destruction through a legal damage-causing animal (DCA) permit.[14]

In the case of Riff Raff, changes within Makalali led to the erection of a two-strand exclusion fence that now separates him from an area that he had formerly used for 15 years. It makes no sense to Riff Raff and the Makalali elephants. This former range has thick riverine vegetation and a large dam within a kilometre of the river, so it is hardly surprising that this man-made boundary is not respected. There has to be a better way to manage elephants that speaks to their biological requirements and responses.[15]

For this reason, Humane Society International (HSI)/Africa advocates the use of alternative, non-lethal methods to mitigate HEC: to curb fence-breaking using early-warning systems facilitated by tracking collars and technology with notification systems. In 2017, HSI/Africa and our local partner, Global Supplies, collared and facilitated the first tusk-bracing intervention on a wild, free-roaming elephant bull: Riff Raff.

Previously conducted on two tame bulls, this novel method involves embedding wire into the elephant's tusk with contact under the lip so that when the elephant fiddles with or attempts to break electric wires, the tusk wire acts as a conductor and the elephant gets a shock. The embedding process causes the elephant no pain and is conducted under sedation. This tool is a conditioning technique to reaffirm the negative consequence of trying to breach fences. As Riff Raff's left tusk had been broken years before, only the right tusk was braced.

The method proved extremely effective and halted all but two of Riff Raff's fence breaches, those having occurred when there was either no or low current on the fence. But boys will be boys, and several weeks after the tusk-bracing intervention, Riff Raff and Charles (another of Makalali's top-ranking bulls) had a battle. This resulted in both bulls breaking their tusks. Neither bull was injured, but each bull's tusk was broken clean through in two places. That ended the tusk-bracing intervention, as Riff Raff's intact tusk broke about 25 centimetres from the lip, causing the embedded wires to fall out. Despite the turn of events, we believe the tusk-bracing technique could save elephants that would normally be destroyed. This is especially crucial where the bulls are dominant, iconic or important for genetic variability.

Another method of control is using a satellite collar to monitor an elephant's location within a 'virtual fence' or safe area in the reserve perimeter. Should the bull breach this, a notification alert is sent to a dedicated number and email. This serves as an early-warning system to proactively prevent fence breaches through a suite of mitigating actions.

As for Riff Raff, pending the red tape of the permitting system and *Norms and Standards*, he is due to be relocated to another reserve, where he will be the dominant bull in a larger land of new, unrelated females: essentially, elephant bliss.[16] Sadly, it's generally far easier to be granted an emergency DCA permit to destroy an animal than to relocate it to a willing reserve. While we await a rubber stamp of approval or rejection to determine Riff Raff's fate, the situation has escalated to a welfare concern. Riff Raff continues to breach a weak zone in the fence and makes nightly forays back into his preferred area. The result is repeated shepherding or chasing him back into the safe area by helicopter.[17]

My heart sinks each time I receive an SMS notification from his collar signalling that he has left the virtual-fence safe area. If only I could send a message back and explain how time is running out for him. Despite some humans having the means to protect these animals and choosing to live alongside them, they do so only as long as elephants behave within the constraints of their (human) ideals – a luxury that serves neither species.

Audrey Delsink of Humane Society oversees the transport of fence-breaker Riff Raff to a new home.

© Waldo Swiegers/AP Images

Elephant transport takes some heavy lifting. This bull is sedated Riff Raff, now approximately 40 years old, being relocated from Greater Makalali Private Game Reserve in South Africa to a larger game reserve.

'In the area of species protection, we should concern ourselves with what is right as opposed to what might be easier, or popular in the short term.'

———————

Dr Richard Leakey

@ Marion Garaï

Jane in the 1990s, with her herd in Venetia Limpopo Nature Reserve, South Africa.

'I'm fascinated by elephants because of the way they treat each other. When you get a look at elephants, you get the impression of them as conscious beings, of figuring things out in their environment, including their relationships with you as an observer.'

Iain Douglas-Hamilton CBE

07

The amazing Jane

A rite of passage and a love affair

Dr Marion E Garaï

This is a story about an amazing elephant named Jane and a gaggle of freaked-out youngsters that she calmed and taught how to be a herd. In the 1990s, young elephants spared from culling operations in the Kruger National Park were being relocated to private reserves. At the time, nothing was known about the effects of their trauma and relocation, and it was this that I made the subject of my PhD dissertation.

A group was relocated to the newly established Venetia Limpopo Nature Reserve near Alldays, close to the Botswana border: hot, dry, bush-covered country. I joined them, camping near their boma (livestock enclosure) along the dry bed of the Kolope River. To my great dismay, on their release, one with a radio collar (which had only a limited range) took off with a companion, and the youngsters effectively disappeared from view in the 30 000-hectare reserve.

In due course, more elephants arrived at the bomas. The first group consisted of four very nervous females aged about five to six: the 'hysterical girls', as I came to call them. A few days later, a further eight arrived, three females and five males, then 10 more young males between 18 months and 2 years old who would otherwise have been shot.

The youngest elephant did not feed. He rumbled the whole day, which is how he got the name Rumble. He still desperately needed his mother.

To satisfy his suckling need, he started sucking the ear of another elephant I named Notch, who was extremely patient. But sometimes it just got too much for him and he pulled his ear away, eliciting screams of frustration from Rumble, which I could hear all the way from my tent.

Rumble wasn't feeding and I was worried he'd die. So, against the rules, I started to hand-feed him through the fence. He soon got the idea to pick up cubes by himself and suckled my hand. He was not yet able to chew lucerne and I was worried he was not getting sufficient food. At his age he should have been fed more often and with milk.

The elephants soon got to know my special call and calmed down on hearing my voice. Only the four 'hysterical girls' remained nervous until they were released. The youngest elephants habituated the quickest. Like most young animals, they had not yet learnt fear and were incredibly trusting. I was grateful for this long boma acclimatisation period, which I also needed to gain confidence.

One day I returned from town to find an adult elephant in a pen. Jane, about 18 years old, had arrived from Zimbabwe, where she'd grown up with humans on a farm, having been an orphan herself from the culling operations at Hwange National Park. The youngsters reportedly had gone berserk with excitement when they saw her!

The following day, five young males were let into the pen with her. A bunch of bullying rascals until then, they were clearly awestruck and were suddenly just five small, helpless elephants. They stood around Jane and five little trunks gingerly came up to touch her, while Jane stood stoically, taking in this new situation of suddenly having acquired five babies at a shot. It was wonderful to watch.

The males did not leave her side and she had to push them away so she could feed. The younger ones tried to suckle, but this she did not allow. The smaller males in the adjacent pen frantically tried to reach and touch Jane through the fence. After a few days, when she'd recovered her composure, she would press against the fence, so they could touch her and suckle; Rumble, in particular, tried to reach her with his little trunk. Jane's arrival spread an air of calm around the bomas.

Jane and the five little males were soon allowed into a larger paddock, to be joined by eight mixed-sex elephants. Jane, excited to be in a larger enclosure, ran around with all 13 little elephants running after her. A big, grey Pied Piper. She had her first – and probably last – nasty experience with the electric fence, and she never touched one again. I never saw any of the small elephants touch the fence, so its danger was quite likely communicated to them by Jane.

After 10 weeks of confinement, Jane was released together with the 13 younger elephants. At first they stayed near the boma, but eventually she realised she was free and walked off, followed by her 13 charges and me, tagging along at a respectful distance. Would I have the courage to follow them? Would I succeed in keeping up with them in the bush? Would they accept me walking behind them, or would they be aggressive? For 5 hours I followed, talking to them and calling my special call to calm both them and myself.

Jane was still nervous, of course, not knowing the area and having the responsibility for 13 young elephants. The next day I spent 3 hours searching for them and eventually found the group on a hill near a dam. Jane suddenly came towards me very quickly, and I was uncertain whether she was happy to see me or telling me to go. But she stopped and then backed off, having satisfied her curiosity.

After that, I lost them again for many days. When at last I found them, Jane again approached, this time coming towards my vehicle – their chosen terrain having made it possible for me to follow by road. I like to imagine that she approached because she was happy to see me. Jane had discovered a dense riverine area, where she decided to stay for a while. The following day, I gathered all my courage to look for them, once more on foot.

Jane suddenly appeared out of the bush, walked up to my empty vehicle and, to my horror, started pushing it. All I could do was take a photograph. Luckily, she soon tired of this game and disappeared into the bush with all the youngsters following, and no doubt suitably impressed. When I caught up with the group, they were drinking at a beautiful dam, all in a row, like Snow White and the dwarfs. I walked to the opposite side and sat watching in awe and taking pictures. Jane was calm, and in due course they started to feed. They were accepting me. This was a dream come true.

For the next 2 weeks, I found Jane and her 13 charges daily in the small riverine valley and near the dam. On the third day, I summoned enough courage to walk into the bush, and a big surprise awaited me: one of the two smallest males, whom I had named Squeak, was suckling, and Jane was standing for him. This was beyond any expectations.

I had named the other, very similar little male Bubble, the only distinction between them being a small bump on Bubble's rump. I hadn't been able to tell which was the younger, but obviously Jane could. She adopted Squeak and treated him as her offspring. Poor Bubble was ignored and he joined the other small males.

In the boma, Bubble and Squeak had spent most of the time near each other, but now Squeak had only one goal in life: to stay near Jane. He was allowed to feed with Jane, slept near her or under her belly, suckled regularly, and received Jane's full attention and care. She touched him gently on the face and lips while he suckled, and constantly looked out for him. Squeak directed all his attention and friendly behaviour to Jane, who wouldn't allow Bubble too close.

One day I found the elephants in the far north of the reserve when they crossed the road in front of my vehicle. Behind Jane were 13, 14 … 21 juveniles! The 'hysterical girls' and the group from an earlier intake had joined their new matriarch.

As time went on, the elephants and I became more confident of each other and I could eventually walk with them, even between them. Jane usually stood guard when the youngsters slept, but I was certain she had accepted me the day she lay down to rest with them.

The next big step towards mutual confidence came the day the elephants panicked, possibly because there were helicopters in the area, which reminded them of their capture. Scared I would unintentionally be squashed in the general pandemonium, I called out to them something to the effect of 'It's me! It's all right! Don't worry!' Incredibly, this had the desired effect and they immediately all calmed down.

Then one day Jane brought the group up to my vehicle on the road, with Squeak and Bloukop, the oldest of the juvenile females, alongside her. She suddenly walloped Bloukop with her trunk. I couldn't understand why, and got out of the vehicle and talked to them calmly. What was the meaning of her behaviour? It happened on three separate occasions, when Jane would approach with the young elephants, and wallop whoever was next to her, while at the same time reassuring Squeak. Then they would go away and feed. It was as if she were teaching them a lesson: 'Behave towards *this* human'. From then on I could walk with them without cause for concern on either side.

Later that year Venetia received another two adult elephants saved from Gonarezhou National Park in Zimbabwe, where devastating poaching was taking place. I was told a male and a female were to join our group. When I arrived back at Venetia from another reserve, the manager told me the bull had aggressively charged him in the boma, nearly breaking the fence in the process, and that I should be careful.

After the newcomers' release, I was understandably nervous. I found Jane and the juveniles and approached them slowly, trying to hide behind the mopane bushes. I soon identified the two new elephants – in fact they were two large females, one of them having been mistaken for a bull, thanks to her exceptionally large head and tusks. I named her Zora after a character out of a children's book about a wild young Hungarian gypsy. The other female had short tusks, and I named her Gona for Gonarezhou.

Gona stayed near Jane, but Zora gave me a bit of a rough time initially. Poor Zora must have had some bad experiences with poachers. She was wary and nervous and had an uncanny way of hiding behind a bush when the others had moved on, and she ambushed me time and again. On one particular day, while the group rested in a small mopane forest, I sat under a tree and waited. After about two hours, I thought it strange that Jane had not moved on. I switched on the tracking radio: no signal. They were gone: 24 elephants had moved off and I hadn't heard a thing.

I soon picked up the signal quite some distance away and began running to catch up with them. Suddenly a huge grey mass rose in front of my eyes, a flash of white, dazzling in the sun. Zora had ambushed me and was coming towards me at full charge. I dived behind a metre-high mopane bush and heard someone shouting: 'It's all right! It's me, don't worry!' Surprised, I realised the sounds were coming out of my own mouth. Zora recognised my voice and stopped. I realised that, under Jane's influence, she had learnt to accept me.

This remarkable relationship with Jane highlights the intelligence of elephants and their amazing ability to communicate ideas to one another; Jane's teaching the others that I posed no threat was evidence of this. Our knowledge of elephant intelligence, cognition and psychology has greatly increased in the last decades. The more we learn about these animals, the more we realise how similar to humans they are. They use tools and have a complex society that is so well adapted to their needs that they have survived for millions of years longer than us – and without destroying their environment.

They have a highly complex communication system, which we are only just beginning to understand. They have feelings and a strong sense of family; they show emotions and empathy and have a knowledge of self, which so far is seen only in the great apes (including humans) and dolphins. Elephants also have a sense of humour and can think complex thoughts and plan ahead. And they mourn their dead. Jane showed all of these abilities and was a truly amazing teacher for the juveniles, and for me. She allowed me insight into her world and understood me without words. The fact that we are decimating these wonderful animals is nothing short of genocide.

'Elephants are simply one more natural resource that is being caught up in human greed on the one hand, and human need on the other. We somehow need people to become reacquainted with nature, or they will have no clue as to the interrelatedness of cause and effect.'

Dr Stephen Blake

08

Last of the big tuskers

For elephant society, the loss of large tusks pales in comparison to the loss of the accumulated wisdom that disappears every time a tusker is felled before his natural time.

James Currie

I hear the ear-splitting crack of a large tree being felled, followed by a resounding thud that shakes the earth. I hold my breath and scan the thick bush, hoping for a glimpse of the animal capable of such destruction, the largest elephant on Earth. Just beyond a small window in the bi-coloured croton understorey, less than 40 feet [12 metres] away, a patch of wrinkled grey skin comes into view, accompanied by an eerie quietude. Even the forest birds have been stunned into stillness, either by the detonation of the freshly broken tree trunk or in reverence to the presence of the King of Kings. Or, possibly, both.

I stiffen as an unexpected blast of wind gusts past my shoulders, picking up the tell-tale molecules of my human scent. The croton understorey in front of me shimmers green and white. And then my window shuts abruptly, concealing any sign of the big bull. I know the wind has betrayed me. 'Now what?' I think to myself. Is he going to charge right at me or walk away? The seconds tick by and morph into minutes. How loud silence can sound in times like these.

Without any warning, I detect movement out of the corner of my eye. A colossal shape is moving slowly towards me, the soft sand underfoot muting his approach. I turn to focus on the elephant looming at twice my height and immediately I'm rendered immobile by the awesomeness of his tusks. Now, just 30 feet away, the magnificent bull stops, his two ivory pillars touching the ground. For what seems like an eternity he stares me down with his trunk outstretched as if acknowledging my insignificant presence.

With considerable effort, the giant elephant lifts his head and shakes it in slow motion. He turns and lumbers to the side, the massive ivory leading the way as he crosses a clearing directly in front of me. The silent forest envelops him once again.

I stand in disbelief, having just shared a moment with one of Africa's legendary big tuskers. At first I am elated as it sinks in that seeing one of these remaining giants is a truly rare privilege. And then a pang of disappointment and guilt grows in my chest as I realise that this is nothing to celebrate.

Mudanda, an exceptional matriarch in Tsavo National Park, Kenya, who died of natural causes aged about 60.

The legendary Isilo from South Africa's Tembe Elephant Park died in late 2013/early 2014, 'officially' of natural causes. However, his body was only discovered 2 months after his death, and was without tusks.

© Tim Driman

If Charles Darwin were alive today, he might feel vindicated by the ongoing elephant crisis. His theory of evolution threw a spanner into the works of the contemporary creationist thinking of his time, which held that God created humans and all living things. Still today, many people find the premise that apes and humans evolved from a common ancestor a very tough concept to grasp. The theories that Darwin proposed – that species evolved through minute changes over many millions of years – were difficult to comprehend. Many detractors of Darwin disputed his theories simply because there wasn't enough solid evidence of real-time evolution to back up his claims. Now enter the elephant crisis …

I was one of the last people to see the largest elephant in the world alive while on a filming trip in December of 2013. His name was Isilo, which means 'King of Kings' in Zulu. He lived in a beautiful mosaic of rare sand forest and marshland in Tembe Elephant Park, located in the far north-eastern corner of South Africa, and home to the last free-roaming herd of wild elephants in KwaZulu-Natal. Not only was he an exceedingly large elephant in stature – Isilo weighed in at a staggering 7 tonnes in his prime and stood 4 metres at the shoulder – but he also owned the largest set of ivory of any elephant alive at the time. Our planet is all the poorer for his untimely loss.

But we are even poorer due to the *circumstances* of Isilo's death, for we will never know the exact record-setting length of his ivory. Nkosi Tembe, traditional leader of the Tembe people and custodian of the Tembe elephants, had earmarked the tusks to stand at the entrance to Durban International Airport as a tribute to the gentle giant. But, shortly after I had filmed him, Isilo disappeared. For months, park rangers looked for him, fearing the worst. His decomposed carcass was finally discovered in March 2014. Although Isilo is believed to have died of natural causes, park officials were alarmed to discover that the 3-metre tusks had been brazenly pulled from the face of this iconic elephant, and, to this day, they have not been found.

Just months after his death and more than 1 600 kilometres away, an equally famous elephant entered the sanctuary of Mount Kenya forest for the last time. Mountain Bull, as he was fondly known, was a wanderer. He roamed the vast area between Laikipia and Mount Kenya, traversing community lands and protected areas alike. Although not nearly as well-endowed as Isilo, and possibly not even a certifiable 'big tusker', Mountain Bull was nevertheless an impressive animal and was the catalyst for an ambitious elephant conservation movement.

He was collared so that conservationists could track his whereabouts. In an effort to further protect him and discourage poachers, his ivory tusks were sawn off. All for nought, though, as several months later, Mountain Bull's carcass was found deep in the forest at the foot of Africa's second-tallest mountain. The corpse was riddled with spear wounds. What was left of his sawn-off tusks was evidently still impressive enough to warrant his brutal death. Mountain Bull's stunted ivory has never been found.

It is sometimes said that bad things happen in threes. And this was certainly the case for iconic bull elephants in 2014. The next victim was a phenomenal Kenyan tusker, without doubt the most famous of the three elephants. Satao was the largest tusker in Kenya and his ivory was probably only marginally smaller than that of Isilo. He lived in Tsavo East National Park and was a truly magnificent animal, with gently curving, almost symmetrical ivory that scraped the ground when he walked. Satao was killed by a poisoned arrow shot deep into his left flank in late May 2014. Ironically, he was killed inside the boundaries of the national park, a place that was supposed to have offered him refuge and safety from poachers.

Three of the largest elephants in the world gone in three months. Isilo and Satao belonged to an elite club of ever-dwindling elephants called big tuskers, animals with tusks heavier than 45 kilograms a side. Just 100 years ago, big tuskers were fairly common throughout Africa. There were literally thousands of them, but then there were also several million elephants. Today, the latest elephant census estimates that there are only around 400 000 African elephants left on Earth.

But perhaps the most shocking issue of the elephant crisis is that, among the remaining populations of African elephants, a strange and visible phenomenon is rapidly taking place. Elephants are losing their 'elephantness'. The gene responsible for large ivory is disappearing, rapidly being replaced by the small-tusked gene and, in some areas, even more alarmingly, by the tuskless gene.

By hunting big tuskers, our species is fast-tracking elephant tusks out of existence. Today, there are few places in Africa where 'hundred pounders' still roam. Elephants, as we know them, are dying out rapidly. As I write this, only an estimated 25 big tuskers remain on our planet. At the same time, entire populations of African elephants are becoming tuskless. It's human-induced and unnatural selection.

Tusklessness in a healthy elephant population lies in the region of 3 to 5%. But we are now seeing the percentage rise to 60% or higher. In Addo Elephant Park in South Africa, for example, over 90% of all female elephants are tuskless. They are all descendants of the 11 small-tusked or entirely tuskless elephants that formed the foundation population, responsible for the 600-plus elephants that exist there today. The initial 11 were the last group spared by hunters in the Eastern Cape in the 1920s.

This is happening in our lifetime, right under the noses of conservationists in some of Africa's best-known parks. According to research published in a 2008 paper in the *African Journal of Ecology*, the percentage of tuskless female elephants in Zambia's South Luangwa National Park and the nearby Lupande Game Management Area increased from 10.5% in 1969 to 38.2% in 1989. This wasn't the result of bad dental hygiene on the part of the elephants; in the 1970s and 1980s, there was a large spike in ivory hunting in Zambia.

In Queen Elizabeth National Park in Uganda, an elephant conservation plan from the early 1990s reported a higher-than-normal percentage of tuskless elephants and deduced that ivory poaching was the main culprit. A 1989 survey of the area revealed that tusklessness in the elephant population could be nearing 25%.

And, lastly, in Gorongoza National Park in Mozambique – where almost 90% of the original elephant population was massacred during the long civil war – almost 60% of all adult female elephants are entirely tuskless and 30% of all the young elephants now bear the same trait as their mothers. The gene for large tusks has all but died out, and the gene for tusklessness appears to be taking over.

Elephants use their tusks for many essential life functions. They are used as tools to lift bark from trees, to dig for roots, lift babies trapped in mud and dig for water. Their functions range from the essential to the seemingly mundane; from serious weapons of defence to simply a resting place for their trunk. And bull elephants also use their ivory to compete for females. While tuskless elephants can survive, they are essentially handicapped. They have lost much of their 'elephantness'.

Many hunters will argue that killing a big tusker does not affect the elephant population in the slightest, that these bulls are well beyond their prime and that they have had many fruitful years to pass on their genes. But science tells us otherwise: elephant bulls only reach their sexual prime at around 35 to 40 years. This corresponds directly with the time when these bulls emerge as 'hundred pounders' or large tuskers. The ivory grows exponentially from this age until the bull reaches the end of his 60-year lifespan. This rapid ivory growth stage is exactly when hunters target the trophy bulls, resulting in the disturbing reality that few are able to pass their genes on to future elephant generations.

For elephant society, the loss of large tusks pales in comparison to the loss of the accumulated wisdom that disappears every time a tusker is felled before his natural death. Elephants are intelligent and complex creatures, and we're only just beginning to figure out the extent of their societal fabric. A fair amount of research has been conducted on the bonds between cows and their calves, the importance of herd matriarchs and the intricacies of close-knit family groups.

But bulls are just as important to elephant society as cows. This is illustrated by the aberrant behaviour of young bulls that are translocated to wildlife areas in Africa without older bulls to act as role models. These teenagers have been known to try to mate with, and even kill, rhinos. They run rampant, tear down fences and attack vehicles and people.

For many years, conservationists puzzled over this, until it was suggested that older bulls be introduced to see if they calmed the youngsters. It worked, and now it's common practice to relocate at least one or two older, more mature bulls to accompany the younger males. Furthermore, older bulls are often accompanied in their final years by several younger bulls called askaris. As the older bull's eyesight, teeth and senses deteriorate with age, these young bulls offer protection and assistance in finding food. It's believed that, in return for protection and guidance, the old bull passes down his knowledge and the finer details of what it means to be a bull in elephant society.

In 2015, this magnificent elephant bull was shot legally in southern Zimbabwe by a German hunter. He was probably the largest bull hunted anywhere in Africa in the last 40 years. Yet many argue that he was still young enough to pass on his genes to subsequent generations.

Many from the hunting fraternity and beyond argued that this hunt was immoral and should never have taken place and that elephants with tusks over 100 pounds should be treated as national treasures, never to be hunted.

Murembo (Beautiful One) was one of a number of Tsavo National Park's great elephants whose tusks each weighed in at 140 pounds (63.5 kilograms).

Today, as I review my films of Isilo, taken before his death in late 2013/early 2014, I'm still in awe of this gentle giant who was revered by so many people, and I'm deeply saddened by his loss. I still marvel at what I see: his massive stature and the incredible pillars of ivory that made him famous. But I marvel even more at what I don't see: almost 60 years of knowledge that would have been passed on to one of his askaris, to live on and grow in this important population.

I realise that Isilo was one of the lucky few to have lived to a ripe old age. We may never recover his ivory, but the elephants of Tembe may still live out his knowledge and wisdom. So many other big tuskers were not so fortunate.

The loss of large elephants like Isilo, Mountain Bull and Satao is depressing. There are pockets of hope, however. The best genes for large ivory currently reside in Tsavo National Park in Kenya and in South Africa's Tembe Elephant Park. Nowhere else in Africa is there such a propensity for up-and-coming big tuskers. These elephant ambassadors need to be protected and nurtured for our children and our children's children.

Some might argue that the human-induced evolution of tuskless elephants is what will save elephants in the long run; that the sepia-toned photographs of enormous 'hundred pounder' tusks are relics of a past era. But I argue that I want my children to see elephants as they are meant to be: large-tusked, healthy and unhindered by human greed.

09

A tribute to the giant tuskers

Portfolio

Colin Bell

Paging through the journals of early African explorers, I'm struck by how much larger and heavier ivory was in those days. This perception is reinforced when I glance through the pages of Rowland Ward's hunting records, where hundreds of trophy elephants are listed, each with combined ivory weighing around 200 pounds. The weight of the tusks of one elephant on display in the British Museum is an astonishing 440 pounds. And that list doesn't record thousands of elephants that were poached or commercially hunted to be shipped to world markets for piano keys, billiard balls and bracelets. Up until fairly recently, a 'hundred pounder' was considered the standard minimum target for the hunting world. As a hunting website site laments:

An elephant with really good ivory is generally considered Africa's top hunting trophy. 100 pounds [45 kilograms] per tusk used to be the magic number, though these days 70 pounds [32 kilograms] is considered very good (or 35 pounds [16 kilograms] for a forest elephant). Finding good, heavy ivory is much more difficult today than it was a few years ago and many sportsmen spend a great deal of time and money in unsuccessful pursuit.

Clearly, hunting and poaching have taken their toll on elephant populations and, in particular, on the size of elephants' tusks. The biggest tuskers get shot first and have been steadily whittled away over the decades, resulting in a rapidly shrinking gene pool. And today tuskless elephants are becoming more and more widespread – living proof that the 'survival of the fittest' mantra is still relevant today, albeit with a slight twist: 'survival of the smallest'.

This chapter pays tribute to the surviving tuskers, last of the elephants with powerful genetics that have produced magnificent ivories – elephants that have been able to evade bullets from trophy hunters and poachers to live beyond 50 years. No one knows how many 100-pound-plus tuskers are left in Africa today, but certainly there are fewer than 50. Maybe that number is as low as 20.

Genetics is a key factor, but so too is environment. Elephant ranges that have poor soil nutrients (like Namibia and Botswana) seldom produce such tuskers, as is evidenced in the Rowland Ward records. Yet, for sheer size (measured at the shoulder), the desert-adapted elephants of north-west Namibia reign supreme. We talk about those special elephants elsewhere in this book. For now, feast your eyes on what was regarded, until recently, as normal for an elephant.

Satao2 with his three askaris in Tsavo East National Park, Kenya. Satao2 died in 2016 from a poacher's poisoned arrow.

The elephants in the Ngorongoro Crater, Tanzania, are almost exclusively male. A study suggested that they may be spill-overs from the surrounding highlands.

There are fewer than 40 giant-tusked bull elephants left in the world; these three, Little Male, Tim and Tolstoy, are among the few. Over a third of them live in Tsavo East National Park, Kenya, where they are under the careful guardianship of the authorities and the Tsavo Trust.

An elephant will grow more ivory in the last 10 years of its life than in the first 10 or 20. The tusks will be used to bully and intimidate less well-endowed males. Also, size really does seem to matter among the females. This is Kamboyo of Kenya.

The story of Tolstoy (left) and Tim (right) goes way back to 1973, when Cynthia Moss's Amboseli research project in Kenya started to document the lineages, behaviours and movement of elephant families of the region. Tim then was 4 years old and Tolstoy just 2. Forty-six years later, it is remarkable that they have been able to steer clear of poachers' bullets and poisons. Tolstoy's right tusk was broken while he was sparring with Tim, and his left tusk was subsequently trimmed (after this photo was taken) to avoid its dragging along the ground and being tracked by poachers.

10

Big trees, big elephants and big thinking

Preserving the species safeguards a suite of critical ecological processes and conserves a whole array of smaller species.

Dr Michelle Henley

People value big trees and big elephants for reasons that are obvious: both are aesthetically appealing, are important for ecosystem functioning and have economic value. But in Zimbabwe and South Africa, the management policy has been that elephants need to be managed to the benefit of trees. So, historically, elephants have been culled.

However, co-existence has always been at the core of ecological webs, so I could never comfortably favour either trees or elephants. It seemed unnatural to observe the contrast on either side of a fence and blame the difference on elephants. It also appeared senseless to set management benchmarks based on how the vegetation looks or should look like in the absence of elephants. Along the extremes of the co-existence of the tree-elephant continuum – no elephants, some elephants or too many elephants – there needed to be a different way of looking at things. There had to be a way of thinking big!

It is increasingly clear that there's no linear relationship between large trees and elephant densities. It's a complex system with uncertain outcomes influenced by many factors, which are constantly in flux. A good number of these are out of our control, simply because of the large space and timescales at which they operate. The longevity of a particular tree depends on many things:

- climate
- fire frequency
- soil type
- topography
- elevation
- surrounding vegetation community
- herbivory from numerous animal species
- spatial distribution of elephants
- physical properties of the tree itself that would determine its palatability
- ability to recoppice when browsed
- regeneration rate from seed
- root and bark structure and hence susceptibility to impact from elephants.

Humans often skew these relationships. We open artificial waterholes, which anchors elephants to an area, then become concerned about the loss of surrounding trees that happen to be intolerant of elephant feeding. To understand the elephant-tree relationship we must look beyond the frame of our narrow spatial view and short timescales. We need to widen the frame. Big trees and big elephants have co-existed since the dawn of the elephant lineage around 55 million years ago. Our hominid line has been around for only about 5 to 7 million years.

Tree-elephant interactions

Big trees provide food, both directly (through the woody plant parts they offer) and indirectly to herbivores. They act as nutrient pumps, bringing micronutrients to the surface and increasing grass species diversity around their base. This is particularly beneficial in the wet season, when elephants are primarily grazers.

Big trees also offer shade – especially important for a creature measuring over 3 metres at the shoulder. It's been found that big bulls prefer areas with high tree cover and low herbaceous biomass, while the opposite is true for females. Generally, elephants tend to cross areas inhabited by people or move into more open areas only at night; big tree clusters offer shelter and security, allowing these animals to conceal themselves during the day.

Big trees are important nesting sites for large birds. Vultures, which play an essential role in cleaning up carcasses, prefer nesting in the upper crowns of large trees. So the interaction between big trees, vultures and elephants, which may bark-strip or fell trees used as nesting sites, is worth mentioning. If the cycle of large-tree replacement has been interrupted, more often by factors other than elephants' ravages, vultures' nesting could be compromised – with negative consequences. In such cases, lowering of the potential diversity of avian or mammalian fauna (such as bats) dependent on large trees as nesting sites or fruit sources could occur.

Elephant-tree interactions

Elephants are ecosystem engineers and major tree pruners. Although they're preferential grazers, they browse during the dry months when they effortlessly break branches or hedge, fell or uproot trees. Depending on the level of impact, these pruning activities may promote plant growth or bring the canopy within reach of other, smaller browsers.

This process has been found to increase the nutrient quality of affected plants, escalate the overall biodiversity of the landscape and promote a mosaic of elephant-impact-tolerant plant species closer to water points. Landscapes used by elephants will have an altered vegetation structure; elephant feeding habits create microhabitats for smaller creatures, leading to a higher diversity of ants, reptiles and frogs.

Elephants are composting machines and fertilising agents. Because they have a low digestive efficiency (around 22%), they deposit large amounts of undigested plant material. With a dung pile being around 10 kilograms, and with an average defecation span of 25.3 hours a day, they are capable of producing 150 kilograms of wet dung a day. Landscapes used by elephants can potentially be significantly enriched by nutrients.

Elephants are constant gardeners. Large bulls can carry fruit seeds up to 65 kilometres from source, making them one of the wild's most impressive fruit dispersers. Not only are the seeds effectively dispersed, but they are deposited in a perfect organic mulch, which aids germination. In so many ways, elephants make their landscape.

Promoting co-existence

The expanding human population is the biggest threat to peaceful co-existence between big trees and big elephants. Africa's average population density in 2000 was 26 people per square kilometre. By 2050, it is projected to be 60 people per square kilometre. This will increase pressure on natural ecosystems: wild vegetation will be cleared for agriculture, compressing elephant populations within their range and competing with them for resources.

Where natural migration cycles still occur over large areas without artificial water points, as in Botswana, elephants can migrate in the summer from heavily impacted local areas, allowing the vegetation to recover. But in other areas we have disrupted the natural cycles of elephant range expansion during the summer months and subsequent range contraction around limited water sources in the dry season. In South African parks, for instance, historical migration routes are fenced off and the landscape is saturated with artificial water points. This increases the encounter rate between elephants and significant big trees, especially those situated closest to water points.

In large conservation areas, elephants have the ability to regulate their own reproductive output in relation to the availability of resources. When water is limited in the dry season and food resources are depleted, females lower their reproductive output. In years with below-average rainfall, females may delay their age of first reproduction, or increase their inter-calving interval. Within family units, weaned calves are hit hardest by adverse conditions, and are usually the first to die on the long treks between water sources or when food or mother's milk is scarce.

With this knowledge, where protected areas are large enough, we can encourage natural ecosystem processes by regulating our own management actions. This would foster more peaceful co-existence between big trees and big elephants.

Over-exploitation of trees and elephants

Big trees are valued as fuel and for construction and furniture. In sub-Saharan Africa, wood fuels around 85% of household energy, with demand growing at 3 to 4% a year. Charcoal use poses one of the greatest threats to natural forests and woodlands because of the preference for harvesting living, old-growth species. By 2025, more than half Africa's population will be urbanised, resulting in a 14% increase in charcoal use for every 1% increase in urbanisation. The combined demand for construction timber, charcoal burning and firewood has led to the destruction of some 68% of forests in Africa. In South Africa and Mozambique, the effect on woodlands surrounding protected areas is starkly visible.

As with big trees, elephants have increased in value as their numbers have decreased. The skyrocketing value of ivory has fuelled a poaching tsunami, resulting in elephant populations crashing over large parts of the continent. In Mozambique there's been a 48% decline in elephants in 5 years and, from late 2015, the Kruger National Park (KNP) has experienced some of the highest numbers of elephant-poaching incidents in its history.

Why protect big trees and big elephants?

The growth of big trees and big elephants takes time. This means that recovery from over-exploitation is slow. Big trees are buffers against climatic imbalance, they are sources of food, nutrient pumps, nesting sites and flood- or erosion-prevention agents. Elephants are keystone and umbrella species. By conserving them, we maintain some critical ecological processes and conserve a whole array of smaller species. They are the trees' constant gardeners.

So, while big trees and big elephants are individually important, they are also collectively vital because of their interacting ecological functions. We cannot place the importance of one above the other, and *we simply must not protect one at the cost of the other*. Our management actions should promote their co-existence.

We have placed big trees and big elephants under terrible threat. It's time for informed strategies and big solutions.

Hlanganini is possibly the largest living tusker in the Kruger National Park today. This potentially puts him at risk, as there is evidence that the poaching tsunami that swept through East Africa will increasingly hammer South Africa's herds.

© Johan Marais

'Animals are more ancient, more complex and in many ways more sophisticated than us. They are more perfect because they remain within Nature's fearful symmetry just as Nature intended. They should be respected and revered, but perhaps none more so than the elephant, the world's most emotionally human land mammal.'

—————————————

Dame Daphne Sheldrick

This image from Hwange National Park in Zimbabwe was taken at Nyamandlovu Pan (Ndebele for 'meat of an elephant') towards the end of the day. Each evening, large herds make their way to the pan for a drink and a wallow.

11

Conserving elephants and biodiversity in Africa's savannas

Providing artificial water in previously waterless regions is the greatest threat to biodiversity in the presence of elephants in large, relatively open-ended wildlife systems in southern Africa.

Dr Richard WS Fynn & Dr Timothy G O'Connor

The iconic African elephant is proving to be a hot potato for conservationists in Africa: there are either too few elephants in some regions because of rampant poaching, or burgeoning elephant populations in other regions, raising concern over habitat destruction. There's been much debate over the potential negative impact of high densities of elephants, both on other wildlife species and overall biodiversity, by causing local extirpation of tree species, homogenisation of vegetation and an associated reduction of habitat diversity. This is a major dilemma for agencies charged with conserving habitats and overall biodiversity. Do we conserve or reduce elephant populations? Is there an acceptable solution for people across the spectrum of viewpoints?

Our position is that there are potential solutions that don't require reduction of elephant numbers. But this depends upon the size of a protected area and its specific mix of habitat types, which forms the environmental template, as well as on water availability across large landscapes. Using examples from several southern African parks, we show that small and medium-sized parks are much more vulnerable to negative elephant impact than large

parks. In addition, we show that water availability across a landscape determines accessibility by elephants to various parts of that landscape, thereby providing refuges for plants against severe impact in the least accessible parts of the landscape.

Apart from the size of parks, we suggest that artificial water provision is the single most important factor determining the outcome of elephant impacts on biodiversity. Finally, we discuss how riparian woodland adjacent to large ephemeral and perennial rivers can persist in the face of large elephant populations, either naturally or with various interventions.

Several studies across Africa have demonstrated that elephants can have major negative impacts on the structure (vertical distribution), abundance and diversity of trees and shrubs. They also show that the extent to which these impacts are expressed differs with park size and the extent of artificial water provision.

In East Africa, Richard Laws summarised elephant impact on woodlands, with examples from Murchison Falls National Park, Tsavo National Park and Lake Manyara National Park. In these areas, elephants have had a severe impact on

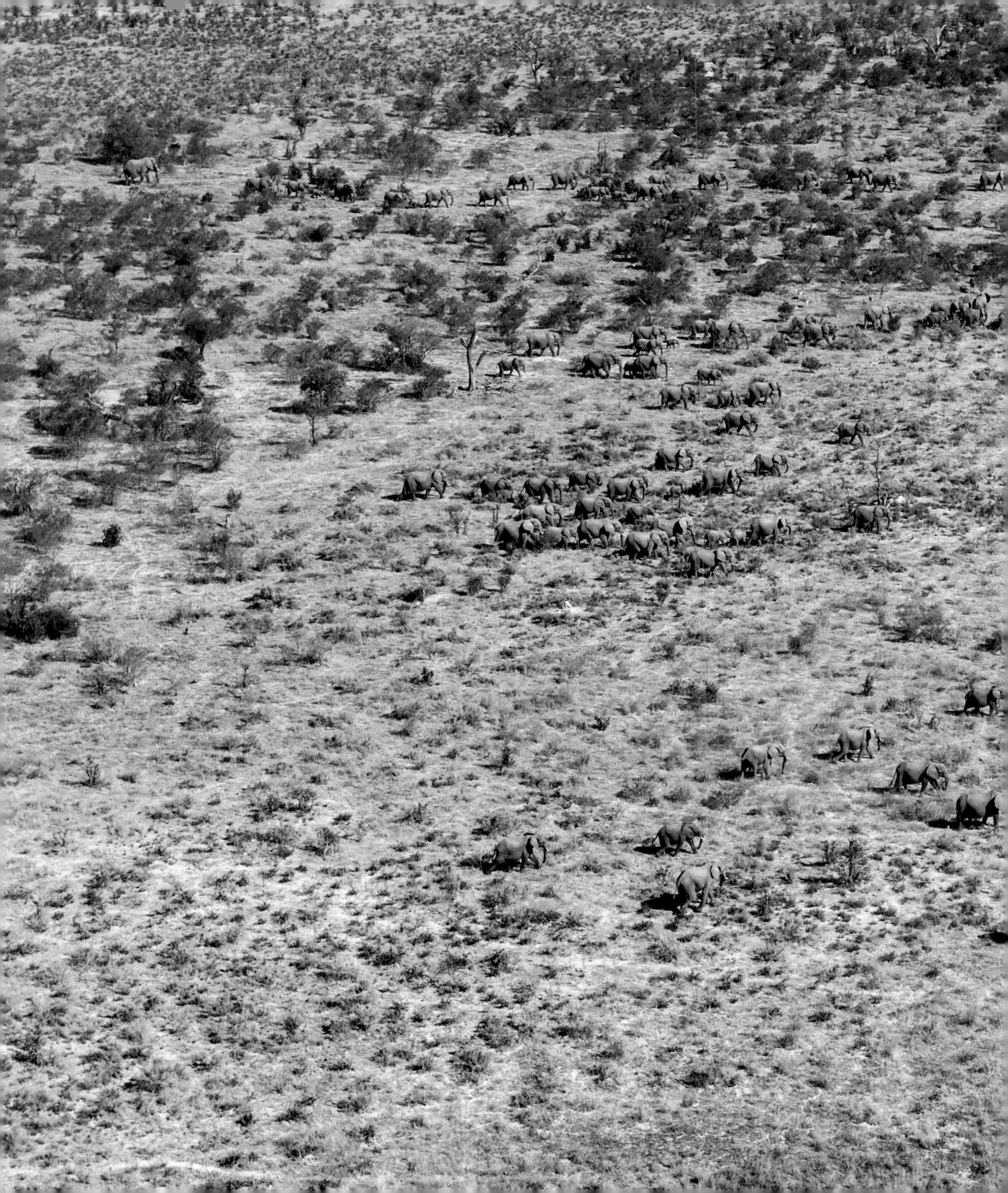

Elephants are the water diviners of the African savanna, and their trails inevitably lead to where the precious liquid can be – or was once – found.

© Colin Bell

These two photographs from the same region in Botswana in dry and wet seasons temper claims that elephants are destructive of vegetation: trees that look hammered in October after being heavily 'pruned' during the dry season are into full recovery by December.

© Colin Bell

woody vegetation, in some cases almost completely transforming woodland and forests to open grassland. Along the Chobe River in Botswana, Lucas Rutina and Stein Moe reported that elephants have had a severe impact on riparian woodland structure and diversity. In the Kruger National Park (KNP), Greg Asner and Shaun Levick noted a six-times-greater rate of tree fall in areas accessible to elephants, with an up to 20% decrease in the number of tall trees. A long-term study by Tim O'Connor in a medium-sized park in northern South Africa showed that there were dramatic declines of 18 tree species, amounting to a loss of more than half their population in 13 years, with one species being eliminated. Simon Chamaillé-Jammes and others in Hwange National Park demonstrated that elephant impact on tree cover was most strongly expressed within 5 kilometres of artificial water points.

Clearly then, large elephant populations can have a severe impact on woody vegetation. But the concern for biodiversity conservation is not that elephants will exert impacts on vegetation structure and diversity in localised areas, but that their impacts will be widespread across large landscapes, reducing habitat diversity and associated biodiversity. For example, studies have shown that the great diversity of mammals, birds and insects have differing vegetation structural requirements, with species' preferences ranging from short scrubland to tall woodland; short, open grassland to tall grassland, or any of a range of structural states between these.

A vegetation state preferred by a species forms part of its habitat requirement, an environment outside of which it does not survive. So conversion of woodland to scrubland, or tall grassland to short grassland across large landscapes can be expected to cause loss of habitat space for some species, and hence a loss of biodiversity. A simplified landscape could also reduce the ability of wildlife to adapt to constantly changing conditions in variable climates typical of African savannas. Loss of tall trees throughout landscapes has obvious negative consequences for birds that are reliant upon them for nesting, or for giraffes, dependent upon them for browse.

On the other hand, creation of shrubland by elephants has been shown by Shimane Makhabu and others on the Chobe River in Botswana, and

Marion Valeix and others in Hwange National Park, Zimbabwe, to benefit browsers such as steenbok, impala and kudu. Thus, if impacts are restricted to zones near water, both shrubland and tall woodland states can persist, increasing habitat diversity for browsers. Similarly, habitat diversity is increased for bird species, with some preferring shrubland and others tall woodland.

Marion Valeix and others demonstrated that elephants in Hwange may also create more open areas with better visibility, which helps reduce predation risk for some large herbivores. Patchy or zoned effects of elephants on vegetation, therefore, can increase overall habitat diversity, favouring biodiversity, but only if these effects are restricted to a portion of the landscape.

Allowing elephants to transform savanna vegetation into scrubland and short grassland, with loss of most of the tall-grass and woodland habitat, would pose a serious problem for overall biodiversity conservation. It appears, therefore, that introduction of elephants into smaller, fenced parks, where they are easily capable of accessing any part of the park at any time of the year, requires maintaining their population at a very conservative number. This can be done through contraception, translocation or culling.

If populations are left unchecked in small parks, they will, over time, attain a high density and deplete their most favoured food types (trees and grasses), and be forced to turn their attention to their next-most-favoured species (known as diet breadth expansion). Given sufficient time, elephants in small parks will eventually deplete their food resources to such a degree that they will face major population crashes during dry periods, leaving a highly-modified, species-poor system relative to what had previously existed.

If elephants possess the ability to induce dramatic, deleterious vegetation change, what approaches might conservation bodies pursue in order to mitigate this? In large parks, several processes may operate to buffer the effect of large elephant populations. One proposal is to ensure that vast areas are available for elephant movement, such as happens in northern Botswana and adjacent parks in Namibia and Zimbabwe. In these parks, herbivores may be less strongly coupled to individual patches of their food resources than in smaller parks, having greater options to move away from heavily affected vegetation. This generally takes place on a seasonal basis, allowing heavily impacted plant species time to recover, and possibly not be revisited by elephants for a long time.

This serves also to increase habitat diversity, which may offer a greater range of food resources, allowing elephants to meet their seasonal foraging needs. This would promote elephant movement between habitats and regions, allowing further time for vegetation recovery.

The value of large areas, of course, depends on the number of elephants. And if the habitat is conducive to life, an elephant population can be expected to grow. This growth will eventually increase impact per foraging area and see briefer periods between revisits to affected vegetation, less time for plant recovery and, ultimately, transformation of the vegetation. Spatial scale is important, but insufficient on its own to mitigate the effects of elephants.

Securing refuges from elephant impact offers the best chance for co-existence between large elephant populations and biodiversity. These are areas to which elephants cannot gain access, such as mesa plateaux bounded by cliffs. Physical refuges are extremely rare, and usually protect a habitat and vegetation completely distinct from the surrounding landscape. They are therefore not a basis for overall conservation of biodiversity in relation to elephants.

Biodiversity can, alternatively, benefit from a virtual refuge, in which the probability of elephant encounter and impact is very low. What circumstances, then, comprise a low probability of encounter? Availability of water is one. Areas far from water are less available to elephants and other large herbivores. Elephants need to drink on an almost daily basis. The potential area of impact by elephants depends on the maximum foraging distances for different social units from a water source. In the case of cow-calf groups, this distance is about 5 kilometres, owing to the inability of young calves to walk long distances. By contrast, mature bulls, which are responsible for most damage to trees, from ringbarking and uprooting, may feed up to 15 to 20 kilometres from water.

The spatial distribution of permanent water in a system is, therefore, a key influence on the proportion of a landscape beyond the maximum foraging distance. Vast areas of current elephant range in southern Africa have a limited distribution of perennial rivers. These include northern Botswana, northern Namibia, southern Angola and western Zimbabwe, an area encompassing northern Zimbabwe and south-eastern Zambia, as well as an area encompassing northern KNP, south-eastern Zimbabwe and western Mozambique.

Under natural circumstances, elephant range expands during the wet season as non-perennial rivers and ephemeral pans are filled by rains, allowing elephants to access feeding grounds other than the winter dry-season areas. Then, as pans and non-perennial rivers dry up, elephants return to their winter feeding grounds. Because elephants tend to concentrate their feeding on herbaceous material during the wet season, woody vegetation within areas accessed only during the wet season is used sparingly. In above-average rainfall years, some large, deep pans or river pools may harbour water throughout the dry season, making elephants more likely to use the surrounding vegetation.

In large natural landscapes, there's a gradient of the encounter probability of woody vegetation, ranging from extremely high to extremely low, indicating a gradient of the effectiveness of spatial refuges across these landscapes. This, in turn, results in a gradient of elephant impact on vegetation from a minimum in some locations to a maximum in others, thereby increasing habitat diversity and, ultimately, biodiversity. So, contrary to general expectations, we predict that large elephant populations will increase biodiversity in these large natural landscapes.

Patterns of feeding

Images of woodland destruction and transformation have become synonymous with elephants throughout Africa, begging the question of why they choose to eat in this way. Although it has long been established that elephants' diet changes markedly over seasons, recent research in south-eastern Zimbabwe by Bruce Clegg has sought to explain this pattern in terms of the feeding value of the choices they make compared with what is available.

Elephants, especially bulls, prefer a diet of herbaceous green grasses and forbs (herbaceous flowering plants) during the wet season because these offer greater nutrients within digestible material than woody foliage at that time. Growth of grasses and forbs during the wet season ensures that large trunkfuls of forage can quickly be harvested. In addition, grasses tend to lack the chemical defences of woody species.

So in order to meet their energy and nutrient needs, it's more profitable for elephants to eat a herbaceous diet. But this wanes as the dry season advances and grasses and forbs dry out. Their digestibility and energy and nutrient concentrations decline, as does their availability. A point is reached where it's more profitable for elephants to turn their attention to the green leaves of shrub and tree species. These remain green longer into the dry season owing to their deep root systems, which can access groundwater well beyond the reach of shallow-rooted herbaceous species.

When most woody species have lost their leaves by the late dry season, elephants turn to the bark and roots of favoured tree species. This has the greatest negative impact on individual plants. These foraging choices during the dry season are neither preferred nor time-efficient for elephants, but are necessary to meet their nutritional needs during the most demanding time of the year.

Thus, under the forage-resource classification developed for herbivores by Norman Owen-Smith, herbaceous material favoured by elephants can be considered a high-quality resource, leaves of trees and shrubs as a reserve resource, and bark and roots as a buffer resource when other food is depleted. As a consequence, woody vegetation within the maximum foraging range of elephants during the dry season is particularly vulnerable to damage by elephants.

A virtual refuge for woody vegetation is, therefore, achieved through a combination of the seasonal spatial pattern of water availability, maximum elephant foraging distance and the seasonal patterns of dietary selection. If the pattern of availability of permanent water is altered, such as by artificial water provision, the security of virtual spatial refuges will be compromised. This has been demonstrated by Simon Chamaillé-Jammes and others in areas with a high density of artificial water points in Hwange National Park.

Provision of water

We predict that large elephant populations will have severe negative impacts on biodiversity in landscapes where all parts of the landscape have almost equal probability of access by elephants all year round, where spatial refuges do not exist or where their effectiveness has been eroded by human activities, such as through the provision of artificial water.

A recent study in northern Botswana by Keoikantse Sianga and others offers strong support for the prediction that effective spatial refuges would buffer large landscapes against homogenisation and loss of biodiversity. Within 5 kilometres of permanent water, elephants foraging during the dry season transformed a woodland of *Terminalia sericea* (a favoured tree in this locality) sandveld into short shrubland, whereas tall, intact *Terminalia sericea* sandveld woodland persisted beyond 15 kilometres, the foraging range of adult bulls from permanent water.

Elephants and other herbivore species clearly created structural and compositional heterogeneity in this landscape, where a spatial refuge for favoured plant species was created beyond 15 kilometres from permanent water.

It is obvious, therefore, that the placement of artificial water points, pumped from boreholes, in landscapes previously inaccessible to elephants during the dry season is a threat to spatial refuges, rendering such areas permanently accessible. Such a programme of artificial water points was initiated in the KNP during the 1970s through concern about access to water by wildlife during droughts after fencing the western boundary of the park had constrained movement to seasonal feeding grounds.

Before management in the KNP began to close these water points, it was not possible to travel more than 6 kilometres from water during the dry season, a distance that consigned almost all parts of the park to an extremely high probability of encounters with elephants. A study by Shannon and others found that elephant impacts on tall trees increased up to 5 kilometres from permanent water, which indicates that the park's landscapes had no effective spatial refuges from elephant impact.

Spatial refuges not only buffer elephant impact on vegetation, but also the impact of other large grazers on grasses, allowing taller, high-quality grasses to persist. Concern about loss of effective spatial refuges, therefore, also applies to tall-grass grazers, a concern that has emerged as a pressing conservation challenge for sable and roan antelope populations in the KNP. Whether decline of these species was a result of the loss of favoured tall grasses or of refuges from heavy predation cannot be ascertained, although both factors probably contributed.

Spatial refuges involving distance from water are also apparently important for the regulation of elephant numbers through their influence on calf and juvenile mortality. Across 13 protected areas in southern Africa, a study by Kim Young and Rudi van Aarde showed that mortality of weaned elephant calves increased with greater daily distance walked between water and foraging sites during the dry season. This finding suggests that calf mortality during the dry season is an important factor influencing the growth of elephant populations.

This effect is readily evident in northern Botswana, where elephant calves experience substantial mortality during the dry season as a result of long treks between water and foraging sites. Calves may die from dehydration and insufficient food of adequate quality, but also because of increased predation by lions owing to their weakened state. This calf mortality has been put forward as the main reason for the size of the elephant population in northern Botswana remaining relatively stable at around 130 000 individuals.

Calf mortality could obviously be reduced by opening up currently waterless spatial refuges through the provision of artificial water. For a period, the elephant population, including calves, would have access to an abundant supply of favoured dry-season foodstuffs and would not have to commute long distances. Not only would calf survival increase, but an already large elephant population would grow.

Simon Chamaillé-Jammes and others showed that, in Hwange, elephants prefer artificial water points where elephant numbers were previously low. This probably relates to better food resources around these less-affected water points and less fouling of water by dung and urine. The effect of an introduced water supply would be transient, lasting until elephants have depleted the food and transformed the vegetation within foraging distance.

A larger population size following the introduction of artificial water creates a greater problem than before because more landscape becomes vulnerable to homogenisation of vegetation structure and loss of biodiversity. Pressure to provide additional artificial water often arises during drought events, but a subsequent – inevitable – drought event would likely result in catastrophic die-off of the elephant population.

For the above reasons, provision of artificial water in previously waterless regions is, in our opinion, the greatest threat to maintaining biodiversity and elephants in the few remaining large, relatively open-ended wildlife systems in southern Africa.

Riparian woodlands

A case for the importance of spatial refuges does not apply to all habitats. The presence of elephants in riparian woodlands challenges the maintenance of biodiversity because these habitats are, by definition, immediately adjacent to water. They occur in almost all instances on terrain easily accessed by elephants, usually harbour biodiversity of special interest on account of their composition and structure (such as tall trees) and, because green foliage is usually available throughout the year, they are generally a preferred feeding habitat of elephants. Taken together, riparian woodlands are possibly the habitat most vulnerable to elephant impacts. How then, might these areas avoid damage over time? Two features appear to be critical determinants.

The first is whether elephants have an alternative source of green forage during the dry season. Of particular value are wetlands and floodplain grasslands that maintain green grasses through the dry season. Elephant bulls prefer green, broad-leafed grasses to woody foliage. If such grasses are available, impact on woody vegetation is lessened.

Contrasting riparian woodlands with or without wetland or tall floodplain grassland in close proximity is instructive. In northern Botswana, riparian woodlands along the eastern edge of the Okavango Delta have not been transformed to the extent of those along the Chobe

River front. These two regions differ in two notable ways. The Chobe River floodplains are dominated mainly by short grasses, whereas the Okavango floodplains support extensive tall-grass habitats dominated by wild rice (*Oryza longistaminata*), hippo grass (*Vossia cuspidata*) and tall sedges (*Cyperus* spp.), providing a bountiful alternative source of forage during the dry season.

The Chobe River front riparian woodlands occur as a narrow strip along the river (which, to the Zambezi confluence, has decreased as a result of human encroachment), whereas there are more than 50 000 islands in the Okavango Delta, each with its own patch of riparian woodland.

So in the Chobe region there's a low ratio of riparian woodland relative to the extensive hinterland woodlands used as wet-season habitat. There are also floodplain grasslands, but these are suitable for short-grass grazers, rather than for elephants. By contrast, in the Delta, there's a much higher ratio of riparian-to-dryland woodland and an abundance of tall floodplain grassland suitable for elephants. During the dry season, a large elephant population from a large area is funnelled into the relatively small Chobe riparian region, whereas such a concentration effect does not occur in the extensive Okavango Delta. These different patterns result in greater elephant impacts on Chobe's riparian woodlands. We predict that, had the extensive tall floodplain grasslands of the Okavango Delta been absent, riparian woodlands in this region would have suffered much greater impacts by elephants.

Knowing the ratio of riparian to dryland woodland is therefore essential in understanding the vulnerability of riparian vegetation. In many protected areas in higher-rainfall regions, such as in the eastern parts of southern Africa and in East Africa, there are sufficient riparian habitats along primary and secondary rivers and streams to provide this balance, and so temper impacts by elephants. But, clearly, the availability of alternative sources of green forage during the dry season, such as tall grass floodplains, is optimal for survival of riparian vegetation under large elephant populations.

In the past, riparian habitats received some protection from excessive elephant impact through human disturbance along rivers, where people tended to settle. For example, the only remaining intact patches of riparian forest along the Chobe River are adjacent lodges (human disturbance) or on steep, inaccessible banks of the river (spatial refuge). This is akin to the 'landscapes of fear', which influences the distribution and impact of herbivores on landscapes, as was observed with the reintroduction of wolves to Yellowstone National Park. At present, elephants have no more fear of encountering humans alongside rivers than elsewhere in parks, and are generally free to forage in riparian habitats.

It would be possible to recreate landscapes of fear in parks where no other options exist (for example, parks without alternative dry-season habitats, such as tall-grass floodplains). This could be done either by limited hunting of elephants only in riparian habitats – a decision likely to be challenged by many – or by creating disturbances in such habitats using drones (which they fear) or harmless shots with rubber bullets.

Alternatively, Lucy King in East Africa has shown that playing the sound of a swarm of bees through loudspeakers will send a herd of elephants fleeing in terror – a method that holds much potential. As a last resort, artificial spatial refuges could be created through single-strand electric fencing to preserve key stretches of riparian habitat, while still allowing free movement of most other animals.

In conclusion, the effects of high densities of elephants on vegetation and biodiversity will differ between small and large protected areas, providing that large areas have extensive effective spatial refuges in their landscapes. Provision of artificial water in previously waterless regions is, in our opinion, the greatest threat to maintaining biodiversity in the presence of elephants in the few remaining large, relatively open-ended wildlife systems in southern Africa.

Conservation of riparian habitat is more complex and depends upon the ecological template, such as the availability of alternative dry-season habitats; for example, tall-grass floodplains and the ratio of riparian-to-dryland woodland. Application of artificial spatial refuges in the form of electric fences or by reintroducing landscapes of fear may assist in conserving key riparian habitats in parks where other options don't exist.

12

Constant gardeners
of the wild

What we call destruction by elephants is
often the exact opposite.

Garth Thompson (photographs and text)

**'Look how these bloody jumbos have hammered
the vegetation!' How sad it is to hear comments
like this, often by safari guides who've been
afforded the privilege of working with these
magnificent keystone animals: people who have
had the opportunity to observe and learn from
the environment and its inhabitants.**

Wild, majestic African elephants don't deserve
the disparaging name of jumbo (referring to a
poor elephant that earned its name by performing
headstands and other degrading acts on a barrel
in some circus). There are few land mammals that
command the presence and have the intellect of
elephants, the largest land mammal with whom
we so fortunately share our planet. There have
possibly been more wildlife books, articles and
documentaries published on elephants than on any
other wild mammal, yet they still rank as one of the
most misunderstood animals.

It's ironic to hear the most destructive creatures
on the planet refer to the action of elephants
in terms of destruction, damage, degradation
and habitat devastation: we who, in the name of
progress, direct bulldozers to clear a virgin forest
that took hundreds of years to grow to make way
for a shopping mall, housing estate or palm oil
plantation. What we call destruction by elephants is
often the exact opposite. Join me on an important

rethink of the many positive effects that these bush
architects have on their environment.

Forming waterholes

Most waterholes in a wild ecosystem were engin-
eered by elephant action over millennia. All
mammals eat soil to acquire mineral salts. Browsers
tend to eat more soil than grazers, and elephants eat
more of it than any other herbivore. The depressions
they create fill with rainwater to form small
waterholes. When these begin to dry up and turn to
mud, warthogs may wallow, expanding the hole and
sealing the bottom off in a process called puddling.
The following year, after the rains, when the dry
depression has increased in size as a result of more
soil being eaten and ongoing wallowing, it may be
'puddled' by buffaloes and then rhinos and, years
later, by elephants.

It's estimated that each adult elephant excavates
about a tonne of soil a year from the various licks
and waterholes – about 2.3 kilograms of silt when
drinking, and two bucketfuls of mud each time
they wallow in, kick and throw it over their bodies.
This sealing process enables the pan to hold water
for much of the dry season and benefits all forms
of wildlife.

Digging wells

At the end of the rains, water continues to move beneath the sandy surface of many dry river beds. Elephants, with their large feet and dextrous trunks, use these tools to excavate to the water table, often over a metre deep. Water seeps into this cylindrical well from which they drink, and which they then leave for the convenience of the rest of the wildlife community.

Providing nutrients

Elephants are consummate herbivores. They're not only browsers and grazers, but they also eat bark, fruits, pods and roots (from up to a metre below ground). Unlike most herbivores, which are ruminants with four stomachs who regurgitate and chew their food thoroughly, elephants have one large stomach and a simple digestive system. After being chewed a few times, their food is swallowed and moves into the belly for a few hours before emerging at the other end, some of the vegetation still intact.

A recently dropped elephant's bolus is not good as compost until other creatures open it up in search of the food it contains. Hornbills looking for seeds and insects are common around a bolus, while people, monkeys and baboons go for marula (*Sclerocarya birrea*) and mongongo nuts (*Ricinodendron rautanenii*), which contain tasty kernels. Wild primates also feed from a variety of other seeds and insects in the dung pile. Butterflies are attracted to the dung for its mineral-rich moisture, and dung beetles eat the finer dung parts. These beetles also dig their dung balls into the ground for their young to feed on after hatching, thus aerating the soil and fertilising it.

Termites and other insects are attracted to the dung – and themselves become prey for many birds. Termites often convert the entire dropping into soil, recycling its coarse vegetation in their fungus farms, turning it into healthy topsoil. When elephants defecate in rivers, pans and lakes, their dung breaks down and feeds the aquatic life.

Left, from top

'Puddling' can be the start of pan creation.

Mining nutrient-rich soil can lead to the creation of waterholes.

A red-billed hornbill feeds from insects in elephant dung.

Butterflies benefit from the minerals in elephant dung.

Elephant highways were Africa's first access roads.

Creating highways

Because of their size, longevity and intelligence, elephants are the main highway builders of the bush. With their expansive knowledge of home ranges, garnered over millennia, they have created U-shaped highways linking feeding areas to waterholes and rivers – highways that other wildlife uses. When flying overhead, it's common to see their trails like spokes in a wagon wheel radiating out from a waterhole.

For hikers, there's no easier way to travel than on an elephant highway. Elephants follow the best contours to each summit and valley, often taking you through tunnels of impenetrable jesse-bush bushwillow (where an encounter with a black rhino used to be a strong possibility). A heavily dung-carpeted highway indicates a healthy elephant population.

Humans use these paths for more than strolling. The tar road from Makuti to the Kariba Dam wall in Zimbabwe, a distance of about 80 kilometres, was tendered for and built in the 1950s by Jim Savory, who worked in the then Southern Rhodesia's Ministry of Irrigation and Water Development. He tendered against big commercial contractors and his bid came in way under their estimates on cost and time. He won the tender and built the road in 18 months (allowing Kariba Dam to be constructed on schedule). This he managed by not wasting time with initial surveys, but following the main elephant paths. He knew that elephants had, over the ages, found the easiest gradients into the Zambezi Valley.

Planting trees

Elephants are often accused of disfiguring and killing the vegetation, but they are, to the contrary, responsible for planting many species of tree. When a ruminant eats a seedpod from an acacia tree, it's regurgitated and chewed finely in order to continue its journey into the next three stomachs. In this way, seeds are often crushed and ultimately destroyed.

Elephants eat large quantities of acacia and other pods. The warmth of their belly swells the seed, and stomach movement rasps and scours it, all of which assists in the germination process. All trees can germinate on their own, of course, but the process is favourably accelerated by passing through an elephant's belly. Germinated seeds, embedded in the bolus, are often dropped in the dung pile a good distance from where they were eaten.

The Lower Zambezi River, on the Zambian side downstream of the Chongwe River, is a good example of this redistribution and replanting process. Tens of thousands of *Faidherbia albida* have grown over the past 30 years, all the way down to the Mushika River, a distance of around 80 kilometres. They are pioneer trees of such areas and proliferate on sand banks and sandy islands.

Their gardeners are elephants, which move to the islands daily to feed off the pioneering sedges and *Phragmites* reeds. While there, they deposit their load, which includes *Faidherbia* seeds from the mainland. As these islands are without baboons and impala, which pluck and nibble off all the young

A scenic grove of cathedral mopane keeping their leaves above the reach of browsers. Vegetation in all wild areas containing elephants looks modified at the end of the dry season, the culmination of a year's pruning. Four months into the rainy season these stark wooden statues explode into green leaf.

shoots, these young seedlings have the opportunity to grow into adulthood, and have produced forests that now line the channels of the river terraces. Over time, the river changes its course and these islands join the mainland, where the trees drop tonnes of pods that feed multitudes of herbivores and omnivores, from tiny gerbils to the next generation of elephants.

Most of these trees have been planted by elephants, yet these animals get blamed for pushing over or ringbarking a small percentage of what they ultimately planted. Seed distribution is what they do, and they do it well. It's not uncommon to see *Hyphaene* palms growing on islands in river systems or on the mainland, far from the closest palm areas, such as the Makgadikgadi Salt Pans. Elephants are constant gardeners and landscapers.

Pruning

In agriculture, pruning is undertaken to remove parts of plants, such as old branches, buds, or even roots. For a while, the plant may not look its best, but pruning encourages healthy regrowth and helps increase the yield or quality of flowers and fruits. In nature, wind, ice, snow and salinity can cause plants to self-prune. This is called abscission.

In all wild areas that are home to elephants, the vegetation looks modified at the end of the dry season, the culmination of a year's pruning. Four months into the rainy season, these stark wooden statues explode into green leaf, making the same trees almost unrecognisable. Elephant pruning ensures that a tree is kept down at browse level for all leaf eaters, from dik diks at the base to giraffes, which shape its canopy.

Of course, most humans would prefer to see a majestic grove of cathedral mopane trees rather than an orchard of scrub mopane modified by elephants. But what can feed from such a high canopy? These trees have grown out of the browse level of even giraffes. So which grove is the more useful?

When elephants push over a mopane tree, the tap root often remains intact. The tree continues to grow on its side, with new branches coppicing towards the sun. This tree is very much alive and offering food to a much wider range of creatures down at this level. So the state of vegetation of a national park or elsewhere should not be judged at the end of the dry season. It's what you see at the end of the rainy season that reflects its true health.

There's further value in elephant action on trees: woody species encroach on grasslands and, if left unchecked, this process can take over grassy areas and dry vlei lines. By pruning and killing encroaching trees, elephants help protect the grassland for grazers, and help to create balance.

Playing God

Ever since colonial times, people have attempted to suppress, rectify and control 'pests' like tsetse fly and elephants, and 'vermin' such as lions, leopards, cheetahs and wild dogs, all seen to compete with livestock and crop farming. Many administrations in Africa sought to control tsetse fly by eliminating hundreds of thousands of wild herbivores. Between 1919 and 1957 in Southern Rhodesia alone, 657 334 wild animals from over 36 species were slaughtered in a mania of fly control. These included:

- black rhino: 374
- kudu: 86 981
- sable: 37 351
- roan: 5 347
- bushbuck: 40 399
- duiker: 184 973
- warthog: 73 146

Following a public outcry, the government implemented an alternative policy: erecting hundreds of kilometres of game fence north of Gonarezhou National Park to separate cattle from wildlife. Thousands of migrating wild animals, cut off from water, died agonising deaths along its length. When this resulted in another public outcry, the government overhauled the Tsetse Fly Department and installed a new director who had acquired tsetse experience in Uganda. What followed is recorded by Dr Colin Saunders in his book *Gonarezhou a Place for Elephants*:

The foundation of [the new director] Ford's preferred tsetse control policy was based on preferred elimination of the preferred habitat of the fly, dense, shady forest and bushy thickets. A large fleet of bulldozers was brought to Gonarezhou and surrounding cattle ranching areas and the wholesale slaughter of wild animals was replaced by the wholesale destruction of riparian woodland, riverine thickets, ancient forests of indigenous timber and any shaded areas in which it was considered that tsetse flies could take shelter from the bright sunlight in which they were unable to survive.

Clanking mechanical monsters worked night and day to clear the verdantly forested banks of rivers and streams, destroying the whole basis of intriguing ecosystems that sheltered an intricate network of flora and fauna, creatures big and small. The bulldozers created a wasteland where previously tens of thousands of centuries-old trees had lined the water courses of a shady paradise.

The Tsetse Control men left behind them a desolate scene of bare and tortured soil, piled high with fallen trees and littered with bushes and vines ripped from the earth. The vegetation was either left to rot or burnt – another stark reminder of the criminally wasteful habits of modern man in his attempt to conquer the earth.

Ford's policy was, in truth, even more disastrous than the method so cruelly used by his predecessors: animals can be successfully trans-located to a wildlife area in a matter of months – it takes generations for the giant riverine trees to grow and for the incomparable riparian forests and shady thickets to become re-established. Many of the senior members of Ford's department were disapproving participants or horrified onlookers as the giant trees came tumbling down before the irresistible might of the bulldozers. One of Ford's senior colleagues stated that 'the bush clearing was a monstrous measure to which I could not reconcile myself'.

The killing of wild animals by the tsetse department ended in 1970. By then, apart from ecological damage, in 38 years nearly 750 000 wild mammals had been slaughtered. All this destruction in only one country.

Elephant culling

In the early 1960s a number of scientists and research officers in a handful of African countries considered it necessary to cull elephants. The dictionary meaning of cull is: *to reduce the population of (a wild animal) by selective slaughter.* This is exactly what happened.

The main states that embarked on this exercise were South Africa, Zimbabwe, Zambia and Uganda. The only accurate figures I have available are for Zimbabwe from 1960 to 1995, during which 50 333 elephants were culled. This does not include those shot by the hunting industry, killed as 'problem animals' or shot for rations by the Department of National Parks and Wildlife. In parks such as Hwange, Matusadonna, Mana Pools and Gonarezhou, between 1960 and 1990, an elephant was killed per week in each of these parks for staff rations – about 9 000 individuals. This brings the total kill by the Department of National Parks in Zimbabwe alone over this period of only 3 decades to nearly 100 000 elephants.

Recent assessment from just one area – Sebungwe, Zimbabwe – in the 2016 report on the Great Elephant Census of Africa states:

There were estimated to be 3 407 elephants in the Sebungwe in 2014.[1]

This represents a decline of 77% during the 8 years between the 2006 and 2014 surveys. A total of 6 907 elephants were culled in the name of conservation between 1960 and 1995.

This decline from 2006 to 2014 was from poaching, Africa's present culling process. In another area, the Zambezi Valley, the 2014 survey found only 11 657 elephants, a decline of 40% (17 569 individuals) between the 2001 and 2014 surveys.

This paints a rather sad and serious picture for the future of elephants in only one African country, which is classified by a handful of scientists, ecologists and purists as having 'too many elephants'.

Value of an elephant

To have millions of hectares of land under cultivation for grain crops, tobacco, fruit, flowers, tubers and exotic timber throughout Africa is considered to be a sign of progress. Who could have a problem with that? It is, after all, how countries fund themselves, feed people and create employment. To do this, land is cleared of woody inhabitants. Let's call this loss 'progress tax'.

We need to remember, though, that tourism is another major revenue stream in Africa, and one of the top three industries for generating foreign exchange in countries like Botswana, Zambia, Kenya, Namibia and Tanzania. In these countries, elephants

Below, left to right

A rich package of seeds and nutrients, ready for dispersal.

A male baboon feeding off uncrushed seeds scavenged from fresh elephant dung. That elephants do not fully digest their food is a bonus to many creatures.

are among the most sought-after animals by visitors. Do elephants destroy more than they conserve?

Think about it this way. While the uninformed may claim that elephants destroy, damage, degrade or cause devastation, could this not be classed as a legitimate tourism tax or progress tax, just as agriculture extracts a tax on the land in order to get the end product successfully to the market? To use Zimbabwean tobacco as an example, it would be ludicrous to believe that all Zimbabwe's elephants could eradicate in 12 months more trees than those used for the curing of tobacco over the same period of time. In the 2016 tobacco season in Zimbabwe, alone, this was just under a million tonnes of firewood.

I leave you with a personal anecdote

In October of 1984 my wife and I moved to Mana Pools National Park to run Rukomechi Safari Camp, situated on the park's western boundary. At the time, the south bank of the Lower Zambezi Valley was home to the largest black rhino population in Africa – around 2 700 individuals. As a guide, I would encounter around five black rhinos a day. On the north bank in Zambia, however, the population had been eliminated by poachers.

The year of our arrival coincided with incursions by poaching gangs from the north. By October 1985, one year later, we were down to seeing one black rhino a week. By 1990 we would get excited at finding a rhino midden. In 1994, 10 years after we had arrived in the Zambezi Valley, the black rhino became locally extinct. This inconceivable tragedy has happened in my short lifetime. Back in 1984, when they were so abundant and I was so young and full of optimism, I could never have conceived of such a tragedy.

What about our elephants? No park is ever quite the same without the opportunity of interacting with these remarkable, magnificent animals. We need urgently to consider the many positive contributions they make to our economies, as well as to their habitat and those animals big and small that live within it. Should we lose this keystone species in the wild, not only will we all be so much poorer as a species and planet, but all we'll have left to show our grandchildren will be caged elephants and a circus jumbo.

Thirty years ago this island in the Discovery Channel of Lower Zambezi National Park was a treeless sand bank. The new *Faidherbia* growth germinated from seeds passed out in elephant droppings.

What looks, to some commentators, like destructive elephant damage in the dry season, makes way in the wet season for ample, lush, more nourishing vegetation for a host of other browsing animals by providing easy access to what would normally be inaccessible foliage.

A bull elephant in Tuli, Botswana.

© Dave Southwood

Rainbow elephants

Virginia McKenna OBE
Cofounder, Born Free Foundation

The elephants are blue
Blue in the evening's indigo
The dawn's first azure light.

The elephants are red
Red from the terracotta soil's
Warm wallowed mantle.

The elephants are silver
Moon ghosts between the trees
And shining through night's veil.

The elephants are grey
Without disguise of light or night
Grey shadows in the forest.

The elephants are gone
Mourned spirits in the air
Haunting our dreams.

13

Ensuring elephant survival through community benefit

Elephant survival depends on our generating appropriate motivation for local communities and their governments to value living elephants. Even if poaching were eradicated, elephant habitat fragmentation and loss would continue.

Romy Chevallier & Ross Harvey

In a remote enclave of northern Botswana, elephant herds migrate in and out of the Okavango Delta: north to Namibia and Angola and north-east to Zambia. Their migratory routes place them on a collision course with local communities: about 15 000 elephants in competition with an equal number of local inhabitants for food, land and scarce water.

Botswana is home to the world's largest remaining population of elephants, estimated at 130 451.[1] Living with or near them can be terrifying. Local communities use the delta's fertile floodplains to plant and harvest crops, and many families depend on these as their principal food source. But elephants are expert crop raiders, intelligently detecting the most nutritious food

sources. When farmers have waited patiently for crops to germinate and grow, only to have them raided by elephants, it can lead to intense human-elephant conflict (HEC).

There are solutions, however. Crop raiding is mainly opportunistic and can be mitigated by elephant pathways and development-free buffer zones.[2] Such buffers help to prevent the relatively frequent occurrences of elephants killing people (and vice versa) simply because they are close to each other. When conflict occurs, it raises the incentive for community members to poach, either for ivory or illegal bush meat.[3]

The Great Elephant Census estimates that Africa is losing roughly 27 000 elephants a year to poaching, mainly because of demand for ivory in East Asia.[4,5] Unless conservation is incentivised, community members living with elephants will value ivory more highly than the animals bearing it, and illegal killing will continue. Outside of formally protected areas, the land-use choices made by local communities will determine the extent to which wilderness landscapes are preserved.[6]

For conflict and poaching to be overcome, natural-resource management needs to be community based. It's essential to galvanise

One of the traditional villages located in the three conservancies sandwiched between Nkasa Rupara and Mudumu national parks in the Caprivi/Zambezi Region of north-eastern Namibia. The buy-in of the local communities is essential if wildlife and conservation are to have a chance of succeeding in the region.

political will among local and national land authorities to implement appropriate land-use planning measures. With elephants, that means building on scientific studies of elephant pathways and implementing effective measures of deterring elephants from human settlements.[7]

Incentives are generated by the social systems, good or bad, that shape human behaviour. The lucrative illicit trade network is one of those systems – including the storing of ivory for speculative purposes – that drives up the *exchange value* of ivory.[8] This undermines the non-consumptive *use value* of living elephants. The fact that there remain legal domestic ivory markets alongside an international ban on its import and export has also driven the illicit harvesting of elephant tusks. This is, however, changing, as the world's largest consumer markets move to shut down their domestic ivory trade.[9]

Nonetheless, simply driving the *exchange* value of ivory downwards does not immediately increase the *use* value of elephants. It's necessary, therefore, to ensure that any economic exchange value lost through a genuine total abolition of the ivory trade is replaced by greater elephant use value, such as increased wildlife tourism. And it's important that a fair portion of any such value accrues to near-park communities as they are critical allies in the fight against poaching syndicates and habitat loss.[10]

As things stand, poaching syndicates are able to co-opt local elites and recruit poachers from near-park communities relatively easily.[11] So conservationists' first line of defence should be to reduce rural poverty, include community groupings in decision-making structures, drive down the price of illegal wildlife products and increase the opportunity costs of poaching.[12]

To ensure revenues are distributed fairly, institutional design needs to improve accountability and transparency within the decision-making structures of communities and governments. Incentives need to be compatible with local values or they will be undermined by competing development priorities.[13] And they require long-term support. Too many donor-funded projects end prematurely, before the programme has become sustainable. What follows, then, is a questioning of community-based natural resource management (CBNRM) as a vehicle for benefit transfer to local communities living with or near elephants.

What is CBNRM?

Community-based natural resource management refers to the management of natural resources by local institutions for local benefit. It has expanded throughout Africa over the past 3 decades as a policy tool for conservation. This is largely in response to the failure of centralised colonial and post-colonial policies to effectively manage natural resources, promote equitable benefit sharing and secure the co-operation of communities in sustainable resource governance practices.

CBNRM can take various forms in different locations. It may place greater or lesser emphasis on commercial or subsistence resource utilisation, rely on consumptive tourism such as hunting, non-consumptive revenue streams such as photographic tourism, or a mix of the two.

If ivory cannot be sold on world markets, conservationists must carefully consider how to increase the use value of elephants. This cannot happen if the value of elephants is driven to zero. In this case, community members are likely to kill crop-raiding elephants. From a conservation perspective, if CBNRM is to work outside protected areas on private and communal land, it must be an economically competitive land-use option.[14] It should be a means of achieving simultaneous conservation and livelihood improvement. CBNRM policymakers need to calculate the opportunity costs associated with particular land uses, and consider how adequately to compensate those who are likely to lose out when conservation is chosen above other options.[15]

The practical implementation of CBNRM initiatives often faces profound challenges. In some countries, such as Zimbabwe and Botswana, resource rights are limited by the central government's decisions. Communities may be consulted but have limited say in decision-making. In most countries, user rights can be transferred for a specific timeframe through tendering, auctioning or other mechanisms. Resource-use rights are often allocated only on the condition that a community-based organisation (CBO) is established with a governing constitution, a resource management plan and audited annual financial accounts.

Policy discourse in sub-Saharan Africa

Some authors have described the colonial and early post-independence approach to natural resource management as state-dominated 'fortresses, fines and fences'.[16] A number of communities in southern Africa were dispossessed of their land and resettled elsewhere, often with little or no compensation. This form of governance has produced resentment towards conservation activities which are seen to undermine community livelihoods.[17]

Today, the discourse tends to be polarised around the best way to generate revenue for communities. Some conservationists support consumptive use, such as hunting, game harvesting and farming, intensive breeding, live capture and sale of game, and the processing of wildlife products.[18] Others support non-consumptive use, including game viewing, photographic safaris, adventure and cultural tourism, the breeding of endangered species for reintroduction into wildlife zones, and the production of forestry and veld products for handicrafts and medicines.

The latter group considers consumptive use as ultimately unsustainable because of its unknown and often unintended negative consequences. For instance, the captive breeding of rhinos is unlikely to satiate demand for rhino horn and could, in fact, exacerbate it.[19] It also contributes little to the preservation of the species in the wild. Moreover, corrupt hunting outfits may kill more than their quota of animals, especially where governance oversight lacks capacity or credibility.

A common point between these polarised perspectives, however, is an emphasis on increased tourism as an option for generating community benefits. Many African elephant range states are exploring new opportunities to expand and market wilderness-related activities, and many are also exploring alternative livelihood opportunities. However, tourism is not a quick win and there are a number of issues to consider.

- Tourism requires infrastructure and services to many areas presently without them.
- Elephants could come to be valued purely for their use value in generating revenue and not in terms of their inherent conservation value.
- Unless partnerships with local communities are well governed and felt to be mutually

beneficial from the outset, both parties are likely to become frustrated.[20] This is because conservation-related tourism generates largely private and unevenly distributed benefits.

- Within some developing countries, little of the revenue generated from tourism is retained and reinvested, and links between tourism and other sectors of the economy are often not well developed.

For these reasons, a CBNRM approach should be supplemented with programmes that raise awareness of the non-monetary value of elephants as keystone species of importance to the complex ecosystems they support.[21] Policymakers need to incorporate both market and non-market benefits into their development choices. This will help to attract investment in conservation rather than in uses such as trade in wildlife products.[22] It may also help to change deep-seated beliefs about the inherent value of wildlife. New beliefs about the importance of conservation cannot be forged through monetary benefit alone.[23]

Challenges of devolution, constraints and capacity

The CBNRM paradox is that it requires the state to confer strong rights over resources to local people – but that it can also withdraw those rights. This requires a delicate balance and the challenge is clearly illustrated in Botswana.[24] Some analysts have criticised the country for not devolving sufficient responsibilities to local institutions, which have become passive recipients of private-sector income rather than active resource managers.[25] This doesn't promote stewardship over wildlife and dilutes the link between responsibilities and rights. It also erodes the incentives for communities to care for the wildlife with which they live.

Besides the challenge of balancing centralisation and devolution, governance presents an ongoing challenge at the central, district and community levels. As with broader economic policies, the design of natural-resource governance institutions in sub-Saharan Africa is often driven not by considerations of technical efficiency, but by an array of personal interests revolving around patronage networks and the exercise of political power.[26] Devolving or decentralising rights over valuable natural resources may conflict directly with such interests and block CBNRM initiatives. Communities also face their own governance challenges, such as revenue misappropriation, poor record-keeping within community trusts and institutions, lack of transparency, poor financial decisions and resource capture by local elites. They also often lack the necessary skills, resources and technical capacity to effectively govern natural resources themselves.

National governments have a legitimate mandate to protect 'public goods' and to ensure the sustainable management of resources in the interest of the entire population. But the role of communities living in the vicinity of these resources is crucial for monitoring and enforcing regulations at the local level. Unfortunately, management rights issued by a central government are often weak, limited and conditional, yet with the government retaining important decision-making control over when and how resources may be used.

At a local level, more benefits need to flow directly to the most affected community members, but these are too often captured by committees or elected representatives.[27] For this reason, some conservationists have suggested moving beyond trusts and CBOs into formal corporate structures – community companies, for instance, based on norms of good governance and accountability.[28]

Key elements of successful CBNRM

What follows are the main elements that provide the foundation for effective CBNRM in southern Africa.

Flexible and resilient systems

Climate change is rendering Africa progressively more water-scarce. This means that places such as the Okavango Delta will become increasingly contested sites as people and elephants migrate into the area. Elephants will also probably move there as they attempt to escape the growing poaching trade in Zambia and Zimbabwe.

CBNRM models will have to be sufficiently flexible to respond to these shifts. They also have to consider wider market developments in both non-consumptive and consumptive tourism. The US and China, for instance, have shut down their domestic ivory markets, as will Hong Kong (which has announced that a ban will be implemented by 2021).

For CBNRM models that depend on trophy hunting, the world is also changing. After 'Cecil the lion' was illegally shot in Zimbabwe, three US airlines moved to ban the shipment of all hunting trophies.[29] A continuation of this trajectory would severely jeopardise CBNRM programmes that fail to adapt to this new reality. Also, with increased human-elephant conflict likely, conservation agriculture will become more important. Cash crops such as chillies can be grown, both as a means of deterring crop-raiding elephants and to generate revenue for local communities.[30] This dynamic approach simultaneously serves the dual ends of elephant conservation and livelihood security.

CBOs with strong governance

Effective community-based organisations require a strong, locally appropriate institutional framework. Governance systems operate within a set of embedded institutions and will succeed or fail depending on their congruence.[31] Ideally, they should be transparent, with built-in monitoring mechanisms to ensure that community trusts distribute resource rents equitably and efficiently. Also, smaller communities tend to work better than an amalgamation of many disparate villages.

Transparent distribution strategies

If CBNRM is to gain traction and benefit elephant conservation, governance challenges around effective revenue distribution must be addressed. CBOs should design their own structures in such a way that they ensure transparency and accountability, and instil a sense of ownership of community investments. Appropriate checks and balances need to be in place to ensure that benefits and decision-making do not become controlled by local elites; it's not merely about how the money is spent, but also about who decides how it is spent. The community trust should adopt governance practices congruent with existing traditional institutions (such as *kgotlas*) which have democratic aspirations of good governance.[32]

Effective channels of communication

Good communication is especially important in contexts of relatively low financial literacy. Even if financial reports are available for public scrutiny (an important requirement), it's not clear they would mean much to ordinary community members. Therefore, the board or an independent third party should communicate clearly about how and where benefits are transferred. This is essential to dispel perceptions in the community that the only people benefiting are trust employees and committee members.[33] Government policy practitioners need to communicate effectively and consult extensively before making major changes. In Botswana, for example, the implementation of a hunting ban in 2014 created uncertainty and risks for communities as well as investors and entrepreneurs involved in conservation, development and tourism.

Direct and indirect value from CBNRM

Sufficient revenue should accrue directly to local community members so they can take ownership of conservation objectives. As Orr points out, CBNRM initiatives must also fare well in comparison with other land-use options, such as agriculture and/or livestock.[34] Communities that bear the opportunity costs of living with elephants should be compensated in whatever way possible if they don't receive direct benefits. Ultimately, conservation should become a driver of development. That way, the conversation can move beyond how to create an optimal mix of non-consumptive and consumptive value for communities. Ideally, the inherent value of preserving the integrity of wilderness landscapes would become the motivating principle behind any development plans.[35]

Biodiversity improvements

Revenue accrued through CBNRM projects is often not used to attain conservation objectives, which creates a complex dynamic. Many communities reinvest wildlife-related revenue into livestock or agriculture, which can undermine conservation goals. For this reason, specific quantitative or zoning limitations need to be imposed to avoid ecologically destructive practices. Communities living with or near elephants should be encouraged to diversify livelihood strategies instead of being overly reliant on CBNRM revenues from wildlife tourism.

Robust systems

In southern Africa, land tends to be allocated for activities that attract the highest expected material utility against competing alternatives. Conventionally, this has resulted in short-term, market-based economic and financial approaches. But natural resources and ecosystems exist far beyond purely financial considerations, and include use values (direct and indirect) as well as non-use 'existence' values. Ecosystems clearly play an important role in alleviating poverty and enhancing community resilience. This needs to inform land-use decisions far more than it does at present. For instance, the contributions that elephants make to preserving ecological integrity and benefiting human welfare in the process are too often ignored.[36]

Conclusion

Elephant survival ultimately depends on generating appropriate incentives among local communities and their governments (at all levels) to value elephants more holistically, the more so because, even if poaching were to be eradicated, the risk of elephant habitat loss and fragmentation would not disappear with it.

The success of future CBNRM programmes will be determined by the extent to which they are able to balance existing wildlife, forestry and fisheries initiatives with new and innovative income-generating activities and alternative livelihood strategies. Having the correct institutions in place is critical to attaining this delicate balance. If we want to secure a future with elephants, we'll have to design more optimal CBNRM institutions that incentivise communities to choose wilderness landscape preservation over the next-best alternatives.[37]

Is there a middle ground for marginal lands?

Communities that live alongside national parks and game reserves are often some of the most impoverished in Africa. They are the ones who most frequently come into conflict with crop-destroying, life-threatening wild animals. They will also often be subjected to poor schooling and health-care facilities, a lack of running water and sanitation, and being miles from shops, even from mortuaries. Life on the front line can be tough and uncompromising. In many of these remote communities, livestock are just about the only source of wealth and prestige.

Yet, there is extraordinary potential value in some of these marginal lands around the Kruger (South Africa), Hwange (Zimbabwe), the Maasai Mara (Kenya), Serengeti (Tanzania), the Okavango (Botswana), etc. If rural communities and the tourism/wildlife industry could combine their assets and skills to create vibrant wildlife conservancies in such places, these zones could rank among the most valuable tracts of land in Africa.

This would mean that the safari industry would have to transform and dig deep to guarantee a steady income for their neighbouring community partners to reward them for converting their land from livestock to wildlife. The payments negotiated would need to be significant and regular, regardless of occupancies – and attention would need to be given to ensure that the women in these communities are not excluded. Ultimately, the visitor would pay more for the privilege of viewing wildlife in pristine, uncrowded environments. But that is a small, fair price to pay to ensure that the next generations from around the planet will have great wildlife-viewing opportunities in Africa.

The image to the left is a scene from the northern reaches of the Okavango Delta. The land is currently utilised for cattle, bushmeat and thatching grass – with smatterings of nomadic wildlife. Yet, these could become prime wildlife-filled Okavango floodplains, whilst the vacant lands further to the north could be transformed into superb cattle-grazing country through the provision of boreholes and water – with seasonal, well-co-ordinated, tightly bunched, night-time corralled, rotational grazing into these wetlands. Money would flow to communities from the lease fees and jobs created by safari camps that could be built here.

And if the partnership is working, an elephant or a lion will be viewed as their asset: a win-win for all.

Cattle regularly migrate into the northern Okavango, which is where human-animal conflict occurs.

This elephant was the victim of poaching and had a wire snare wrapped firmly around its trunk. The stricken animal was located and darted, and authorities removed the snare before it could sever the trunk. The aftereffects of the elephant's experiences with humans resulted in aggression when it was approached.

14

Funding elephant conservation

Be sure to check the motives of those who provide the funds.

Dr Don Pinnock

Elephants protect each other for free. The cost to humans of protecting elephants from ourselves, however, is high and – along with poaching and land encroachment onto wildlands – it's rising. Where that money comes from and the reasons we spend it is central to their survival. Their future on this planet is now in our hands. If we have a plan, what is it? What are the reasons that back it? And, in a rapidly urbanising world, is their survival even important?

This book has answers to the last question. But we need to take careful stock of the other questions because the answers are often not what they appear to be, nor are they entirely clear. Each morning, thousands of conservationists, environmental lawyers, game guards and camp support staff get out of bed and go back to saving bits of the planet, oblivious that they're operating within conservation paradigms of which they are mostly unaware.

In NGO offices, government departments and corporate boardrooms, some managers are making decisions based on calculations far from the passion and commitment of scientists and activists who are saving a natural forest from clear-cutting or a species from extinction. Conservation has a history that, over time, has set its balance on different fulcra

and continues to shift still today. For these reasons, it's important to know where conservation funding comes from and why.

Western notions about conservation of the natural world were based on sustainable *use* and, for the most part, are still there today. These attitudes have an old pedigree. In the 11th century, the English king William the Conqueror created several 'protected areas' – among them what is now the New Forest National Park – to preserve game and forests for noble huntsmen. In the 18th century, expanding cities in Europe needed wood: in Prussia and France, attempts were made to preserve forests from unauthorised cutting by peasants – and for the use of the elite. In early 19th-century India, British colonists, requiring teak to supply the Royal Navy, passed the first formal Conservation Act to prevent wildfires and the cutting of undersize teak trees by local communities.

Since then, protecting nature to ensure a ready supply of commodities or creatures has been the deep keel of most conservation efforts and laws. Land, plants or fish received protection, not because these things had a right to be shielded from human exploitation, but so they could be used in a more orderly, sustainable fashion. Whether it be bison, lions, grouse or tigers, it was generally a hunting lobby that campaigned for the protection of animals.

To industrialising countries of the 19th century, populations of plants and animals on land seemed infinite and, given the vastness of the oceans, any peril to life beneath the waves was inconceivable. Undeveloped forests and savanna were seen as wastelands up for grabs by anyone who could 'develop' them – meaning put them to use. In the United States, however, a group of artists, writers, philosophers and photographers, appalled at the dirty urban wastelands that 'progress' entailed, began to appreciate undeveloped countryside and to value what it offered. Notable among them were Henry David Thoreau, John Muir and Aldo Leopold, each a seminal thinker; together, they would launch a movement that became a counterpoint to the notion of sustainable use. Their view could be described as preservationist: nature protected for its own sake.

The opposing notions of conservation are best illustrated by the differences between Muir, who founded the Sierra Club, and Gifford Pinchot, who founded the US Bureau of Forestry within the Department of Agriculture under Theodore Roosevelt's presidency. Although initially finding common ground with Pinchot on forest protection, Muir's views soon diverged. Whereas Pinchot supported the sustainable use of resources within national forests, Muir believed national parks and forests should be preserved in their entirety, meaning that their resources should be off-limits to industrial interests. They could, he said, be a source of human pleasure but not human use. In 1902, Muir wrote:

We are beyond even the meadows and green fields. We are here alone with nature, surrounded by old primeval things. Tall forest trees, mountain and valley are on the right hand and on the left. Before us, stretching away for miles, is a beautiful lake, its waters calm and placid, giving back the bright heavens, the old woods, the fleecy clouds that drift across the sky, from away down in its quiet depths.

Pinchot, on the other hand, wished 'to make the forest produce the largest amount of whatever crop or service will be most useful, and keep on producing it for generation after generation of men and trees'.

Roosevelt was deeply moved by Muir's vision and, under his presidential watch, created five national parks and established the first 51 bird reserves, four game reserves and 150 national forests.

But the legislative undergirding of US conservation policy – and most countries worldwide – followed Pinchot's use doctrine.

What does this have to do with conservation funding? Still today, what to fund and why exists within a tension between preservation and sustainable use, between welfare and the market.

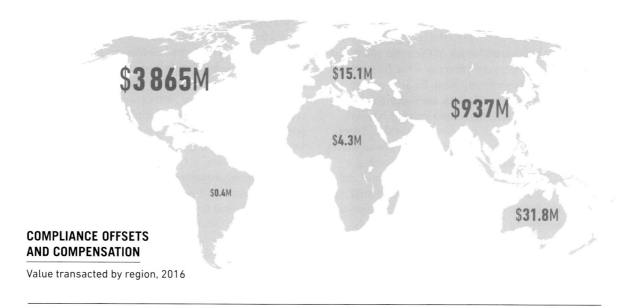

**COMPLIANCE OFFSETS
AND COMPENSATION**

Value transacted by region, 2016

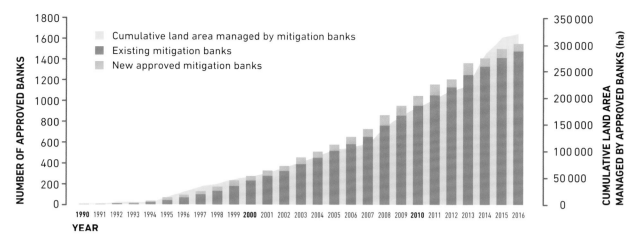

MITIGATION BANKS WORLDWIDE

Cumulative number of approved mitigation banks by year and cumulative land area managed by banks (1995–2016)

Preserving wilderness for its own sake

From about the mid-20th century, conservation was primarily preservationist and driven by national governments. It focused on establishing protected areas, working directly with endangered species and studying biodiversity. In many cases, it was spectacularly successful. At the start of the 20th century there had been only a handful of protected areas in the world, although some had already existed for generations. But by the end of the century, there were nearly 200 000 such tracts covering around 14.6% of the world's land and 2.8% of the oceans.

Conservation biologist Michael Soule underlined the importance of wild places and national parks:

The best current research is solidly supportive of the connection between species diversity and the stability of ecosystems. It has firmly established that species richness and genetic diversity enhance many ecological qualities, including productivity and stability of terrestrial and marine ecosystems, resistance to invasion by weedy species and agricultural productivity.

However, preservationist thinking had a fatal flaw that would come back to bite it. In its terms, conservation required the absence of people. In urging the Washington federal government to declare Yosemite a national park, John Muir urged it to rid the area of 'such debased fellow beings' – its native American residents.

In Africa, those places that colonial administrations declared parks, communities called home. In many cases, the land had been theirs for thousands of years, and they were already living in relative harmony with the now-to-be-protected wildlife.

In 1961, the Worldwide Fund for Nature (WWF) was formed, under Prince Bernhard of the Netherlands, expressly to protect wildlife areas of Africa from being overrun by 'native cattle' and cultivation as the West gave up its colonies. Its founders dreamt of creating a near-contiguous park system from Kenya to South Africa, and one that was under their control.[1] And, in many ways, they succeeded. By 2015, around 14 million people in Africa had been forcibly relocated to make room for wild animals, becoming what Wilfried Huismann, in his book *PandaLeaks: The Dark Side of the WWF*, calls conservation refugees.[2]

In India in 1972, the WWF launched Operation Tiger to save the big cats. Backed by massive publicity, it declared 2010 to be the Year of the Tiger. It failed to mention that between those dates it had virtually forced the Indian government to evict thousands of Adivasi forest dwellers who worshipped tigers and had always protected them from harm.[3]

Saving wild animals, it seemed, had a problematic underbelly, but for most of the 20th century, conservation's main emphasis remained setting aside tracts of land or water and developing ways to protect endangered species for their intrinsic and spiritual worth. In the decades that followed, however, sustainable use began gaining ground again as the primary conservation logic.

Sustainable use reappears

In a series of articles on **Mongabay.com**, Jeremy Hance traced the shift in conservation thinking to the rise of neoliberalism in the latter part of the 20th century. It was a movement that espoused deregulation, distrust in governments and deepening belief in free markets and private enterprise. The idea was that if we could incorporate the dollar value of nature into the current economic system – and convince policy makers and business people to recognise its economic value – we could save the natural world.[4]

Hance called it the rise of new conservation, but it was, in reality, original 'sustainable use' dusted off, provided with a 21st-century coat of paint and given the nod by big NGOs such as WWF International, Conservation International and The Nature Conservancy. The underlying logic was that if it pays, it stays. The unspoken corollary was that if it wasn't useful, there was little reason to sustain it.

This brand of conservation has three defining features. Firstly, in response to complaints by Third World countries that conservationists care only for wild animals and not people, its focus shifted, at least partly, from endangered species towards ecosystem services and poverty alleviation.

The second feature was a shift from government to corporate support for conservation projects, with governments moving from protection of biodiversity within their territory to facilitating what Hance calls neoliberal corporate conservation. Corporations are, by their nature, market-driven; and in return for what, in many cases, are extremely large donations, expect a return in the form of

green endorsements. The companies offering conservation funding certainly need such endorsements: they're often among the worst polluters in the world. Hance lists some of them:

> WWF has partnered with Coca-Cola, Domtar (a major paper company) and megabank HSBC. Wildlife Conservation Society has partnered with Chevron, ExxonMobil, Goldman Sachs and Total. Conservation International has partnered with BHP Billiton (the world's largest mining company), Chevron, ExxonMobil, Monsanto, Nestlé, Shell, United Airlines and Wal-Mart. The Nature Conservancy has partnered with BP, Cargill, Delta Air Lines, Dow Chemical, General Mills, Goldman Sachs, Newmont Mining, PepsiCo, Rio Tinto (a major mining company), Shell and Target. And all four NGOs have partnered with Bank of America, historically one of the largest funders of coal projects.[5]

Kierán Suckling, executive director of the Tucson-based NGO Centre for Biological Diversity, claimed that such relationships tainted many big environmental groups:

> They have started taking many millions of dollars in donations from very big, polluting corporations and putting [the companies' staff] on their boards of directors and advisory boards. Then they started promoting [these] companies' agendas, often by hurting not only the environment, but ethnic and poor human communities.[6]

Conservation was no longer about protecting animals, claims Wilfried Huismann in *PandaLeaks*, a stinging rebuke of the WWF, it's about using them to do business.[7] The NGO gives seals of sustainability to, among other projects, the palm oil industry in Borneo, which cuts down forest, threatens the future of orangutans and practises destructive monoculture; cruel salmon farms in Chile, which pollute formerly pristine fjords; and to companies that have turned the fertile Argentinean pampas into a toxic GMO soy desert.

Some 'sustainable use' conservation NGOs increasingly look like the corporations that fund them, with smart offices, beautiful annual reports, high-impact marketing campaigns and a culture of reproducing super success stories which, on the ground, are less than glossy. Their campaigns alternate between shock/horror warnings and beautiful, cuddly animals that 'urgently need your support'. And it works. The last donation figure Huismann could get for the WWF was $700 million annually, 50% of which went to personnel salaries.

A variation of the 'donate to save a cuddly' marketing pitch is crying wolf when no wolf is in sight. Money is raised, for example, to urgently move elephants 'about to be culled'. An example is three 'problem' elephants that were unnecessarily moved to a reserve in Mozambique at great cost (two subsequently broke out and escaped) amid a flurry of social media praising the NGO that undertook the operation and the US company that funds them. In another case, an elephant

conservation project was started, to great media fanfare but, on investigation, the promised funds never materialised. One needs to ask if these were necessary interventions or conservation theatre carried out purely for marketing purposes. For the animals, nothing is gained.

A variation is the straight marketing scam, where money is raised internationally to solve a crisis that simply doesn't exist. An example is a petition that appeared in mid-2017 on **change.org** to raise money to 'save 80 elephants who will be shot unless we move them' from Atherstone Nature Reserve in Limpopo. But there was hope: 'We have found a home for all 80 elephants! They will be able to roam freely with their families, safe from hunters and poachers.'

The plea was impassioned: 'You can make donations of any amount. The elephants thank you with all their heart. We can't let these beautiful

SPECIES BANKING ATTEMPTS TO MONETISE THE COMPLEX WEAVE OF NATURE

SPECIESBANKING.COM

Ecosystem Marketplace
Forest Carbon Portal
Community Portal
Valorando Naturaleza

HOME   UNITED STATES   INTERNATIONAL   NEWS & ARTICLES   RESOURCES   ABOUT US

International Programs and Banks

Speciesbanking.com initially focused on conservation banking of endangered species in the United States. As initiatives have developed in other countries, we have expanded our coverage to include these programs and provide insights for all involved. SpeciesBanking.com provides international information at two levels: at the Program level and at the level of individual biodiversity Banks.

- Program indicates any law, policy or program that drives biodiversity offsetting, compensation or offset banking for impacts to biodiversity.
- Bank indicates a site, or suite of sites, where biodiversity is restored, established, enhanced and/or preserved for the purpose generating certified credits that may be sold for compensatory mitigation for impacts to biodiversity.

Find a Program [   ] search   **Find a Bank** [   ] search

Offset and Compensation Programs ( ) and Banks ( ) by Region

Mitigation Mail

Don't miss a beat. Subscribe to Mitigation Mail, the Ecosystem Marketplace's eNewsletter on mitigation/conservation banking.

[   ] submit

- Mitigation Mail Archives

New to the Market?

2011 Update: State of Biodiversity Markets
Video Tour of SpeciesBanking.com
State of Biodiversity Markets report
What is SpeciesBanking.com?
US Conservation Banking Backgrounder
US Wetland Banking Backgrounder
Australian Market-Based Instruments Backgrounder

View All Key Resources

Biodiversity PES

Link to our dataset of 51 government mediated biodiversity PES schemes.

elephants die! This herd of 80 elephants is a close-knit, loving family and must be saved. With your help we'll save the fathers, mothers and babies.'

According to Dr Marion Garaï, who chairs South Africa's Elephant Specialist Advisory Group, there *was* elephant overcrowding at Atherstone, but culling 80 was simply not on the cards. The appeal was fraudulent.

A third feature of the new version of sustainable use has been the monetisation of the environment for the purpose of exchange, in line with the neoliberal commoditising of nature itself. A core proposition is that, given appropriate pricing mechanisms and private property arrangements, markets are the most efficient means of distributing goods, services and harm alleviation. The state's role is seen as simply to provide appropriate regulatory and supportive structures for the existence and functioning of expanding commodity markets.[8]

Funds for conservation, under this logic, are generated through compensatory mitigation, an exchange system that requires nature to be valued in monetary terms for the purpose of exchange. It has spawned many names: carbon trading, species banking, biodiversity offsets, biobanking, species offsets, green bonds and climate bonds. At root, these operate through a digital stock exchange – **speciesbanking.com** is a good example – that organises transactions between biobanks (essentially areas or projects doing good things for the environment in one arena) and corporate or state entities (wishing to trash it in another arena).

Monetising the complex weave of nature is essentially not possible, especially over extended timescales, but is nonetheless done by specialist groups that claim the ability. The website of **Absustain.com** explains what it does like this: 'Using Bayesian probabilities and multivariate analysis we translate complex datasets, where causality might lie within a number of different variables, into understandable and actionable insights.'

Wikipedia describes the process more clearly: 'Biodiversity banks rely on existing governmental laws, which forbid companies or individuals [to buy] up land in an area that houses, say, a critically endangered species. An exception is in place [in the US] which allows companies to buy up the land nonetheless, if they also buy a certain amount of compensation credits with a certified biodiversity bank. These credits, which represent a significant extra cost to the company, are then used to provide revenue for the biodiversity bank.'

In other words, landowners with an endangered species can get a permit to harm it if they are prepared to put money into saving a species elsewhere. **Speciesbanking.com** sells credits on everything from vernal pool fairy shrimps and valley elderberry longhorn beetles to tiger salamanders, gopher tortoises and prairie dogs. According to the environmental writer George Monbiot, compensatory mitigation 'subjects our landscape and wildlife to the same process of commodification that has blighted everything else the corporate economy touches'. But it's been hugely profitable in monetary terms, if not for the environment. In mid-2017, green bond issuances reached US$22 billion.

The reason for mentioning this in the context of elephants is that Africa, with its iconic species, is a wonderland of potential biobanks capable of soliciting mitigation funds. But by engaging in such a system, you need to understand the damage being permitted elsewhere. It's an ethical issue quite as problematic as receiving financial support from Monsanto or ExxonMobil.

The mitigation idea landed in South Africa in mid-2017 when the Department of Environmental Affairs published a Draft National Biodiversity Offset Policy. It describes biodiversity offsets as being particularly important in securing threatened ecosystems and critical biodiversity areas, but is less than clear about what damage the offsets would be marketed against.[9]

So where does this leave us? Preservation of the natural environment should, of course, be the task of national governments. But in Africa, conservation isn't an issue that gets you elected. For this reason, it tends to be dominated by non-government organisations funded by corporations, foundations and trusts. They're often steeped in neoliberal attitudes, operating in terms of sustainable use for humans.

Of course, it's important to ensure people are able to live in harmony with the continent's wild places. Rural communities are vital conservation stakeholders. The gold standard of big NGO conservation these days (and it's a good one) is that local jobs should be a priority. That's only acceptable, in my view, if those jobs involve repairing the already damaged environment and not carrying a hunter's rifle in prime wildlife habitats, or clearing forests for a soy or palm-oil plantation.

This goes for environmental workers as well. Increasingly, conservation funding is being dominated by requirements out of step with the ethic that propelled so many people into professions that care for the natural world. The big environmental NGOs are increasingly compromised by corporate association and are thereby often part of the problem. They have the money to do the job, but at the end of the day, who are they doing it for?

To protect elephants, funds are needed and, it seems, big, transnational corporations and trusts are best placed to provide them. Under duress it's tempting to take with thanks from whoever provides. It's extremely easy, sitting in some developing country caring for wildlife and in need of cash, to be co-opted by some distant organisation whose goals are diametrically opposite to yours. So here are some guidelines for conservationists working at the coalface who are needing funds to do their work.

- Before sending off that funding proposal, you need to do some homework on what the funder expects in return. Your good name? Your certification of their actions? Your logo on their masthead?
- Check who owns the corporation or trust; what they do; the size of their green footprint and who else they fund.
- Make sure that they are not using your efforts, name and reputation to 'greenwash' and polish their bad name.
- What say, if any, will you have in their internal environmental policies and they in yours?
- What conditions are contingent on your receipt of funds? Who owns your work? Where will it be distributed? Under whose name?
- Do the promised funds come from a mitigation or offset relationship? And if they do, please make sure that the benefits locally far outweigh the destruction elsewhere.

- Do not begin the project before all the funds are in your bank; delivery often falls far short of promises.
- Be cautious of funds attached to organisations like Safari Club International, which seek legitimacy for trophy hunting in prime wildlife areas. While non-subsistence hunting can preserve marginal areas for wildlife, trophy hunting takes out the best specimens and depletes the gene pool.

The guiding principles for assessing the support you get for the work you do really comes down to your rights and obligations as a citizen of Planet Earth. Does the funder accord with your ethics? These rights are lyrically described by the environmental philosopher Thomas Berry:

Every being has rights to be recognised and revered. Trees have tree rights, insects have insect rights, rivers have river rights, mountains have mountain rights. So too with the entire range of beings throughout the universe.

All rights are limited and relative. So too with humans. We have human rights. We have rights to the nourishment and shelter we need. We have rights to habitat.

But we have no rights to deprive other species of their proper habitat. We have no rights to interfere with their migration routes. We have no rights to disturb the basic functioning of biosystems of the planet. We cannot own the Earth or any part of the Earth in any absolute manner. We own property in accord with the well-being of the property and for the benefit of the larger community as well as ourselves.[10]

Be sure not to become ensnared in that old cliché: he who pays the piper calls the tune. Stand by your integrity even if, at times, it will reduce your funding. Don't become part of the problem that your life's work seeks to solve. And don't ever sell your soul, no matter how much you need the funding!

An elephant basking in the
golden early morning light
at Mana Pools, Zimbabwe.

'There comes a time when
humanity is called to
shift to a new level of
consciousness ...
that time is now.'

Wangari Maathai

Poachers' rifles, gin traps
and charcoal confiscated
by African Parks' rangers.

15

Poaching networks of East Africa

The poaching industry has many links in its chain of cruelty and corruption.

Carina Bruwer

For centuries ivory has been used to make piano keys, artistic carvings, cutlery, daggers and jewellery.[1] It was linked to conflict long before illegal timber and blood diamonds[2], and demand is not decreasing. Africa is losing thousands of elephants to unsustainable poaching to supply mainly Asian demand.[3] Asia has its own species of elephant, but only males have tusks and those of African elephants are much larger.[4] There are also considerably fewer Asian elephants, which means that Asian ivory demand is satisfied almost exclusively by African elephants.[5]

In 1989, after 20 years of intensive poaching, known as the elephant holocaust, the Convention on International Trade in Endangered Species (CITES) banned international commercial trade in ivory. Those 20 years saw nearly half of the continent's elephants being wiped out.[6] Between 1977 and 1987, Tanzania alone lost more than half of its entire elephant population[7], and in the 1980s Mozambique's population shrank from 65 000 to 7 000.[8] This decline was attributed to the legal regulated international ivory trade, causing CITES parties to vote in favour of the international ban.[9] It was thought that doing so would stop poaching. The confusion caused by an international ban in parallel with legal domestic markets in the countries with the highest ivory demand has sustained poaching.

In the 19th century, Tanzania alone had an estimated 20 million elephants.[10] In the following century, elephant numbers began a rapid decline as a result of growing human populations, human-animal conflict, habitat destruction, the ivory trade and, more recently, climate change.[11] Focusing on poaching often removes the spotlight from these other drivers of decline, though they have equally devastating effects. Today the African elephant population is perhaps half a million.[12] Despite this catastrophic decline, elephants are not yet listed as endangered by the International Union for Conservation of Nature (IUCN).[13]

Elephant populations generally grow by 5% a year. If poaching exceeds this, populations decline. Since 2010, more elephants have been poached than those that have died of other causes,[14] and poaching rates have surpassed population replacement rates.[15] Elephants could eventually face extinction if this continues, something bargained on by some syndicates who are stockpiling ivory in order to maximise their profits. Organised wildlife crime is high value and high volume.[16] Wildlife trafficking is the fourth most lucrative transnational organised crime globally, topped only by narcotics, counterfeiting and human trafficking. Trafficking networks are banking billions of dollars a year.[17]

Poaching hotspots

Africa's savanna and forest elephants are currently found in 37 range states covering an area of 3.1 million square kilometres, and most are found in southern and eastern Africa.[18] Not a single range state is unaffected by poaching. Forest elephants in Central and West Africa have been poached to devastatingly low numbers, causing the larger savanna elephant populations to be increasingly targeted by ivory trafficking networks. An estimated three-quarters of Africa's elephants are found in southern Africa in the Kavango Zambezi Transfrontier Conservation Area (KAZA), covering Angola, Botswana, Zambia, Zimbabwe and Namibia.[19] East Africa has 20%, and the remaining populations are found in Central and West Africa.[20] Despite East Africa having only a fifth of Africa's elephants, most poached African ivory destined for international markets is being sourced and shipped from there,[21] with Kenya, Tanzania, Uganda and Ethiopia all implicated in trafficking activities.[22] The East African region has been most affected by poaching, losing around 79 000 or half of its elephants between 2006 and 2016. The decline was mainly in Tanzania, where the elephant population crashed by 60% during this time.[23]

The East African region is extremely vulnerable to transnational organised crime. This has been attributed to the region's porous borders, geographical location, feeble law enforcement, corruption, weak government institutions, poverty and security vulnerabilities in harbours and airports.[24] The areas most targeted by poachers are Tanzania's Selous Game Reserve, Ruaha National Park and Rungwa Game Reserve, and Mozambique's Niassa Reserve. With increasing scarcity of wildlife in Tanzania, and with international attention provoking better law enforcement, poaching has begun to shift south, hammering Mozambique's elephants, which have declined drastically, and moving into South Africa's Kruger National Park.

The transit chain

The container ports of Mombasa and Dar es Salaam are the primary points of ivory export and are where most global ivory seizures have been made or from where seized consignments originated.[25] Most ivory leaves Africa by sea, with consignments often larger than 500 kilograms packed in shipping containers. The size of these shipments indicates well-orchestrated operations by organised criminal networks.[26] Smaller quantities are also trafficked by air and overland. Bole Airport in Ethiopia, which connects Africa to many countries, has seen many seizures.[27]

Kenya, Tanzania and Uganda are the African countries through which the most ivory is moved. Ivory being moved through East Africa is primarily from Tanzania, Kenya, Mozambique, Malawi and Zambia, as well as ivory from forest elephants in Central Africa.[28]

Along these land-based routes, couriers target weak border security, unmanned border areas or 'rat routes' where vehicles and people can cross undetected. Where official border crossings are used, corrupt elements often play a role.

Mombasa

Mombasa is East Africa's biggest port and, since 2009, Africa's biggest exit point for ivory on its way to Asia, followed by Dar es Salaam and Zanzibar.[29] Between 2009 and 2015, 43 tonnes of ivory were seized in, or originated from the port of Mombasa alone.[30] Port security is lax, officials corrupt and the port is geographically well situated for trafficking.[31] Seizures suggest that more Tanzanian than Kenyan ivory leaves from Mombasa, illustrating the ease of trafficking between these countries. The port is perfectly situated for ivory coming from Central Africa and passing through Uganda. With many container trucks travelling daily between the two countries, the risk of getting caught is low.[32]

Dar es Salaam and Zanzibar

Dar es Salaam is another ivory transit choke point, and the second-largest exit point for ivory leaving Africa. In the event of a seizure, the main players are rarely arrested, and there have been few successful prosecutions. The risk of being detected and prosecuted in Tanzania has been dismally low; this, and assistance by corrupt officials, make the trafficking easy. Enforcement efforts have reportedly improved since 2013 when dedicated law enforcement officials started carrying out raids and making arrests.[33] Corrupt port officials have also been removed.[34]

Zanzibar, where elephants are afforded no legal protection in terms of domestic legislation, is a hub for large ivory shipments. Its main port is exploited by Tanzanian traffickers who take advantage of faster cargo clearance than in Dar es Salaam. Zanzibar also links easily with various shipping routes to Asia and has many corrupt officials and ineffective controls.[35]

Illicit trade networks

Because of the clandestine nature of trafficking networks, little is known about their composition. They are as diverse as the countries implicated in the trade and are both fluid and informal. They consist of a web of supplier relationships, with ancillary services and actors, including governments, otherwise legitimate businesses and consumers.[36]

East African networks dominate the sourcing of ivory, as well as its movement through the East African region. They have influence in businesses and social relationships needed to traffic ivory to Asia.[37] They form part of non-hierarchical networks that organise themselves according to existing business, family, tribal and personal relationships, and readjust when there are threats from law enforcement. For this reason, and because these organisations know and understand law, local culture, logistics and politics, and can respond accordingly, law enforcement or conservation strategies are hampered.

Foot soldiers do the poaching, middlemen act as facilitators along the route from Africa to Asia, and buyers and kingpins call the shots. Where poaching was once opportunistic, it is now intentional, organised and commissioned. While many large shipments come from ivory produced by a mass killing, tusks from many smaller poaching events are consolidated into single shipments.[38]

Ivory trafficking networks are often linked to violence against people, but this differs from region to region. When leading elephant conservationist Wayne Lotter was assassinated in Dar es Salaam in August 2017,[39] it highlighted the potentially dangerous nature of networks that do not tolerate obstacles to their profit-making. There is also an element of state-sponsored violence resulting from the militarisation of conservation.

Poachers

Elephant killers can range from opportunistic poachers to skilled hunters, armed militias and military forces.[40] In Central Africa, government forces are often the poachers. In the Democratic Republic of the Congo, the Armed Forces of the Democratic Republic of the Congo (FARDC) have been described as some of the most ruthless poachers in the region, responsible for up to 75% of poaching.[41] Most poachers, though, are people who happen to be living around wildlife and who hunt as a means of income.[42]

In East Africa, they're mostly from communities living near elephant ranges, for whom the high income potential and low cost of harvesting ivory are incentives to poach.[43] It is not seen as a crime, but a means of income.[44] More systematic poachers can travel considerable distances within a country or across state borders. There are also reports of urban-based poachers who poach in an area until local communities push them out.[45]

Elephants are targeted both individually and in groups. Weapons range from traditional arrows and snares to high-calibre hunting rifles and military-style weapons, which complicate conservation efforts and endanger rangers.[46] Sophisticated weapons are widely available and often sourced from conflict zones.[47] Seized weapons from government stockpiles are also leaked to poachers.[48]

CITES has identified poverty as the biggest local indicator of poaching. The 2030 Sustainable Development Goals included combating poaching and the trafficking of wildlife by providing livelihoods.[49] The sustainable use of natural resources and wildlife can benefit both humans and animals, especially in regions with little income potential. Wildlife crime and ineffective conservation policies undermine not only animal populations, they also undermine development, security, good governance, the rule of law and the general welfare of people living with wildlife.[50] It also means that the only groups benefiting from elephants are poachers and traffickers. Local communities are the natural custodians of wildlife and should receive the biggest reward from them. Instead, losing wildlife affects their economic and social development and increases the possible pool of poachers.[51]

Middlemen

Middlemen acting as intermediaries are reportedly mostly corrupt officials or former poachers.[52] They operate in both Africa and Asia and can be local or international. The most prominent are East Africans and Chinese.[53] The latter have been arrested in Kenya, Uganda and Tanzania for ivory-related crimes – organising the procurement, processing and transport of ivory.[54] Other Asian nationals, such as Vietnamese, have also been involved in these trafficking networks.[55]

Middlemen and financiers provide poachers with supplies such as food, weapons, medical supplies and ammunition. They also have access to funds to bribe officials in order to keep the tusks moving. They pay transporters who can move ivory between consolidation points, containerisation points and across the ocean.[56] The more senior middlemen can be government officials, businessmen, diplomats and military officers.[57] Existing shipping companies are used to move the ivory, despite these companies not necessarily being aware they're being used for this purpose. Shell companies are created in order to hide the illicit product.[58] Once ivory reaches Asian shores, local middlemen take over logistics and sell the ivory to buyers or carvers.[59]

Kingpins

Kingpins are the least known connection in trafficking networks and are notoriously hard to track down. The size of shipments, as well as the concentration of poaching in specific areas, indicates a market controlled by a few such people.[60] There are also suggestions of a less formal hierarchy, where kingpins no longer play a central role, so their removal won't necessarily bring down the network.[61] Instead, these networks centre on those who organise logistics and never even come into contact with ivory.[62] This lack of a single kingpin means that there's often no coherent mafia-type organisation able to exercise restraint, but rather a number of entities competing for resources.[63]

There have been few arrests and prosecutions of high-level traffickers or kingpins, and it is this that should be prioritised by law enforcement instead of focusing on impoverished, desperate poachers who are easily replaced.

Finding and removing deadly snares.

© Peter Chadwick

Corruption

CITES has identified weak governance and corruption as the strongest national indicator of poaching.[64] These effects are mostly felt in developing countries where development funds are pilfered, contributing to greater poverty and inequality.[65] Corruption negatively affects conservation efforts where bribery circumvents environmental protection.[66] There are many reports of Chinese-organised criminal networks using their connections with African economic and political elites to facilitate their illicit activities.[67]

Colonial rule stripped indigenous people of their hunting rights, turning them into poachers. This was followed by state monopoly of wildlife, with political elites enriching themselves through wildlife, a practice that continued after independence, and continues today. The prevalence of corruption has been used to justify a total ban on ivory trade. High-level individuals facilitate poaching and trafficking by providing diplomatic cover through documentation and accepting bribes, or by abusing money earmarked for conservation.[68] Rangers can also be bribed to provide information on elephant whereabouts or patrolling patterns, police to transport tusks and provide weapons, or members of the revenue authority to turn a blind eye to containers loaded with ivory.[69]

Various organisations are active in the ports of Mombasa and Dar es Salaam, providing traffickers with a wide range of people who may be open to bribery.[70] The role of shipping companies and clearing agents in these ports is also woefully under-investigated.

Law enforcement

The primary response to the illicit ivory trade has been law enforcement. But this can distract from the need for a multifaceted response involving source, transit and consumer states. A start would be to restore a beneficial relationship between elephants and the communities living among them, as well as addressing the values that drive the demand for ivory in Asia. Understanding illegal markets is hampered by their secretive nature, but is key to implementing a holistic response to all aspects of the trade, and not merely attempting to suppress it through law enforcement. Law enforcement may provide a more immediate solution, but poverty eradication and rooting out corruption should be the long-term goals.

Boots on the ground

In the field

Organised criminals exploit gaps in criminal justice systems and law enforcement.[71] They often poach in areas where there is little or weak law enforcement. In a sense, it's an arms race: the more poachers there are, the more rangers are needed, just as increased trafficking requires increased border law-enforcement.

There are often calls for rangers and officials to employ the latest technology and surveillance, but these authorities are usually under-resourced and lack basic training on evidence collection. Also, rangers receive narrow paramilitary training without understanding the complexity of what they'll be dealing with in the field.

A lack of resources and expertise can lead to frustration, and result in shoot-to-kill policies and the militarisation of conservation, with the danger of violating human rights. Militarised conservation responses are now being employed in many sub-Saharan countries,[72] resulting in anti-poaching units trained to use military techniques, battle weapons, technology and forceful tactics to counter similar tactics by poachers.[73] The high price of ivory has given rise to a situation in which poachers are prepared to take the risk of targeting elephants in monitored and well-protected areas where they are more likely to come across patrols.[74]

The use of force in combating wildlife crimes was encouraged in 2013 by former Secretary General of the Security Council Ban Ki-moon, when he stated that a more militarised approach to wildlife crimes was necessary. Illegal hunting has resulted in more rangers and security companies being employed, new technologies used and many countries actively encouraging the killing of poachers.[75] Military responses run the risk of increasing the prevalence of weapons, endangering rangers, running up costs and alienating communities.[76]

On borders

Ports need to strike a balance between effective operations that continue to raise revenue and profits, while ensuring sufficient security and risk profiling to trace illicit shipments. Traffickers are aware of loopholes, and exploit them. Since 2010, and until recently, tea-carrying containers from Mombasa were exempted from scans, thanks to tea being one of Kenya's main exports.[77] In 2015, ivory coming from Mombasa and weighing nearly 8 tonnes was seized in Singapore and Thailand in tea-carrying containers. After the seizure, instructions were given to scan every container purported to contain tea leaves, but this is reportedly hardly ever done.[78] Basic changes go a long way to securing maritime and land-based entry and exit points, but the continent is short on resources and expertise.

Law enforcement should be supported by strong domestic legislation, with penalties reflecting the seriousness of ivory trafficking. Ivory trafficking is a transnational crime spanning continents; international law enforcement co-operation, in conjunction with demand reduction efforts is, therefore, the weapon most likely to succeed in disrupting the trade.

Snares are used across Africa. They're simple to set and devastating for wildlife.

© Grant Atkinson. Chobe River, Namibia

16

Managing cross-border elephant populations

What is to be done when elephants pay no heed to legal and national landscapes?

Dr Jeanetta Selier

African elephants once ranged across most of the African continent as part of an interconnected population. With the establishment of countries came artificial boundaries, dividing one area from the next, often with no logical geological or cultural features prescribing where these should be.

These boundaries started dictating where people and wildlife could and could not go. In time, growing human populations and rural poverty led to increasing demand for agricultural land, and the subsequent confinement of elephants and other large mammals to protected areas often too small to sustain viable populations[1] and unable to meet the space requirements of these wide-ranging or migratory species.[2]

As most of Africa's protected areas are at or near national borders, the ranges of elephants and many large carnivore species span administrative and political boundaries.[3] Lindsay *et al.* (2017) identified 45 elephant populations straddling the national borders of 34 range states across Africa (excluding Senegal and Guinea-Bissau, which may have no remaining elephants). Of these populations, 15 include more than a thousand individuals. The calculated number of likely trans-boundary elephants (360 499) is more than three times the number of elephants in insular national populations (115 306). This means

that at least 76% of the continent's elephants are members of trans-boundary populations.

This movement across international boundaries has consequences: such as an ad hoc approach to their management – particularly regarding human-wildlife conflict – and an inconsistent consideration of ecological requirements of the species. This results in mismanagement.[4] For these reasons, it's urgently necessary to develop and expand trans-boundary conservation areas as a shared resource.[5]

The term Transfrontier Conservation Area (TFCA) is defined by Hanks (2003) as 'relatively large tracks of land straddling frontiers between two or more countries and which embrace natural systems encompassing one or more protected areas'. TFCAs can encompass varying mosaics of land use and can incorporate private land, communal land, forest reserves and wildlife management areas. They embrace (where appropriate) consumptive use of wildlife.[6] The objective of trans-boundary conservation areas is not only the conservation of biodiversity, but also the economic development of communities within these border regions. One of the aims is to encourage the relatively unrestricted movement of tourists and large mammals through open boundaries.[7] This is especially important for regions with large-bodied, valuable mammals such

as elephants, which could negatively affect both ecosystems and local communities surrounding protected areas.[8] However, several political, legislative and implementation challenges exist that impede the effective management of such animals.[9]

What follows is an assessment of the challenges and opportunities facing conservation and management of elephants in the human-dominated, trans-border landscape of the Greater Mapungubwe Transfrontier Conservation Area (GMTFCA). This population is trans-boundary, overlapping the territories of Botswana, South Africa and Zimbabwe, and ranges beyond designated protected areas.[10] These elephants are exposed to a range of management practices preventing them from using the landscape freely, and are likely to get into conflict with communities on the edges of protected areas.[11]

Elephants are a high-value species for both consumptive and non-consumptive activities and, as a flagship species, attract large numbers of visitors to conservation areas in sub-Saharan Africa.[12] Several tourism and hunting outfits operate within and on the periphery of the current boundaries of the GMTFCA. All of them view or use a single cross-border elephant population that moves freely between the three countries. Selier *et al.* (2014), however, showed that the current hunting quotas for elephants within the GMTFCA are unsustainable, and that there is little or no consultation among these countries in setting hunting quotas. Each country determines its own quota based on restricted subsets of population data.

Photographic tourism is at present the main economic driver within the region and elephants are a primary drawcard.[13] Habituated, viewable elephants, including large bulls with trophy-size tusks, are important to this.[14] Excessive hunting will, therefore, affect photographic tourism in the Limpopo Valley by significantly reducing numbers of big bulls, and could affect the chances of viewing elephants in general.[15] Furthermore, because of its selective nature, trophy hunting will decrease the number of large-tusked bulls.[16]

Activities such as trophy hunting may not only influence the presence of elephants within a landscape, but can also cause animals to concentrate in areas with low human disturbance.[17] This can have a significant impact on an ecosystem, resulting in cascading effects such as the loss of large trees in

such areas.[18] More importantly, when management decisions are not made at the appropriate level for the species or system, a mismatch in spatial scale can occur, with implications for effective management and species conservation.[19] This can have far-reaching implications for cross-border elephants, where the stress effects of hunting may be transmitted to photographic tourism areas within neighbouring countries.[20] Elephants may be forced to use lower-quality resources, or stress can lead to an increase in human-wildlife conflict.[21]

Effective protection and conservation of elephants cannot be achieved by simply increasing the size of protected areas (PA) or improving ecological conditions, but requires the assessment of the local socio-economic conditions within an area.[22] A study conducted in the GMTFCA showed that factors such as GDP/cap[23] poverty and the encroachment of humans on PAs are important in predicting the abundance of elephants.[24] The GDP/cap is a reliable indicator of a country's investment in PA development, and is positively correlated with human welfare, which in turn influences attitudes towards conservation.[25]

Countries with a higher GDP/cap have higher investments in the management of PAs and lower levels of corruption.[26] Burn *et al.* (2011) showed that poor governance is an important driver in the illegal killing of elephants. Further, Smith *et al.* (2003) predicted that countries with a governance score of less than 3.1 would show elephant population declines.[27] Where elephant populations are trans-boundary, high corruption levels and poor governance in one range state have negative implications for neighbouring countries sharing such populations.[28]

Landscape fragmentation goes hand in hand with an increase in human densities, and several studies have shown that elephant numbers are negatively correlated with an increase in agricultural land and human densities.[29] Human presence, especially on the edges of protected areas, is continuing to increase, while state budgets will probably be unable to balance social and PA demands. This will increase human-wildlife conflict along the edges of protected areas and have negative impacts on biodiversity, leading to illegal timber and mineral extraction, bushmeat hunting and more frequent fires. Increased human densities on the edges of these areas may also lead to species extinctions, as can be extrapolated from the high extinction rates of carnivores adjacent to reserves across the world.[30]

The work conducted in the GMTFCA shows that effective protection of source populations in a well-connected system of PAs, where such populations are safe from human threats, remains key to the survival of these species in the developing world.[31] The inclusion of multi-use zones along the edges of PAs will further assist in this and provide for economic incentives through the sustainable use of natural resources. It could also reduce human-wildlife conflict. If we change the way we govern private conservation areas – benefiting landholders and rural communities through a combination of co-ownership and comanagement and policy changes – we could protect wildlife while addressing rural poverty and environmental injustice.[32]

To understand the consequences of management activities such as trophy hunting, and to implement an adaptive quota system based on population trends, long-term monitoring is essential.[33] Where clear objectives are in place, consumptive utilisation is manageable.[34] Monitoring constraints often cited are the cost of monitoring, long-term commitment and planning.[35] However, great advances have been made in developing more cost-effective methods of monitoring offtakes and population trends.[36] And, where there is an economic benefit through consumptive use, a portion of the revenue should feed back into monitoring. But failure to address social issues, such as the inequitable distribution of hunting revenues and lack of involvement of local communities, can undermine the success of hunting operations.[37] Illegal shooting reduces the number of animals available for trophies, impacting not only on the survival of species, but on the revenue to communities who bear conservation costs.

The monitoring of cross-border populations is, however, compounded by the legal landscape in which the species occur, and relies on co-operation between range states, and buy-in from politicians within these countries. Where species' ranges cross boundaries, co-operation and joint management require the development of co-ordinated legislation and policies to improve land-use planning.[38] At present, the southern African legal landscape regarding elephants is fragmented, and there's an urgent need for trans-boundary co-operation

at ground level to address a mismatch between national legislation of the individual countries and regional and international agreements.

Laws applicable to elephants vary, for example, between Botswana, South Africa and Zimbabwe, and the situation is compounded by the fact that not all three countries are party to all international treaties. On an international, regional and national level, elephants are considered a natural resource of great economic potential. It is a view held by all three countries that elephant survival is reliant on their economic value to people. For this reason, the Southern African Development Community Protocol (SADC), the Southern African Regional Elephant Conservation and Management Strategy, and national legislation in all three countries support community-based management programmes that regard elephants as a natural resource,[39] such as CAMPFIRE in Zimbabwe.

Several cross-border legal instruments exist that govern co-operation in the conservation and sustainable use of natural resources; but, at a political level, there seems little actual co-operation between southern African countries managing and using these cross-border, high-value species. Montesino Pousols et al. (2014) warned that if co-ordinated action is not taken quickly to improve land-use planning, a loss of biodiversity is unavoidable.

As the human population grows, demand for natural resources will increase. At present, southern Africa, with more than 50% of the continent's elephants, is considered a safe haven.[40] But with the demand for ivory increasing while the number of elephants elsewhere decreases, pressure from illegal ivory poaching will grow locally.[41] It is therefore vital for range states to collaborate and co-ordinate management actions. Since Africa's rural communal lands cover about five times the area of state-managed forest reserves and national parks, they're an important part of a solution to biodiversity conservation and improved enforcement.[42] For example, enhanced enforcement combined with effective engagement with local communities prevented Nepal from losing a single rhinoceros to poaching between 2011 and 2013.[43] Elsewhere, co-operation between countries, increased landscape connectivity and the ability to generate income from tourism have been shown to work successfully in increasing wildlife numbers.[44]

Conclusion

In making conservation decisions and limiting human-wildlife conflict, it is crucial to understand factors that affect the presence of charismatic megafauna in human-dominated landscapes. The persistence of wildlife is not dependent solely on environmental factors, but also on the legal landscape, a country's investment in PAs, poverty, the encroachment of humans and activities such as trophy hunting.

Solutions require collaboration between countries, the development of co-ordinated legislation, policies to improve land-use planning, plus the development of multi-use zones around protected areas and conservation corridors.[45] If trophy hunting is to take place, it's important for goals between that industry and conservation to be compatible, and that an adaptive framework is used to ensure sustainability.[46] This can be achieved only through the development of an offtake quota, strict regulations and long-term monitoring of offtakes and population numbers. A single, multi-jurisdictional, cross-border management authority regulating the management and hunting of elephants and other cross-border species needs to be established as a matter of urgency.

Immediate action is needed to conserve and protect elephants and to manage the impact of human population increases and activities in their rangelands. We must plan to enhance biodiversity protection, promote sustainable development and improve quality of life for communities. Key to this is the effective management of source populations in a well-connected system of protected areas, where they are safe from human threats in the developing world.[47]

Conflicts can arise when a national park abuts an international boundary and there's no corresponding park or protected area on the neighbouring side. To the south of the Chobe River is Botswana's Chobe National Park. Along the northern banks in Namibia, land is tribally owned and home to impoverished communities, where trophy hunting dominates. Elephants in Botswana cross the river to feed on the lush Namibian grasslands, but soon swim back south to safety. Because Chobe National Park is on the more prosperous side of the border, there's higher employment, better policing and consequently greater safety.

An extraordinary gathering

Dr Paul Funston / Panthera 2017

I don't want to exaggerate, but herds of 500 to 600 have been commonplace at the lagoon at Horseshoe Bend in Namibia's Zambezi Region (formerly Caprivi Strip) during the last two dry seasons. I have twice seen aggregations there of what I estimate to be 1 000 elephants. On these afternoons there are elephants everywhere you look – out on the floodplains, coming and going at Horseshoe and everywhere around. These are truly spectacular events and quite nerve-racking for observers as you can't move for elephants, not all of whom have a positive disposition. It's not only at Horseshoe where we see this aggregation in September/October. Sometimes they come down to a nearby omuramba – an ancient river bed. One afternoon my wife and I were on our boat on the river there and literally hundreds of elephants were swimming across the river to spend the night grazing out on the floodplains.

The historic photograph (**below**) from Peter Beard's exceptional book *End of the Game* (first published in 1965) gives an indication of the size of some of those massive East African elephant herds of the 1960s and 1970s. No one thought they could ever be repeated. But here (**left**) is part of a massive herd of a similar size in 2017 on the Botswana/Namibia border.

Carved ivory for sale in a curio shop in Asia.

17

The illegal wildlife trade

A conversation with Karl Ammann:

'The only hope for elephants and rhinos is if the risk factor is ratcheted up with some of the linchpins ending up in jail. Hit the supply chains.'

Dr Don Pinnock

On the banks of the Mekong River in Laos is a palatial casino named Kings Romans where you can order freshly killed bear cub steak, grilled pangolin, tiger penis or gecko fillet and wash them down with wine matured in a vat containing lion or tiger bones.

The shop offers rhino horn libation cups and bracelets or, for more conventional tastes, religious sculptures and jewellery made from poached African ivory. After a night at the gambling tables, you can pay a beautiful young woman to accompany you to bed. Chinese guests are preferred.

If you're unable to settle your gambling debts, however, you will be locked in the local jail until your relatives pay. If they don't, you could, apparently, be hurled off the roof.

Kings Romans is one of a number of such establishments in the Golden Triangle, thickly forested borderlands between Laos, Thailand, Myanmar and China. It's an area of lawlessness and rebel armies, from which much of the world's heroin and amphetamines come. Similar 'resorts' include Allure, God of Fortune, Fantasy Garret, Regina, Mong Lah and Boten. These destinations lie at the end of a conduit of death for an unimaginable number of Africa's iconic animals.

This information is offered matter-of-factly over a cup of coffee at Cape Town's Waterfront by an unusual Kenyan-based undercover investigator, film-maker and self-confessed troublemaker named Karl Ammann. Unusual because he works alone and digs out explosive information, often at considerable financial cost to himself. A troublemaker because he's uncompromising in exposing wildlife traffickers, as well as governments and respected international conservation organisations when they become part of the problem.

His motivations – he's inquisitive and has a fierce desire to protect wildlife – are often suspect because he has no political or organisational affiliations and doesn't seek to raise funds. He's an elegant, widely travelled, deeply knowledgeable, principled maverick, and delightful company. But how reliable was his information? Corroboration came from a startling report, *Sin City*, compiled by the Environmental Investigation Agency (EIA) in conjunction with Education for Nature Vietnam.[1]

'Laos', begins the report, 'has become a lawless playground, catering to the desires of visiting Chinese gamblers and tourists who can openly purchase and consume illegal wildlife products and parts, including those of endangered tigers.'

Game guards inspect a forest elephant killed by poachers in Dzanga-Ndoki National Park, Central African Republic.

'There is not even a pretence of enforcement. Sellers and buyers are free to trade a host of endangered species products, including tigers, leopards, elephants, rhinos, pangolins, helmeted hornbills, snakes and bears, poached from Asia and Africa, and smuggled to this small haven for wildlife crime. [It is] largely catering to growing numbers of Chinese visitors.'

Ammann began exploring the jungle regions of South-east Asia while visiting his sister and brother-in-law, who ran a hotel in Bangkok and discovered the village of Mong Lah, the last Burmese outpost before China. 'That was 40 years ago', he says wistfully. 'Today, it's another sordid casino town thriving on drugs and prostitution, but then it was beautiful. I made contacts, went on expeditions, met hill tribes.'

On return visits, he realised things were changing fast and he began to document them. Wildlife trading was becoming an issue and he used his connections to probe further, first with questions and, later, with sophisticated button cameras and secret recordings.

'Because of my economic background [he worked in the hotel business], I was fascinated by the changing dynamic from sleepy hill station to illicit marketplace and conduit into China', he says. 'I was able to track changes in the area and thought I could make a contribution to conservation by letting the world know. It became something of an obsession.'

Those changes were to be devastating for elephants, rhinos, pangolins, tigers, bears and many creatures interesting to Asian taste, status, superstition and aesthetic. In the uncontrolled, drug-saturated Golden Triangle, the illicit was profitable, and law the prerogative of anyone wealthy enough to arm and command unscrupulous men. The area was to become, alongside trafficking of narcotics and humans, China's illegal wildlife supermarket. Ammann tried to get information out about what was going on but, he says, nobody seemed interested. The area was a blank on the media map.

The transformation of Mong Lah became a model for the establishment of lawless outposts across the region, catering for largely Chinese customers in search of products and pleasures forbidden in their country. Across the Mekong River, in Laos, a Chinese company acquired a 99-year lease on 10 000 hectares of riverside jungle and built Kings Romans Casino, giving the government a 20% stake. Around 3 000 hectares have been declared a 'special economic zone': essentially a private fiefdom. Clocks there run on Beijing time, trade is done in Chinese currency and businesses are Chinese owned.

According to Ammann, these casino towns make their own rules. Sellers and buyers are free to trade endangered species, and government leaseholders (in the case of Boten and Kings Romans) and a rebel army (in the case of Mong Lah) in and around the Golden Triangle curb any potential law enforcement. According to the EIA report, 'The blatant illegal wildlife trade by Chinese companies in this part of Laos should be a national embarrassment and yet it appears to enjoy high-level political support from the Laos Government, blocking any potential law enforcement'.

Other developments include a private landing dock for boats, a hotel, massage parlours, museums, gardens, a temple, banquet halls, an animal enclosure, a shooting range and a large banana plantation. Unrestrained by – and protected from – any known law, illegal wildlife trade is booming.

Ammann acknowledges the value of reports such as *Sin City* and the integrity of the EIA, but tells me they don't go deep enough. 'You can't find out about these networks the way conservation NGOs do by going around with a notebook logging items. You have to infiltrate', he says, hunching over his coffee and looking the part. 'That means sometimes buying from sellers – and I do that – and working through local operatives.

'The moment money changes hands it becomes much easier. You get information you wouldn't get by just snooping around. So I'm pushing the envelope, which most NGOs have a problem with.

'I send in my guys as bogus sellers of rhino horn. They show photographs and say: "We can get access to this. How much would you offer?" In a contraband investigation that's pretty common, but in the wildlife trade, few people are willing to go to that extent. If I give NGOs this data, they say they need to verify it. But they're not prepared to use my methods, so how can they do that?'

Ammann's methods of tracing networks through secret recordings and a bogus website he set up have paid off. He's traced the circuitous smuggling routes out of Africa and tracked down crooked officials and countless bogus CITES export/import permits.

He found the wildlife trade in Vietnam to be dominated by a handful of key players who are behind the container imports. They have the infrastructure in Africa to get the containers loaded and shipped. They work with retailers, sending cellphone pictures ahead signalling, say, 20 tusks or horns on the way. They've operated with port authorities and key dealers for many years.

In 2008, China and Japan together legally bought 108 tonnes of ivory from African countries in a CITES-sanctioned sale, and on the strength of that, China built the world's largest ivory-carving factory. Two years before, it had listed ivory carving as an Intangible Cultural Heritage. At that stage, according to a report, *Out of Africa*, by C4ADS and the Born Free Foundation, China had 67 registered (and countless unregistered) carving factories and 145 retail outlets. A survey of the outlets found that most ivory items had no identity cards, meaning their source was illegal. In 2013, a seizure of contraband in Guangzhou included 1 913 tusks, meaning almost 1 000 dead elephants.[2]

A 2002 document sourced by the EIA includes a Chinese official reporting the loss of 99 tonnes of ivory from government stockpiles, greater than the amount procured in the 2008 one-off sale. An NGO report in 2013 estimated that 70% of the ivory circulating in China was illicit and that 57% of licensed ivory facilities were laundering illegal ivory.

At the end of 2016, Chinese authorities announced they planned a domestic ivory trade ban by the end of 2017. The State Forestry

Administration, which oversees the trade, ordered all 67 factories and retailers to close. Zhang Dehui, the agency's director for wildlife, said another 27 factories and 78 stores would also have to close by the end of that year. This was celebrated as a game changer for elephants.

However, Ammann is deeply sceptical. 'In all the backslapping which followed I have not seen any reference or comparison to the fact that domestic trade in rhino horn and horn products has been outlawed in China for the last 25 years. This has made no difference and, as the affluence of Chinese consumers rises, so does demand. Since the horn ban, demand has moved from health to wealth, with ever new jewellery and other artefacts carved of rhino horn coming on the market. And poaching in Africa continues unabated.

'So why would eliminating the domestic trade in ivory be any different? I'm not sure to what extent China's enforcement activity is real. It's mostly for Western consumption. They sacrifice a shipment every now and then and that's probably part of the plan. In Vietnam the story is they give the container back to the dealer after six months.

'If traders get a tip-off that the Chinese government is curbing the sale of ivory in China, they send the message down the line saying shift your ivory somewhere else: Laos, Burma, Vietnam. That's where some of the big dealers have set up their operations, places like Kings Romans or the capital, Vientiane. It just means the conduit routes to China are shifting.'

There have been reports that, because of the impending ban, the price of raw ivory is dropping. Again Ammann is sceptical. 'What's hardly ever discussed is the retail price of worked products. On my last trip to South-east Asia, I enquired about the cost per gram of items like an ivory bangle, bead necklace or carved medallion, and the prices quoted were U$2.5 to U$3 per gram. These prices have not changed since I was last there. So if prices of raw ivory have declined as stated, all it means is the carvers and retailers have increased their profit margin.

'And carving is getting cheaper to do. Most items produced for Chinese buyers in countries bordering China are no longer hand-carved pieces of art sold to connoisseurs and investors. They're mass produced. The workshops are equipped with computer-controlled carving devices, in some cases imported from China. The people operating them are not traditional carvers but IT experts who produce eight identical items at the same time, drastically reducing the production cost. The end result might be more traders getting into the game, offering an ever-larger number of new items.'

I shift the conversation to regulation. CITES spends millions of dollars to ensure that the future of endangered species is being taken care of. The backbone of this process is its permitting system, which regulates international trade. It's another Ammann bugbear.

'I've found disturbing gaps in this process', he says, 'and I question the value of CITES mechanisms and the honesty of many of its officials. They're not really conservationists. It's all about not rocking the boat and pretending everything is hunky dory and that they're fulfilling their role. Any information that doesn't fit that picture, they try to hide.

'I've publicly accused them of covering up a wide range of corrupt and criminal activities. I have the evidence. I've asked them to take me to court so I can present information they ignore. They won't. Even with solid evidence, they're actively covering up for China. So where do you go from here?'

An accusation like that needs strong proof to back it, I tell him. Can he supply that? For the next 2 days my inbox pings constantly with documents, reports and photographs that make for startling reading.

According to the CITES website, permits for the transshipment of animals, reptiles, fish or plants are issued in terms of three appendices, depending on the degree of protection a species needs. If a creature is listed as Appendix I, an import and export permit should be issued only if it is not going to be a commercial transaction, has been legally obtained, and the animal's removal is not detrimental to the survival of the species.

For the last point – survival – an exporting country has to provide a 'non-detriment finding' done by the CITES scientific authority, which is meant to be a check on that country's management authority. Wild-caught, Appendix I-listed creatures cannot be exported for commercial or zoo purposes.

For Appendices II and III – species not endangered – export permits, but not import permits, are required and conditions are generally less stringent.

In this process, the devil is in the detail. If a creature is bred in captivity in a CITES-approved breeding facility for CITES I-listed species, then whatever the status of its wild cousins, its 'source code' on the permit is listed as 'C' (captive-bred) and it can be traded.

This has led to what Ammann calls the 'C-scam' and has been used, for example, to illegally export hundreds of wild-caught chimps and gorillas to China from African countries that have no captive breeding facilities. CITES officials in the exporting and importing countries, he says, must know this, but supply the permits anyway and turn a blind eye.

The UN Office on Drugs and Crime acknowledges the complicity of CITES officials in permit scams. 'Corruption', it says 'involves a variety of actors, including the CITES-competent authorities, public officials, villagers, forest rangers, police, customs, traders and brokers, professional/international hunters, logistics companies (shipping lines, airlines), veterinarians, and game farmers, among others.' [3]

There's another problem. Originally all permits had to be routed through the CITES head office in Geneva for inspection. But in 2002, claiming budgetary restraints, the CITES Secretariat unilaterally decided that permits need only be cleared by its officials in the countries concerned and only reported in summary to Geneva.

It also discontinued 'infraction reporting', done when member states were suspected of not acting in compliance with Convention rules and where corrupt and criminal acts might have been committed. For the wildlife traders who hire the poachers, the leaky process was heaven sent.

At the 2002 CITES meeting that year, the host country, Chile, called for a mechanism to urgently limit the circulation of CITES permits to avoid their fraudulent use, but the Secretariat shot down the proposal on technicalities.

According to Ammann, the point was made at the time that countries with poor governance records had resisted exposure to a name-and-shame regime administered from Geneva. So CITES policymakers decided the easiest way to solve infractions was to stop looking for them.

'The philosophy of the Secretariat seemed to be that it wasn't a good idea for people all sitting in a glass house to throw stones', says Ammann, staring sadly over the Cape docks as seagulls scream at us for tidbits. 'It was the end of attempting effective regulation of trade in many endangered species.'

Officials, syndicates and poachers in certain CITES countries with valuable wild species quickly realised that less control meant more opportunities to advance personal interests. Some dealers assembled special wild population capture teams. A CITES official in Guinea told Ammann: 'CITES is the dirtiest of the conventions when it comes to the falsification of permits and fraud'.

By infiltrating networks, Ammann obtained a sheaf of permits from traders involved in a wide range of such transactions. They were using the Middle East and North Africa as transit points, mostly falsely declaring these countries as the points of origin, based on permits stating that the primates were captive-born. When Ammann presented the head of the Egyptian CITES delegation in Geneva with documentary evidence of this, the official claimed it was fake and threw the report into the street outside the conference centre.

'There are worldwide mafia networks of interlinked dealers conducting their business openly', says Ammann. 'They all claim to have good relations with the relevant CITES management authorities and are able to get pretty much any CITES export or import permit they want. In the case of apes or manatees, the standard fee asked for an illegal permit across Central Africa is US$5 000.'

The buyer is then free to stipulate whether the source code is wild- or captive-born, and the management authority will fill in whatever the buyer requires. They don't need a detailed address of a shipment's destination.

'Anybody can fill in pretty much anything with regard to the final destination or the facility sending or receiving them. I analysed over 100 such permits and not a single one had the required exit stamps, or information from the relevant customs authorities about the specific animal or numbers actually inside the crate.'

One permit for two tortoises was used to sanction supply of two elephants. A permit issued for African grey parrots was used to export four African manatees to China. Some animals were shipped from Guinea with a falsified DRC permit.

On at least two occasions, traffickers told Ammann that if the buyer insisted on a proper but falsified permit for apes, they'd make sure their own CITES official would not file the duplicate copies of the permits when they did their annual reports to Geneva. They'd ask the buyer to do the same to ensure the importing authority would not report the imports.

In Japan, the Environmental Investigation Agency (EIA) found that after CITES approved the 'experimental' lifting of the ban on ivory trade in 1999 and 2008, the sale of illegal ivory skyrocketed, as did poaching in Africa. Even after the ban was reinstated, the EIA found that more than 1 000 tusks of dubious origin were being traded in Japan each year, many probably sold on to China.

Between 2010 and 2012, a Chinese husband-and-wife team was caught smuggling almost 3.26 tonnes of ivory from Japan into mainland China, using Chinese nationals in Japan as intermediaries.

The EIA described the trade as loopholes within loopholes: 'Japan is awash with ivory and not a shred of real evidence is required by law to ensure that ivory is of legal origin and acquisition.

'There's no doubt that hundreds of fake and falsified permits are being issued annually', says Ammann. 'In terms of bribes collected, tens of thousands of dollars are ending up with CITES officials in the countries concerned, who see benefit in the trade as a way to increase their personal income, whether legally or not.'

Member governments are obliged to prosecute offending traders and officials, based on illegal activity. They must then confiscate the animals in question and discuss with the countries of origin a possible return of the animals. That's the theory. In practice there appears resistance at every level, starting with the CITES Secretariat, which will not push members to have this enforced.

'The question that needs to be asked', says Ammann, 'is whether this lack of will by the Secretariat to enforce the Convention is a major contributing factor to these illegal transactions. Is illegal trade being actively encouraged by this lack of control?'

Ammann recently had a conversation with a top UN official who knew what had been on Ammann's mind when requesting the meeting. The first question came from the official: 'Are we better off with CITES or without it?' – a question critical observers had asked many times before.

'There's little doubt', said Ammann, 'that in most cases [the official] gets the answer he's looking for: whatever the flaws, we're better off with some kind of regulatory framework than without one. My response at the time was the same. Today I am no longer sure.'

With many recent permit infractions, China seems to be the main beneficiary and appears to have been granted a special hands-off status to do as it pleases. As long as Chinese demand exists and circumventing international conventions has no consequences, the killing and trading will go on.

In a recent trip to Asia, Ammann came across a racket with live elephants. So-called captive elephants in Laos were being bought by Chinese agents and walked over the border into China by mahouts near the border town of Boten, destined for Chinese zoos, circuses and safari parks.

'Many mahouts told me on camera their elephants are captive-bred but have been sired by a wild bull elephant. To avoid stud costs, mahouts in Laos tie captive-bred females to trees in the forest so that they can be mated with wild bulls. Under CITES Appendix I, an elephant with a wild parent in an uncontrolled setting is not considered captive-bred and therefore may not be sold commercially. So this trade is illegal.

'The Secretariat should have recommended the suspension of Laos for non-compliance and lack of enforcement a long time ago', he says, 'but they're a UN body and rocking the boat is not in its nature, even though suspension is a vital enforcement tool at its disposition and re-admission is an option once compliance has been achieved'.

The truth is that the demand for wild animals alive or dead remains high, which is not good news for Africa's animals. 'Wildlife traders are running circles around us', says Ammann, calling for another coffee. 'They're fooling us and most of them are Asian. And most of the NGOs – EIA is the exception – have operations in China, Thailand or Vietnam so they can't rock the boat too much. For an NGO, being banned from a country is a big problem.

'They can be the good cop but can't afford to play the bad cop. I can afford to be that cop. The problem is getting the information out. Where and how can it make a difference?'

The only hope for elephants and rhinos and other creatures, he says, is if the risk factor is ratcheted up, with *some* of the linchpins ending up in jail. Hit the supply chains. 'If the world really became serious about enforcement instead of becoming serious about talking about enforcement, it would be a major step in the right direction. But it will only come on the back of face loss. If necessary, we have to name and shame.

'But for myself, I want to be able to look at myself in the mirror in the morning. So I keep telling the facts and truth as I see them, knowing full well that I will not win any popularity contests.'

Clockwise, starting left

An elephant trunk and bushmeat on sale at a morning market in Cameroon.

Ivory jewellery on sale in Asia. Although some states, including China, have banned the working of ivory, Chinese customers have no trouble accessing it locally or in neighbouring states.

An ivory trader in Hong Kong greets local ivory protestors.

© Martin Harvey

© Kate Brooks

© Todd R. Darling

'For me, wildlife trafficking is
not a wildlife story,
it's a crime story.'

Bryan Christy

A poacher captured in
KwaZulu-Natal, South Africa.

18

Arms and elephants

Elephant poaching is no longer exclusively a conservation issue. 'Blood ivory' and 'ivory-funded terrorism' have become popular appeals used by defence agencies, major foundations and NGOs as a call to action for foreign military intervention. But more guns is bad news.

Kathi Lynn Austin

In September 2014, I travelled to an air force base in South Africa on the outskirts of Pretoria to attend the biennial Africa Aerospace and Defence Exhibition (AAD),[1] Africa's biggest arms show and South Africa's top-grossing event.[2] A side theme of the arms trade fair was anti-poaching efforts, which had brought me there for two reasons: firstly, to see for myself the cutting-edge weaponry and technology that was being marketed by the world's leading arms companies for use in Africa's so-called poaching wars; secondly, to meet wildlife protectors serving on the frontlines so I could learn as much as possible about poacher weaponry.

The first goal was easily achieved. As I made my way through hangar-sized exhibition halls and cavernous showrooms, sales representatives were keen to point out which of the military-grade armaments and dazzling technology products on display best suited anti-poaching operations. Beyond the usual firepower, I was introduced to the world's most advanced drones with briefcase-size data-processing systems; sophisticated electronic and infrared detection equipment; and aircraft that had been specially adapted for spying over wildlife range areas.

A side event sponsored by a business-oriented lobbying group, the US Chamber of Commerce, targeted an audience already predisposed to viewing poaching and its dismantling as nothing short of warfare, and poachers as the 'enemies' in counterinsurgency terms. A *Top Gun* meets *James Bond* affair, it brought together army scientists, defence firms and counter-terrorism experts to discuss the most promising, if costly, military-grade surveillance tools to counter South Africa's cross-border poaching threat, a new defence expenditure terrain.[3]

Among the trade show's countless exhibits of state-sponsored military hardware, I discovered private military companies' wares also on display. They, too, were hoping to cash in on the anti-poaching industry growth spurt. The promotional websites and brochures of some suggested that, while the firm might have sprung up nearly overnight, their staff nonetheless had gained military experience in places like Afghanistan and Iraq, or during South Africa's apartheid wars. These private companies advertised everything from the deployment of combat-style security forces and intelligence-gathering services to training aimed at turning rangers and dogs into battle-ready forces prepared to kill any would-be poachers on sight. Market conditions appeared to be in their favour.

'The illegal wildlife trade threatens not only the survival of entire species, such as elephants and rhinos, but also the livelihoods and, often, the very lives of millions of people across Africa who depend on tourism for a living.'

Yaya Touré

© Peter Chadwick

With attractions for the whole family, the event could not have done more to signify the growing trend towards the militarisation of conservation. Sunglasses-clad ranger dogs jumping from helicopters, mimicking a first-responder anti-poaching operation, entertained parents as much as their kids. Schoolchildren were only too eager to line up and pay for the opportunity to place their rainbow-painted handprints on the exterior of a tank in a show of anti-poaching solidarity.

The star attraction of the 4-day event was taking selfies in front of a three-storey metal transformer with the body of a legendary armoured vehicle and the faux head of a rhino. Featured as Africa's largest superhero robot, it took over 600 hours to build in order to tower above the exhibition halls and convey a specific message: 'The defence industry is in a unique position to strengthen conservation efforts. We have technologies and equipment that are making a real difference.'[4]

The 2014 trade show certainly proved that a military-style arms race with poaching syndicates was in full swing.

New era, new assets

The use of military assets and strategies on behalf of conservation efforts is not new.[5] Modern militarised forms of anti-poaching stretch back to the 1970s and 1980s, when elephants last faced wholesale slaughter. The 1989 international ban on ivory is largely credited for bringing the species back from that particular brink of extinction. But by then, militarised tactics promulgated at the height of the elephant bloodbath had taken strong root: government-enacted shoot-to-kill policies, the deployment of national defence forces to protect wildlife reserves, and state decisions to forcibly remove poachers and suspect villagers from national parks and conservation areas.

The most recent chapter in the militarisation of conservation began about 5 years ago in response to the epic upsurge in elephant carnage that exploded across Africa following the one-off bulk sales of ivory from certain African states permitted by CITES.[6] While the different contexts and actors involved throughout elephant range states make it difficult to generalise about today's anti-poaching

methods, what is clear is that they have become progressively more militarised.[7]

Some African countries and private wildlife protectors claim they have been left little choice but to ratchet up their front lines of defence, deploying more boots on the ground and bigger guns in situations where well-armed poachers have adopted more aggressive and violent tactics. When armed Sudanese raiders with assault rifles reportedly swooped through the territory of Chad and killed up to 450 elephants in Cameroon's Bouba Ndjida National Park, the country deployed its national army alongside rangers.[8]

Armed intervention, focused on guarding against incursions and stopping poachers, is a fairly common fallback position. In critical situations, heightened weaponised responses can achieve short-term conservation goals. Outside of war zones and their peripheries, however, the costly entrenchment of the military option at the expense of addressing wider political, security and sustainable development issues may ultimately doom the elephant species.

Some experts argue that, since poachers are merely replaceable foot soldiers, 'whack a mole' approaches treat only the symptoms.[9] The motives of poachers are diverse and complex, a reality that the militarised spin in popular media tends to obscure and dehumanise. Poachers may be driven by poverty or a desire for otherwise unattainable upward mobility. A great number have found themselves vulnerable to coercion by organised wildlife traffickers in league with corrupt leaders and officials. Still others find themselves lashing out at policies banning them from age-old subsistence hunting grounds to make room for sport hunters, safari operators and wealthy foreigners on holiday.

More anti-poaching firepower may result in more interdictions or more poachers being killed, but approaches designed to battle only poachers skew solutions in the wrong direction. The real problem is a demand-driven, globalised black-market trade fuelled by powerful transnational crime organisations, which militarised anti-poaching initiatives alone are incapable of solving.

Beyond incorporating reactive anti-poaching strategies long in use, this modern militarised era has been boosted by innovative trends that continue to gain popularity. Chief among them are the use of emerging technologies, the deployment of foreign forces, a hefty reliance on private security contractors and the transformation of conservancies into quasi-militarised protection zones with all sorts of new bells and whistles.

During 3 years of frontline research on the poaching crisis, I found that these new assets reveal as much about what is being tried as about what is missing to protect both vulnerable elephants and humans from the twin dangers of industrial-scale poaching and wildlife crime.

Drones and surveillance systems[10]

Drones and their integrated data management systems, a sign of the times, are among the most touted potential anti-poaching 'game changers'.[11] Pioneered by military contractors looking to increase market share, drones are one of the earliest high-tech military applications converted to civilian use for conservation. Experimental models are being tested in wildlife reserves from Kenya to Namibia and South Africa. But since drone programme costs are prohibitively high for most cash-strapped conservation agencies, international donors and tech companies largely foot the bills.[12]

Anti-poaching entities have limited human and financial resources, which is one reason for their common refrain: the goal is not to be strong everywhere but rather strong where it counts. As eyes in the skies, drones offer enhanced data acquisition and monitoring, particularly over wider and more inaccessible areas. Still, most high-tech experts agree that there are a number of practical challenges facing the mainstreaming of their deployment: price tag, training, terrain suitability, payload and endurance constraints, and the adequacy of ground-to-air tracking and surveillance systems integral to their use.[13]

Conservation stakeholders point to broader concerns they consider equally important, such as regulation and privacy issues, social impacts and fit with other conservation priorities – not to mention the fear drones engender in local communities that view them as 'killing machines' capable of striking targets by remote control.[14]

High-tech giants from Silicon Valley and Washington Beltway firms, along with national agencies, continue to pursue state-of-the-art technological conservation approaches. The goal described to me by one iconic brand is a 'transformative planetary event': harnessing speed, accuracy and cost-saving automation to provide coverage in a class of its own. Since it takes more than drones and artificial intelligence to unseat ivory kingpins, to be truly groundbreaking, the high-tech product must produce data simultaneously in sync with enforcement imperatives.

Foreign troops and mission creep

Elephant poaching is no longer perceived exclusively as a conservation issue. 'Blood ivory' and 'ivory-funded terrorism' have become popular appeals used by defence agencies, major foundations and non-governmental organisations as a call to action for foreign military intervention in African conflict areas.[15]

Over the past decade, wildlife crime has steadily climbed the rungs as a global threat. In Africa and elsewhere, this lucrative scourge – with an estimated annual value between $7 and $23 billion – has fuelled corruption, inflamed conflict and undermined the rule of law. Although ivory crime typically starts with the poaching of elephants, those responsible for the illegal supply chains and trafficking of tusks are transnational criminal organisations: mafias, cartels or syndicates. For this reason, conflating conservation and modern warfare strategies without attention to law enforcement tools does little more than buy time, if that, for endangered elephant populations.

Co-ordination and assistance is, of course, important when wildlife protectors and foreign military stakeholders overlap in war zones and insurgent hotspots. This is particularly true where rebel groups, regional militias and rogue military units threatening international peace and security are involved in illegal ivory harvesting and smuggling.[16] In places like the Central African Republic and Mali, UN peacekeepers have provided backup and training to anti-poaching units, helping them protect elephant herds under attack from illegal armed groups.[17] But when preparing more forceful conservation methods in response to the worst crises, every precaution should be taken to ensure that rangers and local civilians already overburdened with the weight of armed conflict are not put at even greater risk.[18]

There is also a tendency for counter-terrorism rhetoric to overshoot. Sceptics point to the paucity of evidence that violent extremist groups like Al-Shabaab or Boko Haram are significantly involved in the illegal ivory trade. They note that foreign militaries tilt local biases against outside interference and draw attention away from long-term elephant-saving requirements.[19] The latter includes the strengthening of African criminal justice systems, more local community engagement initiatives and reducing demand in Asia.

Even well-disposed foreign military operations have their limitations when trying to outgun poachers more familiar with the terrain. An example is the long-running fight against the Lord's Resistance Army (LRA) in the Democratic Republic of the Congo. Both the United Nations and CITES called for action against LRA poaching.[20] But US special forces-backed operations in and around the heavily poached Garamba National Park were unable to make a dent in the wildlife slaughter.[21] There was even suspicion that one of the American-funded Ugandan military units deployed in the fight used its military helicopter to both poach elephants and transport their ivory off the battlefield.[22]

In nearly every conflict I have covered over 25 years, I have found corrupt troops using the cover of foreign operations to pillage what's within easy reach.[23] I've also documented far too many cases of military-grade weaponry brought in under the guise of foreign operations à la *Lord of War* style, only to wind up in the wrong hands, stoking more conflict and poaching. Foreign militaries are expected to comply with mandated rules of engagement as well as international human rights and humanitarian law, and effective oversight and monitoring can prevent abuses. The same cannot be said of private military and security companies that have come to dominate anti-poaching operations outside of actual war zones.[24]

Private military and security contractors

Private military and specialist anti-poaching firms post-9/11 are fast becoming a cornerstone of pachyderm protection both on state and private lands.[25] These agencies do essentially the same work as state-employed rangers, but without the same governing structures, constraints or moral duties. The dark side of this engagement has opened a Pandora's box of opportunistic bad apples.

Quite a number of anti-poaching intelligence and security businesses and non-profit organisations have paltry vetting or oversight structures. Their operatives range from unemployed soldiers repurposed as private anti-poaching warriors to revamped mercenary firms cloaking other motives, such as countering Islamic extremism, intelligence collection and industrial espionage for paying customers.[26] Together, they have raised millions in the way of charitable funds, some aided by hype, emotional website appeals and flashy bravado-style fundraisers far removed from the complexity of the problem. Others that have been caught misappropriating funds, severely mistreating interdicted poachers, or bragging about their kill rights have, once exposed, merely changed names or moved shop to other countries.

Part of the trouble is the security vacuum that unmoored, private firms inhabit. Another is the way their operations are incentivised by donors. I have encountered instances where contractors are under official investigation for planting and falsifying crime scenes; still others stand accused of the murder of people they fraudulently set up as poachers. From a rule of law and human rights perspective, any potential for 'scalping' to reap rewards from otherwise well-intended international conservation funds must be eliminated.

Grave abuses aside, the insensitive use of militarised tactics by outsiders may accelerate local tensions and alienate neighbouring rural communities vital to long-term conservation success. Accurate or not, perceptions that armed white foreigners are gunning for local black poachers has stirred a few hornets' nests.[27] Further eroding effectiveness is the fact that fewer than a handful of the anti-poaching privateers I've come across could properly handle forensic evidence, execute arrests or build criminal cases necessary to deter poaching in the first place.

The explosive growth of private anti-poaching firms – whether they are morally committed to saving elephants or simply using the crisis as a door-opening business opportunity – requires better safeguards, standards and regulatory processes, especially in remote areas where contractors are empowered to use deadly force.[28]

As with any lethal industry for the public good, a high bar should be set to reduce susceptibility to harm. Refining screening and accountability measures can stem the risks of extra-judicial killings, the excessive and disproportionate use of force and other human rights abuses. They may also reduce donor liability and complicity. The reputation of individual security service providers depends on their own actions, but voluntarily signing on to international standards can go a long way in shoring up their image.

Fortress conservation

Advertisements for Africa's national parks and wildlife reserves often feature elephants and tourists in safari vehicles surveying each other amid the backdrop of golden-hued savannas, verdant deltas or mopane bushveld. It's hard to imagine that sometimes, hidden from this storybook view, is a shadowy world. In certain Kenyan, Zimbabwean and South African parks, this world contains fortresses termed 'intensive protection zones' where iconic species are guarded with more robust firepower and military resources than outlying areas.[29] In hard-hit regions, entire national parks are fast turning into fortress-like spaces, with beefed-up armed responses that are threatening to spiral out of control. The Kruger National Park's recent decision to deploy grenade launchers where hundreds of thousands of visitors and wildlife herds roam is but one example of a desperate measure for desperate times.[30]

Rangers on the front lines are at greatest personal risk as a result of wildlife agencies militarising the conflict with poachers. Across Africa, over a thousand rangers have been killed in the line of duty in the past decade.[31] Most rangers I interviewed in South Africa told me they originally signed up for nature conservation jobs, only to find themselves pressed into becoming paramilitaries. South Africa's minister of Environmental Affairs made this clear when, on World Rangers Day in July 2017, she stated that almost all of the nation's ranger corps had been converted to anti-poaching units.[32]

Poachers have followed suit by better equipping themselves and using more violent tactics. The cycle of violence has bled donors and state treasuries while putting more rangers, poachers and elephants in the crossfire – and caused mortalities. In the dead of night, hunters often cannot be distinguished from those being hunted, which is why national parks forces in South Africa have died from so-called friendly fire.

Although the militarised horse is already out of the fortress barn, hope lies with reining it in and setting a course in a more humane direction. Rather than escalating an 'us versus them' strategy, a 'winning hearts and minds' approach may yield longer-lasting results. No matter how much battle responses are hyped and glamourised, more often than not they fail to prevent poaching, and are contributing to a rising death toll on both sides. Local community initiatives and enhanced law enforcement measures deserve a better try. The goal should be a more just planet where humankind and elephants living in close proximity can both feel safe.

The other side: poachers' arms

Almost my entire professional career has focused on the role that guns play in armed conflicts. Since it takes two to tango, I have made it my mission to look at the weapons used by both sides before proposing solutions to enhance peace and security. Which brings me back to my attendance at Africa's largest arms trade show. From the anti-poaching agencies present, I went looking for details about the poachers' guns.

I failed to find them. There was hardly even tacit recognition of the need to tackle the gun and ammunition supply chains that are a linchpin in the ivory syndicates' vast criminal networks. Caught up as they are in the day-to-day battle against the foot-soldiers, anti-poaching forces tend to be reactive rather than preventive. They seldom build up the expertise and tradecraft – or have the mandate – to proficiently tackle highly organised crime machines, let alone their weapon sources.

So what is the poacher armoury? There are a variety of ways they kill elephants. Cyanide or other industrial chemicals may be put into elephant watering holes. Arrows may be dipped in native plant-based poisons and shot from bows.[33] Spears and snares may be employed, but most of the elephants targeted by poachers are shot with small arms. These range from the ubiquitous AK-47s and shotguns to semi-automatics and powerful hunting rifles that can bring down a charging elephant with a single shot.

Rogue security forces, military units and police tend to use weapons from government arsenals. Subsistence and opportunistic poachers typically rely on stolen firearms or those left over from previous bush wars and colonial conservation departments. Militias often barter ivory for guns with smugglers or unscrupulous businessmen.

Wildlife-poaching syndicates, however, are in a league of their own. They depend on steady flows or bulk supplies of weapons to consistently pull off high levels of poaching. The arms they rely upon may be newly manufactured or from ageing stockpiles. Either way, they're typically sourced from dedicated international gun-supply chains and veteran traffickers.

Follow the guns

A well-known approach for uncovering criminal organisations is to follow their money. This is often hard to do in Africa, where ivory is cash-traded at source (though the approach has netted several Kenyan kingpins).[34] What can be done more readily is to follow the guns, a technique overlooked by the conservation community.

Gun serial numbers are unique identifiers that enable investigators to chart a gun's path throughout its entire supply chain, from its manufacture and export to its local importation, sale and distribution to end-users. Serial numbers also provide irrefutable forensics. The numbers don't lie, their memories don't fade and they can withstand the strongest scrutiny in court.

By trailing guns, I've built dossiers against some of the biggest traffickers in the world, like the convicted Russian merchant of death, Viktor Bout; the Western European and African smugglers of high-calibre 'CZ' hunting rifles; and the rhino and elephant syndicates operating between South Africa and Mozambique.

But before anti-poaching entities can do this, they need systematically to record data on the guns, bullets and spent ammunition linked to poaching crime scenes. Accumulated data such as this can reveal patterns and unifying factors, in much the same way a crime scene holds information enabling investigators to go after a serial killer. Solid detective work will bring the criminal kingpins into sharper focus far more than military-grade weapons rerouted from the war on terror.

Disarming

Elephants are facing their darkest days. The highest aim of preservation efforts should be to reverse the trend of fatalities filling both human and elephant graveyards. A way to halt the elephant butchery before the species goes extinct is to curtail the inflow of illegal firearms, depriving syndicates of the means to shoot their prey. Choking illicit gun and ammunition supplies would not only help reduce the number of people dying on both sides of the poaching wars, it would de-escalate the arms race between the two opposing forces. In short: Reduce the guns. Reduce the killing. Reduce the risk of extinction.

Links to Kathi Austin's documentary exposé on poaching kingpins *Follow the Guns* can be found on CAP's website at **www.conflictawareness.org**

Opposite
The Singita Grumeti Fund in Tanzania combines cutting-edge technology with vehicles and boots on the ground to combat poaching.

'Despite the misinformation put out by those who stand to profit from trade in wildlife products, CITES trade bans can and do work. This has been proved for both rhinos and elephants. There were periods in the 1970s and 1980s when up to 9 000 rhinos and close on 90 000 elephants were being poached annually throughout Africa. The elephant slaughter was stopped in its tracks in 1990 when the ivory markets were shut down by the uplisting of all elephants to Appendix I. Rhino poaching was halted in just one year around 1993 when all the rhino-horn consumer countries implemented the full CITES trade ban regulations. With no market and no trade, poaching dried up. For endangered species trade bans to work, there need to be full international and domestic trade bans without any loopholes and no mixed messaging. Sadly, today's CITES regulations are riddled with exemptions and loopholes that criminal syndicates exploit to drive demand and poaching.'

Colin Bell & Dr Don Pinnock

'People who don't know about CITES think it's a conservation group. It's not. It's a trade-based organisation.'

Dr Cynthia Moss

'What started as a conservation treaty regulating trade is on the verge of becoming a trade treaty regulating conservation.'

Judith Mills

19

CITES and trade: is this the organisation to save elephants?

CITES ought to be the organisation that best safeguards the lives of elephants. But does it?

Adam Cruise

CITES (the Convention on International Trade in Endangered Species of Wild Fauna and Flora) is an international agreement among governments whose sole aim is to ensure that international trade in endangered specimens of wild animals and plants does not threaten their survival. The Convention currently regulates the international trade of about 35 000 wild species between its 183 member countries – called Parties[1] – making it the largest, and potentially the most effective, conservation body in the world.

Elephants are undeniably the Convention's flagship species; an elephant is incorporated in the CITES logo and more time and effort is spent on elephants than on any other species. In fact, one of the key criticisms of CITES is that elephants take up the bulk of proceedings, leaving thousands of other listed species with little or no attention and funding.[2] CITES, therefore, ought to be the organisation that best safeguards the lives of elephants. It certainly is able to but, because of a number of inherent flaws within its framework, it doesn't. Not even close. Worse, the Convention has arguably been implicated as a major cause of the recent poaching crisis, which has witnessed over a third of Africa's elephants wiped out in the past decade.

How does CITES [not] work?

To understand this apocalyptic anomaly specifically in terms of African elephants, one has to delve into the complex working apparatus of CITES.

Fundamentally, as its name suggests, the Convention functions as a trade, and not a conservation, convention, per se. The organisation does not determine *in situ* conservation measures for elephants or any other species, nor does it consider habitat loss or anthropogenic environmental degradation. It allows Parties largely to self-regulate and participation in the Convention is voluntary. Each Party may designate one or more management authorities to administer a licensing system for exporting or importing a listed species, as well as one or more scientific authorities to advise them on the effects of trade on the status of CITES-listed species.[3]

Paradoxically, countries that tend to have the highest elephant populations are also the least equipped and financially resourced to protect them. These poorer countries face difficulties carrying out accurate scientific research and are generally disadvantaged in controlling poaching and the illegal ivory trade; as such, they are susceptible to corruption. One has only to compare the under-resourced customs and scientific officers in Liberia and the Democratic Republic of the Congo with those in the USA and Germany to understand the problem.

It's important to note, too, that although CITES is legally binding once Parties decide to participate, its regulations do not take the place of national laws.[4] For example, even though Parties prohibit trade in ivory multilaterally or bilaterally, they are allowed to trade ivory commercially at a national level as CITES has no capacity to address trade within a country's borders. All it can do is 'urge' or 'recommend' that Parties restrict or close down their markets. Almost every Party from Austria to Zimbabwe has a legal domestic ivory market in some form.

Also, punishments for poachers and criminals caught trafficking vary widely from country to country. For example, while implicated in large-scale ivory poaching, a country like Mozambique is notoriously lax in prosecuting offenders; and Laos, a country renowned for moving large volumes of ivory, has not brought a single known ivory trafficker to book in spite of clear evidence of such crimes. Again, CITES lacks the framework to address these issues.

The three appendices

International trade in wild species is categorised in Appendices I to III and a corresponding permitting system is instituted appropriate to the threat posed to those species by trade.[5]

The Appendix I listing provides the highest possible protection given to any species under the Convention and encompasses those threatened with extinction. Commercial trade in wild-caught specimens of these species is illegal, unless for non-commercial purposes, such as sport-hunted elephants for trophies, or under exceptional circumstances, such as for scientific research, captive-bred species, certain trade in live animals and educational purposes. In these exceptional cases, international transactions may take place provided they are authorised by the granting of both an import and an export permit.

Appendix II includes the largest body of threatened species under CITES regulations: 21 000 versus only 1 200 in Appendix I and 170 in Appendix III. These are not necessarily threatened

CITES delegates in the main debating hall at CoP17 in Johannesburg, South Africa, 2016.

© Colin Bell

with extinction, but may become so unless trade in specimens of such species is subject to some form of regulation. Such 'protection' still allows hundreds of thousands of global species to be traded annually.

Generally, all that is required under Appendix II regulations is a non-detriment finding – evidence that trade will not harm a species – by the exporting scientific authority and an export permit granted by the exporting Party. No import permit is necessary, although some Parties, like the USA, do require import permits as part of their stricter domestic measures.

Appendix III includes species that are not necessarily threatened with extinction and is distinct from Appendices I and II, since a Party is entitled to make unilateral amendments to it. These species are listed after one member country has asked other CITES Parties for assistance in controlling trade in a species.

Conference of the Parties, Secretariat and Standing Committee

The Conference of the Parties (CoP) is the supreme decision-making body of the Convention and comprises all its member states, as well as other United Nations agencies and international Conventions and NGOs. All are allowed to participate in discussions, although only Parties may vote.[6]

Species can be added to or removed from Appendices I and II, or moved between them, only by the Conference of the Parties, either at its regular meetings or by postal procedures. Parties agree on a set of biological and trade criteria to help determine whether a species should be listed in Appendices I or II (species from Appendix III do not require CoP permission and may be added or removed at any time by any Party, unilaterally).[7]

At each regular meeting of the CoP (usually every 3 years), Parties submit proposals based on those criteria to amend species listings within the first two appendices. The amendment proposals are discussed and then submitted to a vote. In order for a species to be listed, up-listed or down-listed, a two-thirds majority of Parties present and voting must be achieved.

The administrative and procedural duties of the Convention are conducted by the Secretariat,

which is administered by the United Nations Environment Programme (UNEP) and is located in Geneva.[8] The Secretariat – apart from playing a co-ordinating, advisory and servicing role, arranging committee meetings between CoPs and providing recommendations on which way to vote on proposals – has limited capacity for enforcing regulations. When informed of a contravention, the Secretariat will notify all other Parties. It will then hand over the responsibility to the Standing Committee, which is made up of representatives of Parties from all six continents. The committee generally sits in session between CoPs and, importantly, provides policy guidance to the Secretariat concerning the implementation of the Convention.

At a Standing Committee meeting in January 2016, 27 countries were dealt trade suspensions for non-compliance, 16 of them in Africa. The committee will usually give any offending Party time to respond to the non-compliance allegations, while the Secretariat may provide technical assistance to prevent further infractions. Failing that, all the Standing Committee and Secretariat are able to do is to recommend – not enforce – trade sanctions for any country that fails to comply with its regulations.[9]

Implementation of the recommendation to suspend trade depends entirely on each individual Party, although it was noted that about half the member Parties have not enacted internal laws for properly implementing the existing CITES system.[10] Mozambique is a prime example of this. The country has few national laws in place for arresting and charging poachers as criminals. Although this is slowly changing, the Mozambican government has traditionally regarded poaching wildlife as a legitimate means of subsistence.

CITES and elephants: a catastrophic history

Perhaps the biggest criticism of CITES is that it functions as a negative, or inverted, model. Instead of taking a precautionary stance by assuming species generally to be Appendix I-listed and then down-listing particular species once scientific studies prove their eligibility for trade, it takes the opposite stance: unregulated trade in all species unless proven otherwise. The agreed-upon CITES biological criteria for an up-listing for elephants was, and still is, 'a percentage decline of 50% or more in the last 10 years or three generations, whichever is the longer.' [11]

This upside-down approach immediately proved calamitous for elephants. In the decade during which CITES came into force, Africa's elephant populations, which by 1977 were listed only in Appendix II, dramatically crashed. Between 1978 and 1988 the number of African elephants declined by more than half from 1.3 million to around 600 000.[12] It was estimated that over 90% of the ivory in international trade was from poached elephants. Kenya's population suffered the worst, showing a decline of some 85% since 1973. In protest against the ivory trade, Kenya decided to burn its national stockpile of ivory in 1989, the first demonstration of its kind, and the world began to take notice. In response, the US Congress passed the African Elephant Conservation Act, which banned the import of African elephant ivory into the US for commercial purposes.[13]

1989: All elephants on Appendix I

That same year, a proposal was presented at the 8th Conference of the Parties (CoP8) in Lausanne, Switzerland, to list African elephants in Appendix I. The biological criteria for an up-listing were certainly applicable, but it was not without resistance. There was an outcry from some Parties, especially those from southern Africa and from some Asian countries. The subsequent vote only just reached the requisite two-thirds.

The decision, though, proved remarkably effective. It generated much global publicity. This was the one and only time CITES demonstrated its ability to protect Africa's elephants. As soon as the ban came into force, most major ivory markets began to shrink and then close down, particularly in Europe and the USA, and African elephant populations slowly started to recover.[14]

However, throughout the 1990s, the southern African countries led by Zimbabwe and South Africa continued to fight against the up-listing through legal challenges. They claimed elephants within their borders were well managed and not seriously challenged by poaching. For just under a decade, their demands were held at bay.

1997: The spectre of split-listing elephants

In 1997 – by cleverly using the biological criteria that had been agreed upon at the 1989 CITES meeting – Botswana, Namibia and Zimbabwe were able to show that their populations were far from the 50% decline necessary for Appendix I.

At the 10th Conference of the Parties in Harare that year, with Robert Mugabe famously opening the proceedings by declaring that it was time elephants in Zimbabwe paid for their survival,[15] Parties voted to transfer the African elephant populations of Botswana, Namibia and Zimbabwe to Appendix II, albeit with an 'annotation'[16] – the three countries were not permitted regular international ivory trade for commercial purposes. The annotation seemed to defeat the purpose of a down-listing except that ...

1999: The one-off sale experiment

The down-listing opened the door for the CITES Standing Committee to permit a one-time, experimental export of 49.4 tonnes of government-stockpiled ivory from the three southern African countries to Japan in 1999.

In 2000, CITES agreed to the establishment of two systems to inform Parties on the status of illegal killing and trade: Monitoring the Illegal Killing of Elephants (MIKE), and Elephant Trade Information System (ETIS).[17] Essentially MIKE and ETIS were established to prove or disprove any causality between ivory stockpile sales and poaching levels. The primary aim of the experiment was to flood the runaway illegal market with legal ivory, but it merely exacerbated the problem.

Between 2004 and 2006, more than 40 tonnes of illegal ivory – almost the same quantity as the legal stockpile – were seized. It was only the tip of the iceberg. Between August 2005 and August 2006, it was reported that as many as 23 000 elephants may have been poached.[18]

However, the CITES monitoring systems saw it differently. From 1999, following a 5-year analysis from MIKE-ETIS, a decline in the volume of illegal ivory was registered. CITES hailed the experiment a success. Even so, Steven Broad, Executive Director of Traffic, an IUCN-backed organisation that compiles the wildlife trade data on behalf of

MIKE-ETIS, admitted that it was unclear 'whether this was cause and effect or a coincidence … we don't know,' he said.[19]

2008: Another one-off sale

In 2002, buoyed by the perceived success, the Standing Committee, with support from a number of prominent non-governmental conservation organisations including – most notably – the World Wildlife Fund (WWF),[20] approved a second export of 60 tonnes of government-stockpiled ivory from Botswana, Namibia and, this time, also from South Africa (the latter having been added to the split-list proponents earlier that year). Trading partners had yet to be approved by CITES. The export did not take place until 2008 but when it did, it came in the wake of fierce opposition at the 14th Conference of the Parties at The Hague in 2007.

Many African elephant range states were strongly opposed to the split-listing. In 2006, 19 African countries signed a declaration in Accra, Ghana, calling for a ban on international trade in ivory. At CoP14, Kenya and Mali, supported by Ghana, Chad, Democratic Republic of the Congo, Niger and Togo, submitted a proposal and a working document to CITES proposing a 20-year moratorium on ivory trade and urging range states not to submit down-listing proposals during this period.[21] Both of these submissions were rejected.

The CITES Standing Committee responded to their concerns in a spectacularly contrary style. Not only did the committee add Zimbabwe to the list of approved sellers again, it almost doubled the amount of government-stockpiled ivory from 60 to 108 tonnes, and added China to Japan as an approved buyer. As a small concession, CITES agreed not to approve additional one-off sales – a term that had become something of a mockery – at least until 2017.

In November 2008, 102 tonnes of ivory were bought, most of it by the Chinese government, which then sold it on to merchants for a profit. The legal ivory, which was poorly tracked by the authorities, provided a slippery conduit for illegal ivory selling below market value to flow into China via seaports in Malaysia, Singapore, Thailand and Vietnam.

2008–2016: Carnage

While the first CITES-sanctioned one-off sale in 1999 proved controversial, at best, the second in 2008 was an unmitigated disaster. The influx of ivory into China reinvigorated the massive government-approved ivory-carving industry, which had been waning since the 1989 ivory trade ban. As the largest ivory market in the world, it was no coincidence that as soon as China entered the scene, poaching across Africa began again in earnest. A study in 2016 found that the 2008 sale corresponded 'with the abrupt increase in illegal ivory production, and a possible tenfold increase in its trend'. An estimated 71% increase in ivory smuggling out of Africa further corroborated the finding. The study suggested that partial legalisation of banned goods did not reduce black market activity, as was hoped by CITES.[22] Furthermore, the message to Chinese and Japanese consumers that ivory was off-limits was all but erased the instant the two legal sales were approved by CITES.

Back in Africa, the Great Elephant Census of 2016, a pan-African survey of African savanna elephants, revealed elephant populations had crashed by a third of the total population – or 144 000 elephants – in the 7 years after the one-off sale.[23] A separate analysis in 2016 by the IUCN African Elephant Specialist Group, which included a survey of forest elephant populations, revealed a similar pattern.[24] Even the Appendix II nations, with their so-called well-managed populations, felt the effects of poaching.

Sales of wild live elephants

As the carnage of elephant populations across the continent became apparent, Zimbabwe, unable to benefit from any further one-off ivory sales, began selling off live elephants. In 2012, the country announced plans to ship some 200 elephants. Most of them were to be juveniles that would be violently separated from their families in Hwange National Park and flown to Chinese zoos, circuses and 'safari' parks. Three shipments, in 2012, 2015 and 2016 of some 80 elephants, have so far taken place. Unaccustomed to the transportation and alien environment, many of these elephants have been traumatised, injured and some have even died.[25]

According to an annotation on the Appendix II listing, CITES allows for 'trade in live animals to appropriate and acceptable destinations'. The Convention defines 'appropriate and acceptable destinations'[26] as: 'The Scientific Authority of the State of import is satisfied that the proposed recipient of a living specimen is suitably equipped to house and care for it; and the Scientific Authorities of the State of import and the State of export are satisfied that the trade would promote *in situ* conservation'. [27]

This annotation is constantly violated. Helicopters are used to cause herds to stampede and juveniles, unable to keep up, are darted, roped and bundled into trucks. The violent means of snatching them in the wild often causes trauma, injury and death of baby elephants. It is clear that the elephants are neither suitably housed nor cared for, nor does the action benefit *in situ* conservation of those elephants left in the wild, especially the mothers and family units of the snatched calves.

In January 2016, CITES authorised the sale of 18 wild elephants from Swaziland to the United States. Elephants are listed Appendix I in Swaziland but an exemption allows the sale of live elephants as long as a non-detrimental finding from both exporters and importers has been secured; and if both the importing and exporting countries issue permits.[28]

The CITES biological criteria, as with Zimbabwe's case, account only for statistical data. While the sale of 18 or even 200 elephants does not detrimentally affect the overall numbers, the criteria fail to consider the traumatic effect on those individual animals removed from the wild, as well as the distress of the family herds that are left behind. Ultimately, given the nature of the current poaching crisis, the CITES approval of the sales of live wild African elephants is both callous and misconceived.

September-October 2016: Last chance to save elephants at CoP17

By the time the 17th Conference of the Parties (CoP17) took place in Johannesburg at the end of 2016, it was obvious that neither the split-listing, the one-off sales experiments nor the sale of live elephants had worked in conserving elephants.

Support to reinstate a blanket Appendix I listing that had proved so effective in 1989 had been growing. Twenty-nine African countries, calling themselves the African Elephant Coalition, put their weight behind a five-point proposal that included a call for an up-listing of the four southern African countries and a cessation of the sales of wild live elephants.[29] Zimbabwe and Namibia, which had both submitted counter-proposals, were calling for unrestricted trade but South Africa, as conference hosts, remained largely reticent, despite a tacit approval for legal trade. Botswana's government, however, had secretly begun to reconsider its former support for split-listing. The stage was perfectly set for a change of heart within the Convention.

Yet, once again, CITES failed to protect elephants. MIKE and ETIS, incredibly, were still denying any proven link between the 2008 sale and the poaching crisis, while the Secretariat gave a misguided recommendation to all Parties ahead of the vote *not* to adopt the proposal. And then, a flawed annotation, which had been implanted in the CITES mechanism way back in 1983, suddenly surfaced, which all but scuppered any chance of all African elephants being placed back in Appendix I.

This strange recommendation against an Appendix I listing resulted from a curious clause buried deep in the appendices text. Largely due to threats from Namibia and Zimbabwe's eagle-eyed delegations and their subsequent threats based on this clause, the Secretariat became worried that if the Appendix I proposal were adopted, then the current Appendix II listing and its 'preventative' annotations would be removed. The clause states that if a change in listing occurs, the Convention allows any Party to enter a reservation within 90 days. Bizarrely, CITES stipulates that any country that enters such a reservation would automatically not be treated as a Party to the Convention.

As a result, a Party entering a reservation against a vote to include all African elephant populations in Appendix I could commercially trade in ivory to any other Party that also entered such a reservation, and do so without violating the provisions of the Convention. So as soon as an up-listing to Appendix I was adopted, a country like Zimbabwe could take out a reservation. If China did the same within 90 days, the two countries could trade in ivory with each other to their heart's content.[30] This reservation clause completely undermines the listing process, and therefore renders the entire CITES system futile.

Also, by recommending that Parties not adopt an up-listing, many countries – say in South America – that have no elephants and couldn't care less one way or another, would simply follow the Secretariat's directive and vote accordingly.

The Gaborone amendment

A second issue – equally ludicrous – rendered proceedings at CoP17 superfluous. In 1983, in Gaborone, Botswana, Parties voted for an amendment to the text of the Convention that allowed regional economic blocs to accede to the treaty.[31] It was a strange amendment, given that it didn't mean anything for 33 years until the 2016 Johannesburg conference when the European Union participated for the first time as a full Party with all 28 member states (plus one extra vote as the EU) in one voting bloc.

The EU all but upended the way voting traditionally occurs at CITES. By wielding 29 guaranteed votes, the European Union held all the voting cards. The pattern at CoP17 was consistent: whichever way the EU decided to vote, on whatever species, the outcome always went the way the EU voted. With such power, the EU essentially overrode the CITES system.

EU policy – decided by unelected Commissioners – supported the outdated view that elephant populations of the four Appendix II countries were healthy, well managed and did not meet the biological criteria for an up-listing. Consequently, all EU countries voted to maintain the status quo.

But that was not the only reason they voted along these lines. The EU is the world's largest exporter of 'antique' or pre-Convention ivory, acquired before CITES came into force in 1975. This is another of the CITES trade exemptions and is legal as long as a certificate 'proving' each item's age is submitted.

According to the CITES trade database, during the past decade, EU countries legally exported more than 20 000 carvings and 564 tusks.[32] It has proved another foil for the laundering of illicit ivory. This was highlighted by numerous seizures of so-called pre-Convention ivory from recently slaughtered elephants in Belgium, the United Kingdom, Germany, Czech Republic and Spain.[33]

Some individual countries like France and Luxembourg went against the grain in the lead-up to CoP17, vociferously supporting the African Elephant Coalition and an Appendix I listing, but were forced to toe the line of the bloc's dictate at the conference. Assenting voices unwittingly became dissenting ones. The EU effectively forced some of its member states to vote against their sovereign mandate.

As a direct result of the EU vote, despite a last-minute appeal from Botswana to support it,[34] the proposal failed to gain the necessary two-thirds majority for an up-listing and the fate of elephants was once again left hanging in the balance.

Conclusion

Despite the remarkable success in the wake of banning trade in ivory in 1989, CITES has not only failed in its task of protecting African elephants, but has hastened their demise through a series of inherent flaws that have all but rendered the organisation's conservation ability redundant.

The continued survival of elephants instead relies on the will and action of individual countries independently of CITES. Some countries have risen to the challenge. On 30 December 2016, China announced its commitment to shut down its domestic trade by the end of 2017, essentially closing the world's largest market for poached ivory. The USA has effectively done the same thing, so has France and more countries are set to follow.

Unless CITES can go beyond just recommending courses of action and become more effective with enforcement of its regulations (among a long list of other internal overhauls), the organisation may as well be tossed onto the scrapheap of other failed international organisations. In the final analysis, the Convention on International Trade in Endangered Species of Wild Fauna and Flora is in dire need of reform; and it must happen soon, before it does any further damage to Africa's fast-dwindling elephant populations.

A heavily sedated young elephant is being cajoled, pushed and dragged into its padded crate to be transported from South Africa's Kruger National Park to its new reserve.

© Daryl Balfour

20

Translocating elephants: are welfare and conservation in conflict?

If we don't bring welfare into elephant management, we will continue to see problems on private reserves.

Dr Marion E Garaï

In many ways, South Africa's approach to wildlife management is unique within Africa. For the past 100 years, the focus has been on restoration rather than on the maintenance of functioning ecosystems.[1] It's also the only country on the continent that allows private ownership of wildlife, including elephants, which are confined within electrified fences (even the national reserves have elephant-proof fences). Also permitted is the commercialisation and use of wildlife, with protection and management of elephants the responsibility of the owner.

These factors pose problems different from those of any other range state, and also present challenges to management. Any animal population confined within a space will eventually expand beyond what's optimal for that space. One way of dealing with rising elephant numbers in the last century was to cull surplus individuals. This posed ethical issues and, eventually, translocation to other reserves seemed an ethical alternative.

Because of open ownership possibilities, many private game reserve owners bought elephants as tourist attractions. Initially, only same-aged juveniles were translocated in small groups. But there's a downside. The long-term effects of translocation on elephants' behaviour are only

now emerging, and are raising questions: is this management option still tenable and ethical, and what are the implications for elephant psychology and social wellbeing?

In South Africa, welfare and conservation fall under two different government departments. This causes problems for the ethical management of elephants on smaller reserves. It also creates a dilemma. Rules and regulations at governmental level are based on ecological principles, but these often don't correlate with the sociological requirements of elephants. What, then, are the implications of translocating elephants, and is ecological management feasible without welfare considerations?

Initially, there were no regulations on the number of elephants owned or on how to manage them, resulting in many small, unviable populations. The Elephant Management and Owners Association (EMOA) then developed a policy on translocated elephants, which was subsequently incorporated into the National Norms and Standards for Elephant Management in South Africa.[2] Data on how many elephants live on private properties was first collected by me in the early 1990s and a database was developed by EMOA in 2004.[3] The Elephant Specialist Advisory Group of South Africa (ESAG) recently updated the database.[4]

A team of vets, rangers and field assistants from Kruger National Park manoeuvering a young, sedated elephant into its crate for transporting to a new home.

For purposes of studying migration patterns, this bull elephant was fitted with a radio tracking collar to monitor its movements in the Kruger National Park and surrounding reserves. The sticks kept the elephant's breathing channels open while it was under sedation.

In a vast open system, elephants can migrate and, although the ecological balance will fluctuate, it will restore itself and elephant populations will stabilise.[5] In a fenced-in environment, for animals that cannot fly or creep under the fence, migration is not possible. At some stage any entrapped animal population will grow and progressively destroy its own food sources. So a fenced game reserve, private or national, has to be managed. How it's done, and for what reason – such as to prevent biodiversity loss, curb specific tree loss or manage for high-population increase – is the owner's decision. Many reserves with elephants opt for retaining biodiversity, meaning that animal numbers have to be managed. There are then few options: translocation, contraception or culling. In all three, we humans decide which individuals are removed or not allowed to breed.

History of translocation

From the 1960s to 1990s, wildlife was managed to keep numbers at a previously agreed carrying capacity; about one elephant per square mile (2.6 square kilometres). Anything above that number was culled. In Zimbabwe alone, this equalled about 46 000 elephants.[6] In 1967, the Kruger National Park (KNP) instituted an annual elephant cull to keep its population stable, leading to the deaths of nearly 17 000 elephants between 1967 and 1994. Initially, entire family units were culled, but from 1978 onwards, the juveniles were captured alive and sold to any bidder.

The first translocations of juveniles began with reserves such as Pilanesberg and Hluhluwe-Imfolozi.[7] In the early 1990s, demand for elephants increased, both from the private sector and from overseas zoos. Because the new owners were unaware of the difficulties in raising young orphaned elephants, and with no information available, mortality was high.[8] Deaths occurred mainly among the very young orphans, the main causes being stress, pneumonia, malnutrition, infections, salmonella, snake bite, falling victim to lions, or lightning. Following the culling moratorium in 1994, however, the technique of translocating entire family groups was refined.

Eventually, though, the market for elephants within South Africa was saturated and the demand diminished. Today it's difficult to find a reserve still able to take elephants, with many reserves looking to reduce their populations.

Elephants on state and private reserves

By 2015, there were over 28 000 elephants in South Africa, distributed over more than 80 reserves. More than three-quarters live in the KNP and the private reserves against its unfenced western boundary.[9] Of these 80 reserves, 17 are state-owned at provincial or national level, the rest are private.

The numbers on many private reserves are small – 20 reserves have from two to 10 elephants and only 19 have more than 100. According to Indian elephant specialist Raman Sukumar, a population size of between 100 and 200 elephants can be regarded as viable, meaning that the many small groups in South Africa are not viable and don't contribute to the overall conservation of the species.[10] These reserves are effectively large zoos.

Effects of translocation and defragmented populations

Since the initial days of removal, we have learnt that translocated elephants living in small, socially disrupted groups behave differently from a 'normal' free-roaming population, and require intensive management. Some of the many issues to be considered are discussed below.

Reproduction rate

A 1994 study showed that 68% of South Africa's elephant population in smaller reserves increased at a much greater rate than in a stable, natural population.[11] The reason is that females in confined areas conceive at a much earlier age than in an undisturbed population. Their inter-calving interval was in some cases 2 years or less, with one young female conceiving at 5 years. This corresponds to the reproduction rate of elephant groups in zoos.

In the case of nine African elephant females in Europe, the first age of parturition was 7 to 12 years

and inter-calving periods were from 26 months upwards.[12] In 10 wild populations, the average age of the first parturition was 8 to 12.8 years and the general inter-calving rate was 4 years.[13] Although this initial rapid reproduction increase on the smaller reserves surveyed has more or less normalised (due to the age structure and the females being older at conception), it still remains high.

Early onset of musth

Young bulls in smaller reserves were found to be entering their first musth period much earlier than the normal mid-20s, and some were siring offspring in their early teens.[14] This is similar to bulls in zoos, which were siring offspring from age 9.25 years.[15]

There's a strict dominance hierarchy among bulls, and when older males are lacking, young bulls enter musth prematurely, before they have had the time and opportunity to learn how to deal with it psychologically and physically. The physiological state of musth greatly increases testosterone levels and makes males aggressive.

During adolescence, males gather social and ecological information from older bulls, which is crucial to stabilising the society, and the presence of older males suppresses the onset of musth in younger bulls.[16] In the 1990s, there were reports of orphaned elephants being aggressive towards other large game, such as buffaloes, hippos and rhinos. Following the killing of at least 30 rhinos by elephants at Pilanesberg National Park, and well over 50 at Hluhluwe-Imfolozi Park, a study found aggression towards other game was due to the absence of larger bulls.[17] Once larger bulls were introduced, the killing eventually ceased.

Adult bulls

Between 1998 and 2000, the KNP responded to a demand for adult bulls, and a total of 84 bulls were translocated to other reserves. What followed were a number of breakouts by some of these translocated animals aged 30 years or more. There were several reasons for this: inadequately built bomas (acclimatisation enclosures); the presence of only one or two bulls (lack of a male hierarchy); older bulls (40+) being translocated, or bulls with a homing instinct who broke the fence and started to walk in the direction of their original home in the KNP.[18]

Social disruption

Translocation of a few individuals, even an entire family unit, leaves the relocated group and the remaining elephants socially disrupted. A study on the social development of orphans showed that, although the young elephants attempted to reorganise a social unit by mimicking a family group, they remained extremely nervous for many years and always kept close together, avoiding areas with human activity.[19] They never joined up with the family units that were later introduced to some of the reserves.[20] Many translocated groups lack a matriarch and not every young female is capable of taking over this demanding role.[21] As was suggested by elephant researchers Lee and Moss, personality traits of individual females underlie leadership ability.[22] This was evident in the orphans: some young females did quite well in leading their group, others lacked the ability.

Social disruption has many other effects; for example, subsequent mothering behaviour is either overprotective or very slack. Females that have not learnt mothering skills may display aberrant behaviour towards their calf or another's calf, even killing it.[23] Infanticide by either the mother or another female is also well documented among zoo elephants.[24]

Elephants have a long period of adolescence, which corresponds with a second major stage in brain reorganisation.[25] This means orphaned elephants that lack normal family structure will not develop certain specific strategies or social skills. This leaves them less equipped to deal with unpredictable situations and stress, such as large-scale poaching, which can cause severe social disruption and changes in behaviour patterns and genetic structure.[26]

Many elephants that have gone through severe trauma, such as separation from family members, are often socially less competent.[27] They may also show poor physical condition.[28] Highly traumatised zoo and circus elephants are often much smaller than their wild equivalents.[29]

Translocated elephants that have experienced separation from family members decades before often respond inappropriately to calls of other elephants. Some don't know to bunch together defensively and remain vigilant, nor do they respond to a 'safe' call.[30] In a socially undisturbed population, on the other hand, matriarchs were found to be capable of distinguishing between up to 100 different calls.[31]

Impaired social skills were found in a population at Pilanesberg National Park, which had been translocated in the 1980s and 1990s and then experienced a severe fire.[32] Although the fire took place in 2005, a long time after their translocation trauma, the elephants were found to be less experienced in dealing with the stress.

Matriarchs in undisturbed populations build their knowledge within an expansive social network consisting of dozens of other elephant families. An example of such a network was in Kenya's Amboseli National Park, where one family was seen in aggregation with no fewer than 50 other families.[33] Translocating family units out of these extended networks leaves the matriarch without her social environment and support. She has to familiarise herself with and adapt to the new ecological environment and try to integrate into any existing elephant society. She must find the best forage, water and safe places, a responsibility causing her great stress. This can have serious health and psychological effects.[34]

Incidents involving elephants on reserves

A survey undertaken in 2012 by the Elephant Specialist Advisory Group (ESAG) showed that over 80% of South African reserves reported problems of some form, involving over 100 elephants.[35] The main issues were breakouts by males and females, attacks on vehicles and damage to infrastructure such as water pipes, electrical poles, carports and buildings. Although these are seemingly minor events, if perpetrated repeatedly they can become bigger issues that may result in the responsible elephants being euthanised or hunted. The survey also reported five human deaths, although incidents involving people are generally under-reported so as not to threaten tourism. On one reserve a matriarch and baby were killed by another female elephant.

Many of these issues can be traced back to early social disruption, as well as to management's ignoring the need for a stress-free social environment. This includes stress caused by intense tourism (high volume of vehicles, self-drive tourists, even game rangers driving too close[36]), human ignorance, hunting, repeated translocation, high density of elephants or nutritional stress due to fence restriction. Some attacks in the KNP were mostly due to ignorance by tourists who did not understand or heed the warning signs given by the elephants.[37]

Conclusion

You cannot simply remove elephants from a group without causing possibly severe and long-term problems. For this reason, their welfare must be part of the solution.

In South Africa, animal welfare (applied to animals in captivity or the pet trade) falls under the Department of Agriculture, and conservation falls under the Department of Environmental Affairs. This unfortunate division creates several problems, especially concerning small populations of large mammals on private reserves.

So how applicable are welfare standards to elephants on private, fenced reserves?

Among the many similarities between a socially disrupted group of elephants in a small reserve and elephants in close captivity is the lack of freedom of choice. In a small population there may be only a few young bulls, so the females cannot choose their preferred partner, nor do they have much choice over where to roam, on what to forage, or with which other family groups or individuals to associate.

If welfare is important for zoo elephants, it's just as important for elephants kept on private reserves. Conservation may deal with ecological systems, but managing a population means dealing with individuals. For this reason, welfare at reserve level is important.

From the moment a group of elephants is taken out of an original population and translocated, trauma, stress and social disruption will result, having lifelong effects and potentially persisting through successive generations.[38] Ecological manipulation, such as water control, fencing of specific areas or plants, roads, buildings and tourism will all affect the elephants.[39] For these reasons, elephant welfare *must* be considered in order to give such elephants the best and as-near-to-natural-as-possible environments. This is not significantly different from managing a zoo population.

The South African government needs to develop welfare standards, not only for those elephants in captivity, but for all elephants, as no population in South Africa is totally free ranging or requiring no management. We've erected fences and are interfering with their normal lives; we *manage* them, and management cannot be successful without welfare considerations.

I believe that, because welfare has been left out of the deliberation, with decisions based on ecology only, we are experiencing problems with translocated elephants. There clearly needs to be a paradigm shift in how we manage populations on small reserves. There are two scenarios:

1. We need to understand that a smaller, fenced reserve limits the environment and sociality of elephants, so thinking in only ecological terms will result in problems.

2. In larger reserves, we must prevent disruption and promote restoration of natural social structures. Natural dispersal processes of habitat and sociality must be reinstated as far as possible. As Rudi van Aarde *et al.* point out, 'We need to tackle the artificial manipulation of the limiting resources'.[40]

The first situation means managing the elephant population as one would a captive population. We must include welfare issues and focus on social factors as much as ecological ones. The second option could be achieved by creating a network of corridors linking conservation areas. This would allow for meta-population management.[41] Ideally, corridors should link seasonal vegetation areas so the elephants can move, for example, from summer- to winter-foraging areas or from wet- to dry-vegetation areas.[42] Although the second option should be the long-term goal, it's unlikely to occur in the near future. Therefore, the first option ought to be our primary focus until the second can be realised.

So are welfare and conservation in conflict? In some instances they appear to be, but this need not be the case. A fine balance is required between the two. Without bringing welfare into elephant management, however, we will continue to see problems on private reserves. If we want to preserve elephants, these magnificent ambassadors of animal evolution, we need to pay much more attention to social factors.

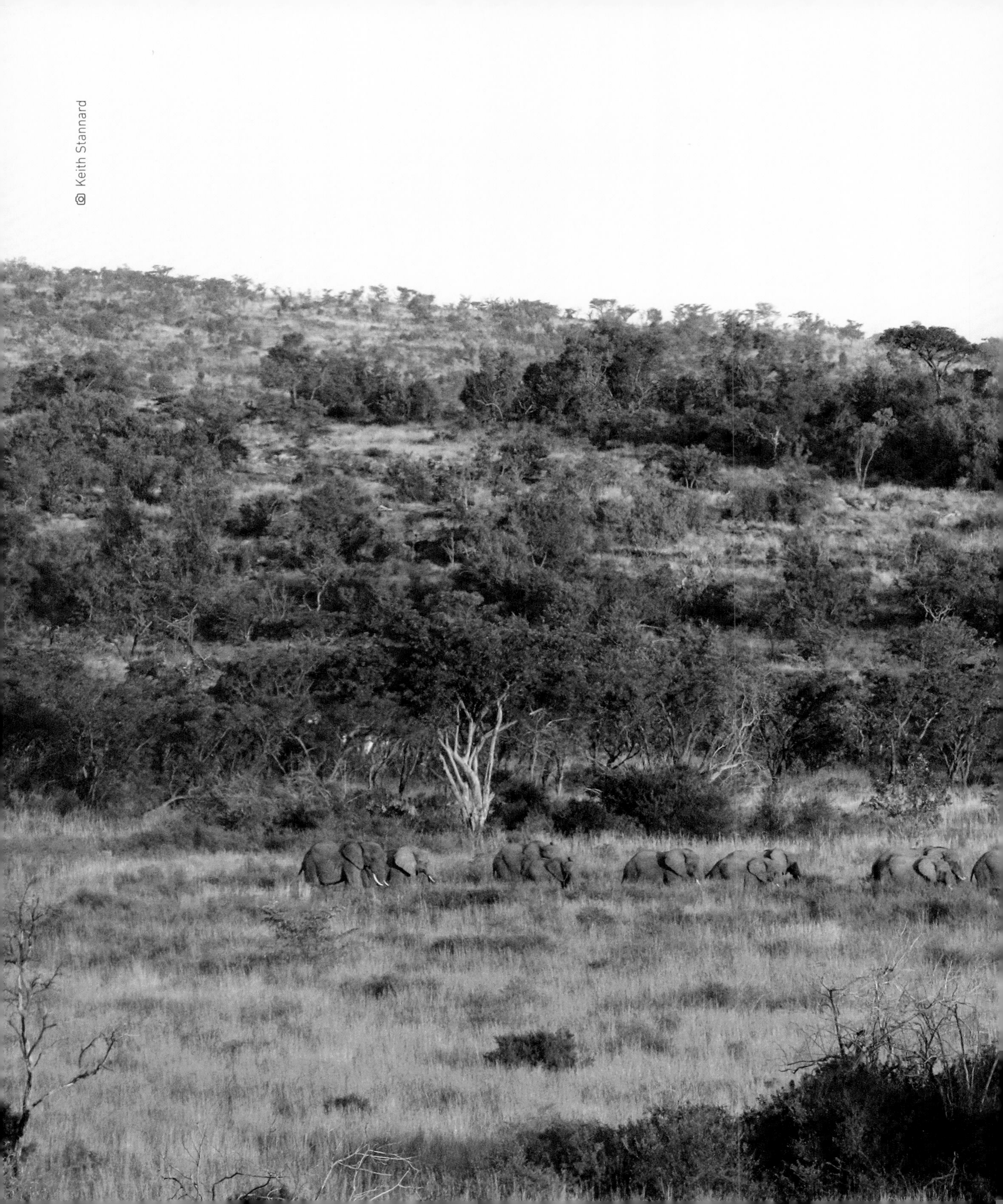

© Keith Stannard

In the late 1980s this was a cattle farm in South Africa. Over a 6-year period, a game reserve was created from these marginal lands by amalgamating neighbouring farms, removing fences and re-introducing wildlife. Today Welgevonden Game Reserve is a flourishing wildlife sanctuary of close on 70 000 acres that is home to the 'Big 5', forming part of the UNESCO-recognised Waterberg Biosphere Reserve. Welgevonden was one of the first reserves in South Africa that translocated whole breeding herds of elephants, thus keeping families together.

21

Just Addo

Portfolio

John Vosloo

John Vosloo is a city lawyer, but with one foot firmly entrenched in elephant country. The house he lives in abuts South Africa's Addo National Park, where he can on some days view elephants from his study. Over the years, John has spent countless days photographing elephants, mostly in Addo, and has amassed a stunning array of elephant images.

But Addo has not always been home to relaxed elephants. Turn the clock back to the early 1900s, when the early settlers were expanding into the region, planting crops and making homes for themselves. Addo's elephants were considered 'rogue', roaming far and wide throughout unfenced bushveld, leading to many deadly confrontations between settler and animal. In 1919, the decision was made to destroy all elephants throughout the region. By 1931, only 11 Addo elephants remained and the national park was proclaimed to protect these last elephants. Over time, Addo was increased in size and four large bulls were introduced from Kruger National Park to improve the genetics. Today Addo National Park is well over a million acres, covering five of South Africa's seven biomes, stretching all the way from the arid Karoo inland through to offshore islands and everything in between. The park is now home to around 600 elephants and, in places, has some of the densest concentrations of elephants in Africa, offering some of the best viewing. Enjoy a few of John's tributes to these magnificent animals in this chapter and elsewhere in this book (including the front cover).

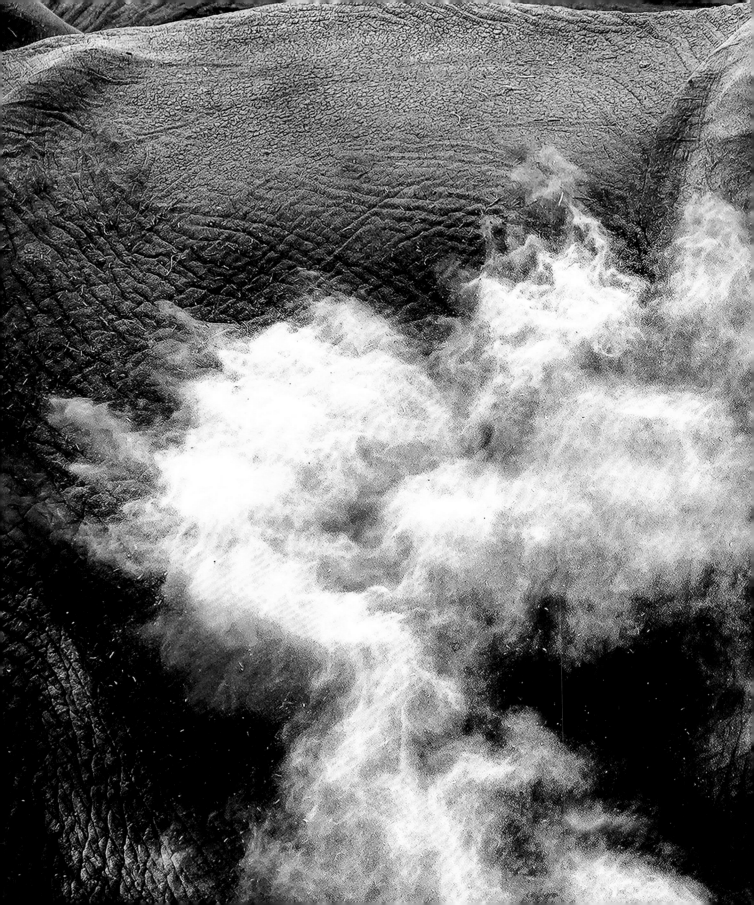

22

Making a safe haven

Botswana has become a refuge for elephants in a time of crisis. But the country's neighbours now need to open protected corridors so herds can return to their traditional ranges.

Colin Bell

With around 130 000 elephants (and potential home range for up to 200 000 elephants), Botswana is home to more than 33% of Africa's savanna population – as opposed to less than 3% some 50 years ago. Plummeting elephant populations elsewhere and the safety provided by Botswana has attracted a steady migration from neighbouring countries, where expanding rural communities and rising poaching levels are putting regional elephants under increasing pressure and stress.

'Elephants clearly have a cognitive ability to understand where they are threatened and where they are safe', says Mike Chase of Elephants Without Borders. 'In this case they're seeking refuge and sanctuary in Botswana where they are well protected.'

It was not always like this: Botswana's elephants have had their share of hard times. In the decades leading up to the 1980s, ivory was a sought-after international fashion product. Pendants, artefacts, furniture and jewellery from elephant tusks was the norm. Locally, just about every safari guide had an elephant-hair bracelet and many safari guests wore ivory bangles and necklaces. Even wildlife reference and field-guide books had elephant-hide leather covers. Throughout Africa, elephants were being slaughtered in great numbers for sport and for their by-products. Some years close on 100 000 were killed, equating to an elephant poached every 6 minutes. Currently, it's every 15 minutes – but still catastrophic.

During much of this time, elephants were listed as CITES Appendix II, which permitted commercial trade. The result was an extremely lucrative international market for elephant products, which the criminal poaching syndicates and commercial traders exploited. As security was non-existent along Botswana's porous, unfenced borders and in the country's wildernesses, poachers had free reign. Elephants in those days were jittery, reclusive and most often very aggressive.

The 1980s ushered in escalating poaching levels, resulting in even angrier elephants. Many campers had their tents trashed and vehicles smashed. One new start-up safari company had its Land Rover welcomed on arrival in the Moremi Game Reserve by an aggressive elephant that thrust its tusks into their radiator. The vehicle and its startled guide and passengers were shoved ferociously for over 50 metres on the first day of their inaugural safari. The company closed shop the very next day. Researchers tell stories of how

This apocalyptic photograph of a hunting camp in Botswana in 2016 documents the 2 years preceding Botswana's big-game hunting ban. This is a glimpse into the future of all elephants should we choose to ignore the ongoing slaughter of these highly intelligent, emotional and sentient beings.

© Daniel Dugmore

skittish elephants and their families would hide in forests during the day, sprinting to waterways in the late evenings for a quick drink before dashing back to the safety of the trees.

But the situation was turned around in the early 1990s when, as a result of the carnage, CITES uplisted all of Africa's elephants to Appendix I, creating a worldwide ban on all ivory trading. This embargo ushered in the 'golden years' for elephants. Demand for ivory was largely eliminated and syndicates that fuelled poaching moved to more lucrative products and activities. Elephant populations across Africa stabilised and began increasing for the first time in many decades. During this time Botswana's elephants began to relax as they learnt that people in vehicles no longer posed a danger.

Botswana's new tourism policy

In the early 1990s, Botswana revolutionised its tourism policies. The country turned its back on mass tourism, which had become the norm in many parts of Africa, and actively put in place policies and processes that would target the high-tariff/low-volume/low-impact tourist: visitors who would create jobs and training at scale for rural citizens. The country's enormous hunting concessions were carved into smaller concessions (on average 100 000 hectares) and tendered out to tourism operators, usually on an exclusive basis. Some areas were designated solely for photo tourism, others for sport hunting and some a mixture of photo and hunting tourism.

Safari companies invested in building small, personalised, high-end lodges, and many new jobs were created in rural areas. The positive spin-off from the dramatic change in tourism policy was significant, directly benefiting around 40% of the people living in northern Botswana. Thanks to the thriving tourism industry they could put food on their tables. This reduced the need to poach in order to survive. The many more eyes and ears traversing these concessions also deterred poachers. The elephants calmed down even more. Safari vehicles were no longer chased by stressed, aggressive elephants.

However, the resumption of elephant hunting for sport in 1996 and the downgrading of

Botswana's elephants by CITES to Appendix II changed much of that. Regulated international commercial trade in ivory was now legal. Furthermore, hundreds of elephant-hunting licences were issued annually to Botswanan hunting companies. One elephant that was collared moved 70 kilometres in 24 hours away from a hunting area after the first bullet was fired at the start of that hunting season. A safari guide who was driving alone in an area that abutted a hunting concession had his vehicle overturned and crunched by an aggressive elephant on the second day of a hunting season. He survived but was bruised and shaken. Everyone knew that game-viewing etiquette around elephants suddenly had to be much more circumspect.

Then, in September 2013, Botswana made the boldest move yet, banning all hunting on state lands and game reserves. Elephants were largely safe again and have calmed down progressively ever since – although there will always be grumpy individuals. Tooth abscesses, drought, heat, musth or even the loss of a family member are some of the reasons why these insightful animals may have bad days.

Years of breathtaking game viewing, with regular, close-quarter, high-quality elephant sightings, has helped enhance Botswana's tourist reputation, driving up demand while creating even more jobs in the rural areas. The wildlife and tourism sector is now the second-biggest contributor to the Botswana economy, after diamonds. And the government knows that diamonds are not forever. The core of Botswana's safari industry is based around its vast, pristine wilderness areas and its magnificent elephants.

Beyond the country's borders, however, pressure is mounting. Even though its poaching policies are strict and backed by the defence force, Botswana is starting to feel the effect of the poaching tsunami sweeping across much of Africa. Initially, criminal syndicates targeted the great elephant herds of northern and eastern Africa. But with elephant populations in East Africa now at less than 15% of what they were, the syndicates are switching their attention southwards. Botswana's long, unmanned, unfenced border with Namibia and Zimbabwe (with Zambia close by) is difficult to monitor. Poaching is rising despite the country's stringent anti-poaching laws. Even the Kruger National Park in South Africa

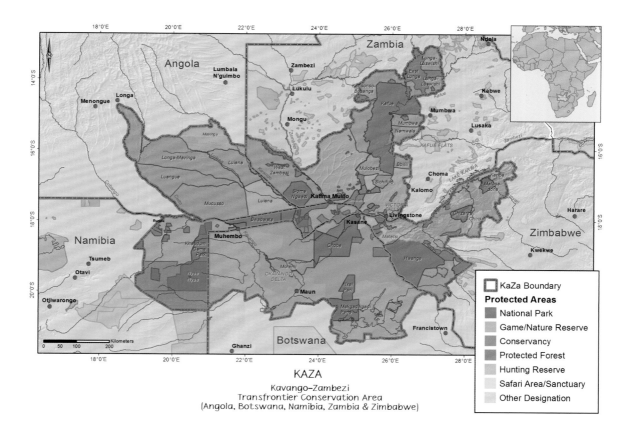

KAZA

Kavango–Zambezi
Transfrontier Conservation Area
(Angola, Botswana, Namibia, Zambia & Zimbabwe)

KaZa Boundary
Protected Areas
National Park
Game/Nature Reserve
Conservancy
Protected Forest
Hunting Reserve
Safari Area/Sanctuary
Other Designation

is starting to feel the effects: after decades of almost zero elephant poaching, South Africa recorded a record 68 elephants poached during 2017.

Just one poaching incident changes elephant behaviour. Late in 2017, one of the safari concessionaires in northern Botswana noticed that their normally relaxed elephants were suddenly jittery and aggressive. A few days later, the bodies of a number of poached elephants were discovered nearby and the cause of their aggression was established.

The KAZA potential

What lies in store for Botswana and its elephants? Let's begin with the numbers. In 1973, Botswana was estimated to have around 11 000 elephants. The results of the 2016 Great Elephant Census placed Botswana's elephant population at around 130 000. That census mirrors the trends picked up in the 2002, 2010 and 2014 calculations. The figures are dramatically down from the 2012 census, which recorded over 200 000 elephants, although the 2012

tally has been discredited, its methodology having been found to be riddled with errors. Today, the overall birth rate of elephants in Botswana roughly matches their mortality rate and the consensus is that Botswana's elephant population has levelled out at around 130 000 and will stabilise somewhere between 130 000 and 150 000. In areas of extremely high elephant densities, it appears that calving intervals have increased: elephants appear to know instinctively what's good for them.

Contrast these numbers with those for Angola, which has fewer than 4 000 elephants in a fertile country almost double the size of Botswana; and with Zambia, which has only around 22 000. Yet both Zambia and Angola have vast tracts of perfect elephant habitat lying dormant and unutilised.

Can Botswana sustain its current elephant population? Will the terrain be overrun by elephants to the detriment of other species? Will elephants destroy the vegetation on which they depend? What about human-wildlife conflicts? These are many of the questions that Botswana authorities continually have to grapple with.

This is where the Kavango-Zambezi Transfrontier Conservation Area (KAZA) initiative comes into the picture. KAZA is an agreement signed by Angola, Botswana, Namibia, Zambia and Zimbabwe, binding them to create a grand transfrontier 'Peace Park' straddling these countries and covering an area roughly equal to the size of France or Texas. KAZA's territory includes a major part of the upper Zambezi Basin, the Okavango catchment area and the Okavango Delta itself. The centre of KAZA is at the confluence of the Chobe and Zambezi rivers, where the borders of Botswana, Namibia, Zambia and Zimbabwe meet. Chobe National Park, Hwange National Park, the Okavango Delta and Victoria Falls are contained within it.

The goal of KAZA is 'to sustainably manage the Kavango Zambezi ecosystem, its heritage and cultural resources, based on best conservation and tourism models for the socio-economic wellbeing of the communities and other stakeholders in and around the eco-region through harmonisation of policies, strategies and practices'. Its vision is 'to establish a world-class transfrontier conservation and tourism destination in the Okavango and Zambezi river basin regions ... within the context of sustainable development'.

These are lofty goals. If they can be fulfilled, this entire region and its people will benefit. The crux will be whether KAZA stakeholders have the political will to open up ancient elephant migration routes and safe corridors to connect Botswana's elephants to those well-stocked food pantries waiting for them to the north in Namibia, Zambia and Angola. If these corridors can be recreated, it will be a win-win situation for all member countries and their rural communities. Millions of euros have already been spent on the project, but much of it on expensive, mainly foreign, consultants who have produced reams of important-sounding documents and beautiful maps. Sadly, there have been few tangible changes on the ground and some of the consultants' work has been impractical, even disastrous.

An example of the latter can be seen in the land-use map created for one of the most crucial KAZA hotpots in Namibia, directly opposite and adjacent to Chobe National Park (see page 217).

This area is community land; it should be able to host abundant wildlife and is a vital corridor that could link Botswana's highest elephant and buffalo densities to the vacant lands and reserves to the north in Namibia and Zambia. This potential corridor would start on the north bank of the Chobe River near Kasane.

Towards the end of the dry season, in particular, large concentrations of animals could pour across the Chobe River into these productive floodplains and beyond – but are unable to do so. With no corresponding national park across the river in Namibia to afford protection, and no well-structured photo-tourism industry delivering tangible benefits to villagers at scale, livestock takes precedence over wildlife. Any elephants that venture across the Chobe come into contact with hunters and local villagers, who are understandably concerned for their own safety. They beat drums and harass elephants, sending them scurrying back across the Chobe River into the safety of Botswana.

How these consultants could produce such an impractical and poorly considered land-use plan and map is beyond comprehension. Did they really expect small pockets of hunting to be compatible with a prime, neighbouring photo-safari attraction along a busy river and a wide-open floodplain used by game-viewing tourists? Many Chobe River cruise tourists have unfortunately been witness to elephants and buffaloes being shot for sport. These incidents, caused by poor planning and inappropriate zoning, damage the good standing and tourism brands of both Botswana and Namibia in the international tourism markets.

The solution would be to develop a practical plan that includes a number of smaller, more easily achievable, step-by-step milestones, working with governments, the local communities and their leaders to implement these stages. Wildlife corridors could open up on either side of the Kasika settlement. Communities would need to be fully consulted and compensated and, in time, benefit from tourism money and jobs generated along such corridors. Well-positioned, but more isolated hunting concessions could open up at a safe distance from the Chobe River – to the north of these corridors and out of earshot and sight of the droves of photo tourists who enjoy the game-viewing pleasures of the river. Some time back, the local chief and his headmen were consulted and were fully supportive of the idea. Without these safe corridors,

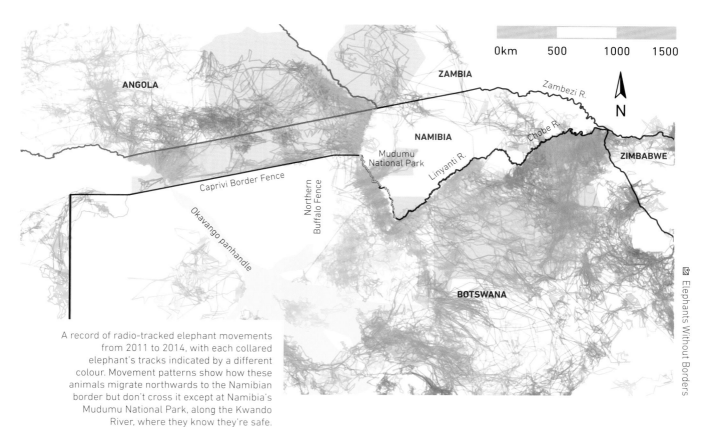

A record of radio-tracked elephant movements from 2011 to 2014, with each collared elephant's tracks indicated by a different colour. Movement patterns show how these animals migrate northwards to the Namibian border but don't cross it except at Namibia's Mudumu National Park, along the Kwando River, where they know they're safe.

the KAZA ambition of spreading and sharing wildlife populations will remain a daydream.

Right now, many of Botswana's elephants are scared of Namibia. The EWB map (above) charts their radio-tracked movements over a 4-year period. They migrate northwards to the Namibian border but don't cross it much except at Namibia's Mudumu National Park, where they know they're safe. Elephants quickly work out what is good for them!

The people equation

Both Namibia and Botswana have been successful pioneers in community-wildlife conservation. Initiatives like Integrated Rural Development and Nature Conservation (IRDNC) and Community-based Natural Resource Management (CBNRM) have been applauded for their ground-breaking work, which has integrated communities into, and given them a stake in, the wildlife and tourism industries.

The Botswana tourism model, formulated in the early 1990s, ensures broad-based and equitable earnings generated by the private-sector tourism industry. Each year, every lodge or concession has to pay neighbouring communities, local land boards or government an annual lease fee. During the year, further payments of up to 19% of gross turnover go to government structures through a combination of resource royalties, training levies and VAT. These combined fees may amount to 100% of a lodge's net profit in a bad trading year, or up to 30% in a good year. The net result is that government (and sometimes communities) receive at least 30% of the profits generated by the tourism industry without having to invest anything. In return, safari companies have the privilege of operating sustainable businesses in one of the most interesting wildlife areas on Earth.

Those hefty payments help incentivise government (and sometimes communities) to look after their wildlife and their environment. Botswana's thriving rhino reintroduction programme is world-class and would not have been successful without the full support of government.

In an attempt to limit local corruption and inefficiencies, the government recently set up a controversial, centrally controlled tourism Land Bank, which aims to regulate the issuing of all tourism leases and licences, as well as the collection of many of the fees generated by the tourism industry. One of the reasons was to help stem local land board and community corruption. It was found that influential community members were often earning a greater share of the lease fees paid by the safari companies, at the expense of their community. Only time will tell whether this Land Bank system will work, or whether it will create antagonism at grassroots level and stimulate bushmeat and ivory poaching.

With all its experience in community development, it's surprising that Namibia has not been able to make its Chobe River conservancies more viable and better-functioning assets for the benefit of their communities and the country. The result is that essential wildlife corridors, which KAZA needs to entice Botswana's elephants back to their former more expansive ranges, are blocked. A few new lodges have opened up on the Namibian banks of the Chobe, and most of the houseboats that work the river are Namibian-owned and based. But they spend most of their time viewing wildlife on Botswana's banks of the Chobe River.

© Michael Poliza, Okavango, Botswana

always be mutterings and there will always be some poaching. However, Botswana's elephants have been able to expand their range further afield within the country, enticed by the Selinda Spillway and the Boteti and Nhabe rivers that have been holding water all year round after several seasons of good rains. Some elephants have been able to move out of their traditional home ranges of the Okavango, Chobe and the Linyanti into places as far away as the Central Kalahari Game Reserve. Some have migrated even further south to Khutse Game Reserve and the Makgadikgadi Pans National Park. Other elephants increasingly migrate between Chobe and Hwange National Park in Zimbabwe. And the Tuli/Mashatu area in the south-east of the country is now home to around 900 elephants.

Since China banned all trade in ivory at the end of 2017 and shut down its ivory-carving industry, there is real hope that poaching levels will decline as demand for ivory products subsides. But first, Hong Kong and the rest of the world will have to follow China's lead and ban all domestic trade in ivory. The internet has become an effective channel for criminals to advertise and sell illegal, endangered-wildlife body parts.

If the world wants to stop illicit wildlife trade, then internet trading platforms like TauBau and others must do their bit. Dodgy yet legal sites offering loopholes, such as 'mammoth tusk' sales, must be shut down, as they become avenues for poached ivory to be advertised and traded on the sly. The giant Chinese Alibaba internet company was able to implement processes that successfully stopped the selling of counterfeit luxury goods on its platforms after the Kering Group (the owners of Gucci, etc.) threatened to take them to court in New York. The same efforts must be applied to stop the selling of all endangered-wildlife body parts over the internet. And CITES must uplist all of Africa's elephants to Appendix I, and permanently close down all the trade loopholes.

There is hope that KAZA corridors will then begin to open up and elephants will spread again into their former ranges. More southern African states will then be able to create a viable, vibrant, sustainable, inclusive tourism and wildlife industry that follows an expanding, migratory elephant population.

Looking ahead

The hope is that the KAZA initiative develops healthy momentum, so that local people and wildlife within the region can prosper. There are few areas in Africa with the tourism potential of the Chobe River's Namibian bank. Yet most people there live in poverty, surviving hand to mouth. For now, wildlife on the Namibian side of the river is mostly transient, perhaps still testing the safety of ancient migration corridors to Namibia, Zambia and Angola.

For the rest of Botswana's elephant range, matters are almost as good as they realistically can be. No system is ever perfect and there will

Kabulabula, Kasika and Impalila Communal Co...
and
Surrounding Areas under Conservation Mar...

Legend

- Airstrip
- Settlement
- Conservancy or Community Forest Office
- Crushpen
- Medical Facility
- Kraal
- Tourism Establishment
- Water Point
- Joint Venture Enterprise
- 3 km Buffer Line along Border
- River
- District road
- Main road
- Track
- Trunk road
- Communal Conservancies
- National Parks
- Community Forest
- Forest Reserve
- Country Border

Exclusive Wildlife: No Disturbance

Multiple Use: Livestock Priority

Settlement & Cropping Area

Exclusive Wildlife: Trophy Hunting Only

Settlement & Cropping Area

Kasika

Multiple Use: Livestock Priority

Exclusive Wildlife: Trophy Hunting Only

Exclusive Wildlife: Trophy Hunting Only

Exclusive Wildlife: Trophy Hunting Only

Multiple Use: Tourism Priority

Exclusive Wildlife: Trophy Hunting Only

Exclusive Wildlife: Tourism Only (No Hu...

Kabulabula

Kabulabula

Settlement & Cropping Area

Kasika

King's D...

Exclusive Wildlife: Tourism Only (No Hunting)

Chobe Savanna Lodge

Botswana

Kazuba
Mourkandi

Slowe

Mpalsa...

Milvere

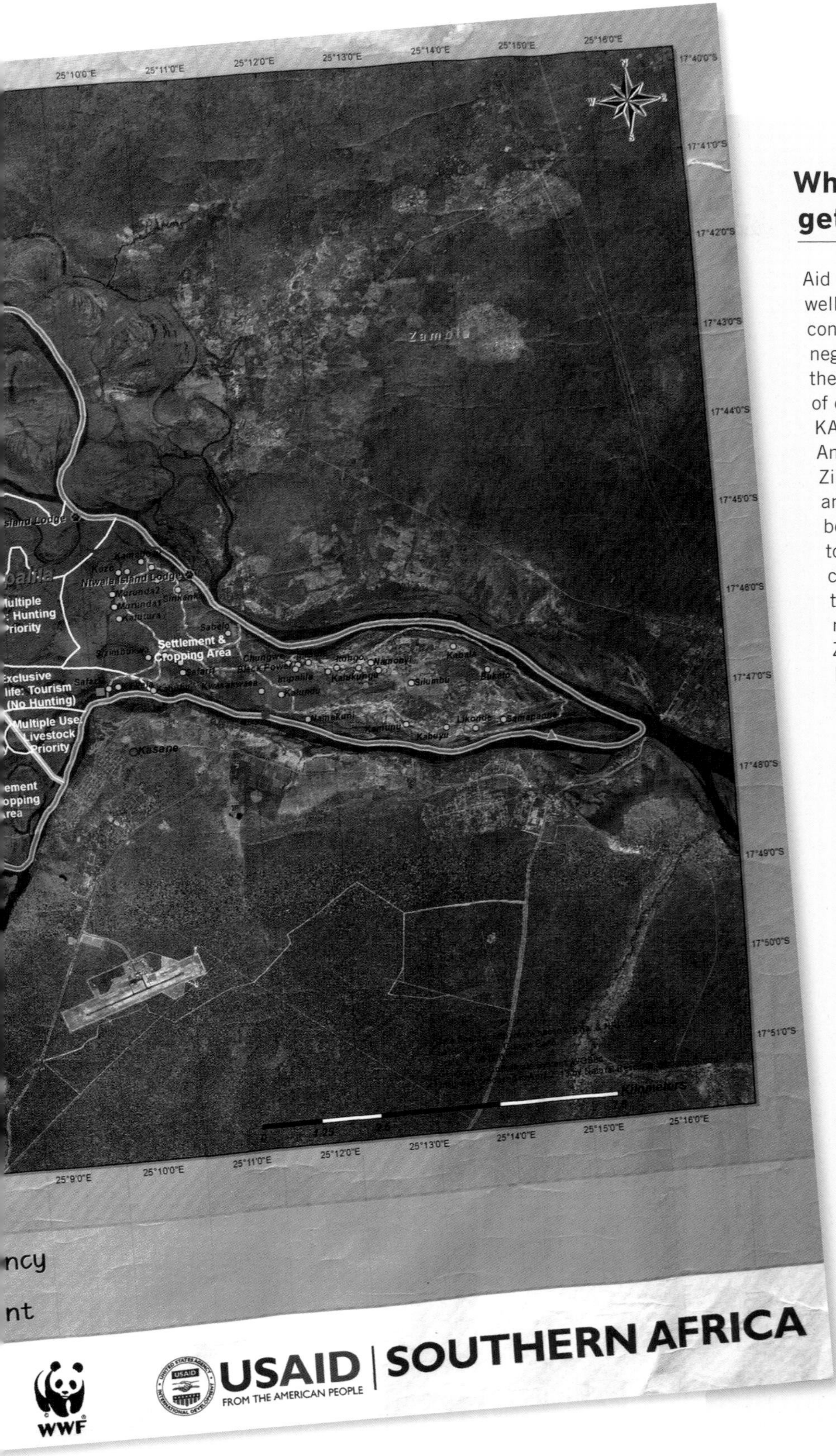
Zambia
Settlement & Cropping Area
Exclusive
life: Tourism
(No Hunting)
Multiple Use:
Livestock
Priority
Kasane
USAID | SOUTHERN AFRICA
FROM THE AMERICAN PEOPLE
WWF

When consultants get it wrong

Aid can be a double-edged sword. Too often, well-meaning donor money comes with consulants and conditions attached that negatively impact on the effectiveness of the project. This 2013 map is an example of consultants getting it totally wrong. KAZA is the ambitious agreement by Angola, Botswana, Namibia, Zambia and Zimbabwe to expand wildlife corridors and enable communities to become beneficiaries of their wildlife. For it to work, safe wildlife transits need to connect Botswana's large elephant herds to the abundant food supplies and national parks to the north in Namibia, Zambia and Angola. That conduit bottlenecks along the north bank of the Chobe River.

This map of the Namibian lands across the Chobe River and directly opposite Chobe National Park was funded by well-meaning, well-respected donor organisations. Small hunting pockets have been arbitrarily drawn and scattered throughout the corridor, often very close to photo-safari game-viewing areas, tourism lodges and villages. Apart from blocking elephant transits, what would happen if a bullet missed its target and killed a villager or tourist? Many photo-safari guests have crossed paths with and witnessed distressing episodes of elephant or buffalo hunting, which negatively affects Namibia's brand and its reputation in international tourism markets.

Practical, informed and effective zoning that has the community's full buy-in is essential for any land-use programme to be successfully implemented on the ground.

Fortunately good sense prevailed and this chaotic land plan has been quietly shelved.

23

Botswana's sanctuary

Tracing the unusual trajectory of the world's most dedicated and successful leader in elephant conservation

Kelly Landen

Blue skies, warm sunlight, all quiet but for the rumble of our Landy bumping across a thick sand cutline in northern Botswana's Chobe Forest Reserve. It's the best time of year: the wet season, when the vast landscape is lush with green vegetation and littered with rain-filled pans. I can see only the horn tips of six buffalo 'dagga boys' (old bulls forced out of the herd) above golden grasses waving in the cool breeze. A few hundred metres away, a large family of elephants is weaving through the Zambezi teak, snatching browse along the way. A small calf, hidden below its mother's belly, stretches up to drink from her warm teat. It's the way I always imagined wild Africa should be.

The story of Botswana's elephant population has been a bright spot in a sometimes bleak landscape. Outside of Botswana, the African elephant is in crisis. Habitat loss, competition for resources, civil unrest, terrorism, human-wildlife conflict and escalated rates of poaching for ivory threaten their very existence. In some areas, populations have decreased by as much as 70% in less than 10 years.

While it's tempting to despair over the relentless flow of images of elephant slaughter, Botswana has made steady and determined progress towards protecting and sustaining its elephant population. The recovery of this population, once on the verge of extinction, is one of the continent's greatest conservation achievements. Botswana now hosts the world's largest free-ranging elephant population, which at times has amounted to over 50% of all Africa's savanna elephants. It's one of the last places on Earth where elephants still follow ancient migration routes unhindered by development. It is also home to spectacular populations of wildlife whose seasonal migrations are among the longest in Africa.

This didn't come about by accident. It took leadership from lawmakers and foresight from policymakers, coupled with the vision of one Motswana doctoral student. This chapter is about how these people together sparked real change. And it's about a vision of elephants as partners in promoting a culture of peace, developing ecotourism and improving rural livelihoods. Elephants are an umbrella species. They're ambassadors for conservation, helping locate and connect key conservation areas in Africa's largest wilderness area.

Botswanan elephants follow well-used paths on a daily trek between their feeding grounds and water.

© Colin Bell

A man with a plan

Two decades ago, a young Motswana conservationist named Mike Chase hatched a plan. Driven by his steadfast belief that every elephant life matters, Chase proposed an ambitious and never-before-attempted study of the elephant migration routes in northern Botswana and adjacent countries.

Chase, a PhD student, was convinced that arming himself with hard data was a key plank in his fight to preserve the elephants, not just of his homeland, but of the entire continent. He believed then – as he does now – that elephants are a leading indicator of the health of a country's ecosystem, the proverbial canary in the coal mine. When a country's elephant population shows signs of assault, chances are its entire ecosystem will suffer the consequences. Though his affection for the world's largest land mammal was obvious, it was his insight that saving elephants could provide a path for a broad-based conservation agenda that made his impact on Africa's wildlife so important. Counting and tracking elephants was just the beginning in transforming Botswana's approach to conservation.

In 2004, Chase established a Botswana-based NGO, Elephants Without Borders (EWB), to gain support for his research. By the time his doctorate was completed in 2007, EWB had fitted satellite collars on nearly 50 elephants in northern Botswana, Namibia and Zambia, making it the largest telemetry study of elephants at the time. The study provided new and unexpected information, shattering long-held assumptions about elephant movements and migrations, particularly the idea that they stayed mostly in one area. Annual ranges were found to vary considerably from 910 square kilometres for a 45-year-old bull along the Chobe River to the home range of another young bull which, at 24 828 square kilometres, spanning four countries, was the largest ever recorded for an African elephant. This was highly significant, considering that the average home range of elephants in the rest of Africa is around 3 000 square kilometres.

These telemetry studies, coupled with population assessments and trends brought to light in EWB's aerial survey across the region,

Mike Chase (front left),
Kelly Landen (right)
and other spotters.

were used to delineate the boundaries of KAZA, the Kavango-Zambezi Transfrontier Conservation Area. Under this agreement, Botswana, Namibia, Zambia, Zimbabwe and Angola formally acknowledged that wildlife conservation transcended national boundaries. Elephants could no longer be studied in isolation – they were telling us where and how far they needed to range and what territory should be protected if they were to flourish. This resuscitated the argument for re-establishing and preserving transnational wildlife corridors.

In 2010, during a meeting with Kitso Mokaila, Botswana's Minister of Environment at the time, Chase was told that the government would be willing to help fund a thorough aerial survey of wildlife throughout northern Botswana. He submitted a proposal for Elephants Without Borders to take the responsibility and was granted the permission and support.

Flying along straight transect lines a mere 100 metres above the ground, having to contend with heat, bumpy winds and high-flying birds, takes immense skill and concentration. Each day for 4 months, the team woke before dawn and were in the air at sunrise for 5 hours of flying. They traversed Botswana's rich biodiversity and ecosystems, from rolling grasslands to sandy ridges, wetlands, pans, river systems and extended woodlands. The exercise unveiled a magnificent panorama of the country's unique wilderness areas.

What the survey revealed was surprising. Over the previous 15 years, numbers of some wildlife species in certain areas of northern Botswana had fallen substantially. However, population estimates from this survey, plus nine government aerial surveys, showed that northern Botswana's elephant population had increased during the early 1990s, but had remained consistent and stable from 2004 onwards.

The end of trophy hunting

Data was the most compelling ambassador Botswana's elephants could have. Though humans have long been fascinated by the species – mythologising them, even revering them – without hard evidence of their decline, there was no pressure on governments to step forward in their defence. But in 2012, just 2 years after Chase's aerial survey, the Department of Wildlife and National Parks conducted their own survey, which echoed the reduction in wildlife populations across the northern region.

Botswana's response was swift and remarkable. In August 2013, the Environment ministry announced that hunting would be banned across the country from 1 January 2014. From then, no quotas, licences or permits would be issued for hunting of game animals listed in the Wildlife Conservation and National Parks Act. It was a remarkable moment of leadership in the country's history of conservation.

The government explained that the temporary ban was because of the decline of several wildlife species. This was due to 'a combination of factors such as anthropogenic impacts, including illegal offtake and habitat fragmentation or loss'. Ministry spokesperson Caroline Bogale-Jaiyeoba said there had been significant declines, and mentioned also tsessebe, sitatunga, lechwe and springbok. 'Of particular concern is the fact that all the surveyed species except elephant and impala declined in at least one protected area.'

The ban was unexpected, and interpreted by some as a direct challenge to the wildlife policies of other southern African nations, where hunting was widely practised. However, after nearly 20 years of hunting being conducted in Botswana, there was hardly any evidence to support a widely held view that the hunting industry was contributing to improving the quality of life for all members of the communities. Funds were channelled through community trusts with little, if any, accountability and did little to benefit the individual community members at a household level.

The decision placed Botswana at the cutting edge of conservation. It committed the country to protecting its wildlife heritage by providing wildlife with a safe haven. It also created a new model of African tourism – low impact/high value – minimising the human footprint on the natural world. It turned the country into a prime tourist destination, bringing in far more revenue than it would lose from the hunting ban.

Tackling the illegal ivory trade

While Chase's work tracking Botswana's elephants had yielded success locally, he knew that to maximise his impact meant aiming even higher. His research had revealed that migration routes and movements were even more expansive than most ecologists had previously thought. Not content with the impressive but nevertheless incomplete picture he'd assembled, he began to percolate another plan. What if, Chase thought, we could count every remaining elephant in Africa?

In December 2013, Botswana hosted the first African Elephant Summit. It was opened in Gaborone by President Ian Khama, and hosted representatives from key countries along the ivory-trade chain. This included elephant range states Botswana, Gabon, Kenya, Niger and Zambia, ivory transit states Vietnam, Philippines and Malaysia and the two primary ivory destination states, China and Thailand. The summit adopted 14 urgent measures to halt and reverse the trend in illegal killing of elephants for the ivory trade.

The summit was an opportunity for Chase to announce EWB's most ambitious effort yet: a Botswana-led initiative called the Great Elephant Census (GEC). It remains the largest and most accurate elephant and animal survey in Africa to date. The announcement was applauded and embraced by most of the African elephant range states, eager to commit to participating in the ground-breaking initiative.

As the survey rolled forward, in 2015 Botswana hosted the Kasane Conference on the Illegal Wildlife Trade (IWT). Thirty-two countries plus the EU and nine international NGOs met to review progress on the London IWT 2014 Conference Declaration. The result was the Kasane Statement, which contained 15 new commitments for action on issues such as trafficking, demand reduction, tougher law enforcement and ensuring local communities benefit from conservation. It committed participants to improving co-operation between countries on trading routes, while strengthening prosecution mechanisms.

Botswana's Minister of Environment, Tshekedi Khama, appealed to all African nations to commit to saving endangered species, despite financial pressures. 'I'm fully conscious that our economies are struggling to recover from the recent economic downturn', he said, 'but we cannot use this as an excuse for inaction as there will be a bigger price to pay in future'.

At the IUCN World Conservation Congress in August 2016, the much-anticipated final result of the GEC surveys was announced just as the scientific peer-reviewed paper was published[1]. The results were shocking and painted a bleak picture of the pan-African situation: fewer than 400 000 savanna elephants survive in Africa. During the previous 7 years, ivory poaching had reduced elephant populations across the continent by 30% – by 144 000 individual animals. Poaching, human-elephant conflict, climate change and habitat loss are so intense that elephant populations are shrinking by 8% each year, continent-wide; that's about 28 000 elephants a year, 76 elephants a day, one death every 20 minutes.

Chase's hard data highlighted the elephants' struggle, offering a better understanding of their plight, and spurring international action. It was now clear: many elephant populations faced the threat of local extinction.

Up-listing elephants

Three weeks after the IUCN Congress, Botswana took the floor at the 17th meeting of the CITES Convention of the Parties (CoP17) in Johannesburg, South Africa. With an unexpected twist, Minister Tshekedi Khama said, 'We are fully aware of the serious poaching crisis facing elephants across much of Africa. We … support an up-listing of all African elephants to Appendix I. If we do not take decisive action now and wait until the next CoP in 2019, the results will be catastrophic. Many more thousands of elephants will have been poached for their ivory, many more rangers will have lost their lives, and the criminal

syndicates and corrupt officials who are enabling the trade to flourish will continue to profit from this destructive trade.

'This will send the clearest signal yet that the ivory trade will not be tolerated and will allow us to divert resources into the things that matter: supporting local communities, maintaining habitat connectivity, managing human-elephant conflict and tackling the criminal networks that have caused so much destruction for our continent's proudest emblem – the African elephant. Botswana now supports a total, unambiguous and permanent international ban on the ivory trade, including domestic markets, and supports the inclusion of all African elephant populations on Appendix I.'

Although the proposal was rejected, many governments, including the 39 countries of the African Elephant Coalition, stakeholders and NGOs, welcomed and commended Botswana's stand. It had set the stage for a complete international and domestic ban on trade in ivory.

One of the key arguments at CoP17 against a unified listing was that the chief criterion – significant population decline – was not occurring in three of the Appendix II countries: Namibia, South Africa and Zimbabwe. According to this argument, these elephants were not threatened and the countries should be allowed to have the option of trading ivory.

However, a peer-reviewed scientific paper published in the journal *Biological Conservation*[2] in 2017 showed that 76% of Africa's elephants are in transboundary populations. The authors make the strong case for harmonising the protection of elephants under CITES by applying its criteria across the species' range as a whole, rather than to individual countries. Using data from the Great Elephant Census and IUCN/SSC African Elephant Status Report 2016, the study shows that six trans-boundary populations numbering almost 250 000 elephants, or 53% of the total African population, are currently 'split-listed', meaning they're shared between Appendix I and Appendix II countries.

In January 2017, Botswana took another bold step by banning elephant-back-ride tourism, influencing neighbouring country facilities to do the same.

Postscript

Shortly after the two chapters on Botswana were written for this book, there were sweeping changes that are radically transforming the country's politics and, in turn, its wildlife and elephant policies.

President Ian Khama reached the end of his two terms as head of state in March 2018 and handed over the presidential reins to his vice-president, expecting his handpicked successor to continue with his policies. President Mokgweetsi EK Masisi is now firmly in charge of the country and has surprised the Khamas (and just about everyone) by rapidly reversing many of the previous administration's revered wildlife policies and oversights.

While Khama's anti-hunting policies created a photo-safari boom, one of his biggest mistakes was not to fill the voids created when he banned hunting at the end of the 2013 safari season. For some unknown reason, most of those ex-hunting concessions have been lying dormant ever since, even though offers were made by photo-safari companies to lease many of these areas.

Communities that received meat and cash from hunting were left with empty stomachs and drained bank balances. Furthermore, a number of rural Okavango communities had the revenues they earned from tourism channelled away from them into the new Land Bank that was set up and controlled by central government. These communities are understandably discontented at their losses and in an election season their voices get heard.

There has also been little clear policy direction on how communities can effectively deal with roving elephants that destroy crops and harass – even kill – people. These incidents are reported in the press and reach the ears of politicians and the pro-hunting lobby, which has been relentlessly exploiting them to persuade the Botswana government to reverse the ban on sport hunting.

Khama's ruling BDP Party failed to win an outright majority in the 2014 elections, garnering just 46.5% of the popular vote. But they retained power because opposition parties were fragmented. The next elections are in October 2019, giving President Masisi time to consolidate his and his BDP Party's position. But to win elections he needs to make wholesale changes to address matters not dealt with sufficiently under Khama's period in office.

The future of elephants has been debated in parliament and at local level around the country, including whether Botswana should reintroduce sport elephant hunting (and trophy hunting in general) and whether it should change its commitment to upgrade all elephants to CITES Appendix I.

As this book went to press, the nation – and indeed the world – was awaiting the outcomes of these deliberations. Would Botswana reverse the positive 'Brand Botswana' gains made, the jobs created and foreign exchange revenues earned from photo-safari tourism since the hunting ban? Could its bold no-hunting stance, which has elevated the country as arguably the most prestigious wildlife safari destination in Africa, hold? Or will short-term politics for votes win the day?

By pure coincidence, as the new president took office, the ivory-poaching tsunami that has slowly been moving southwards through Africa arrived in Botswana (and South Africa). For decades, both Kruger National Park and northern Botswana have been relatively free of ivory poaching. But those golden years are over and both countries are now having to deal with sophisticated, highly trained ivory poachers well connected to international criminal syndicates.

Sadly, Dr Mike Chase of Elephants Without Borders was caught up in the politics of the day when he reported on the sudden unexpected surge in elephant poaching, evidence of which he discovered during a 2018 aerial elephant census survey across northern Botswana. As a scientist, he reported what he saw, but it was not the message the new Gaborone administration wished to hear and he came under sustained personal attack.

Questions being asked are whether President Masisi's wildlife policy changes are out of genuine concern for the wellbeing of rural communities and elephants, out of embarrassment at the unanticipated surge in poaching under his watch – or simply to consolidate votes. It is not clear whether he (or his advisors) fully realise that the high regard in which Botswana is currently held by the international safari tourist market could be irreparably damaged by opening sport hunting of elephants.

According to the World Travel and Tourism Council's 2018 Economic Impact Report, Botswana's tourism sector's contribution to the country's GDP will have increased from P13 billion in 2013 (the last year of hunting) to P23 billion in 2018 – a remarkable 70% increase in just 5 years since hunting was abolished. And advance bookings for 2019 and 2020 are the strongest on record. Many of Botswana's supporters around the world are hoping that the new administration is not considering jeopardising these extraordinary gains (not forgetting the new jobs created and the VAT and other taxes earned through photo-safari tourism) for a handful of rural votes and the lobbying of a few disgruntled sport hunters.

Time will tell …

Dr Don Pinnock

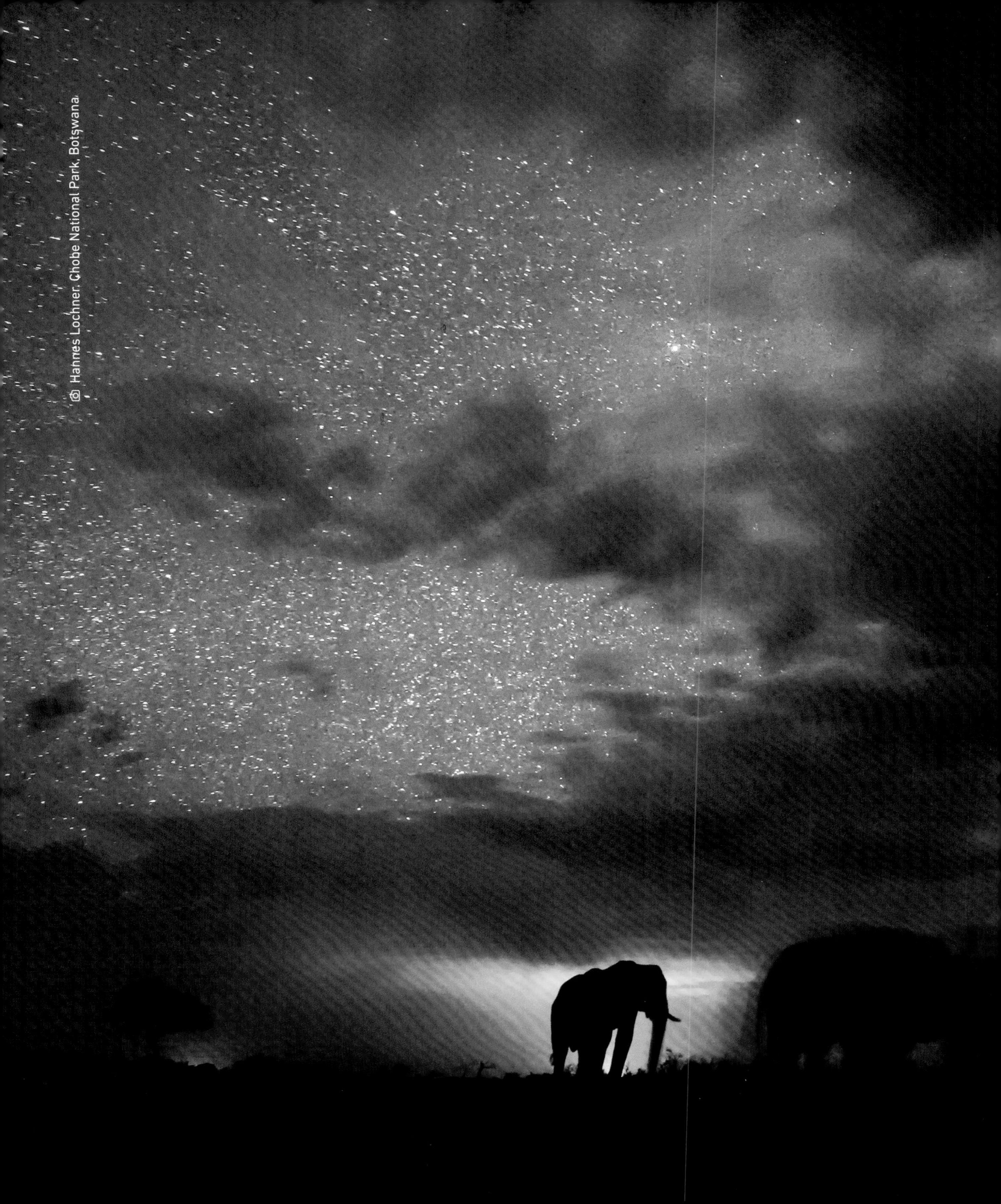

© Hannes Lochner. Chobe National Park, Botswana.

In the tracks
of giants

Dr Ian McCallum

To walk in the wake of elephants,
to be small in a world of giants,
to learn the spoor of silence,
and the deep rumbling eloquence
of kin.

To move in the skin of elephants,
to feel the alliance of sand,
the contours of land
and the far-reaching pull
of water.

To be alive to the sway of elephants,
to remember the songs of seasons,
the ancient lines of migrations,
and loosen your reasons
for fences.

To wake up to the web of intelligence,
to the wild origins of sentience,
to find your voice and raise it,
that others may raise theirs
for elephants.

24

Learning from Zimbabwe's 'presidential elephants'

This intelligent being, gifted with conscious thoughts and emotions, was clearly thinking. She may not have been able to speak my language, nor me hers, but she'd chosen to commune with me nonetheless.

Sharon Pincott

Zimbabwe's presidential elephants roam a relatively small slice of unfenced land in western Zimbabwe's Hwange Estate, their huge footfalls etched daily in the Kalahari sand. I arrived in March of 2001 to work with them, a fearless young woman from Australia, eager to embark on a new life working alone, untrained, unpaid and self-funded.

The elephants were said to inhabit just one section of land and to be specially protected from hunting, culling and other ills by a 1990 presidential decree. Although the decree was probably well intentioned at the time, the realities on the ground were worrying.

Few elephants remained just on the Hwange Estate. Hwange National Park – some 100 times larger than the estate – beckoned, with a mere railway line separating these two areas. The elephants, I also discovered, didn't belong to just one herd. Which were presidential elephants and which were not? They were all said to be presidential when (and perhaps only infrequently) they were within the estate's boundaries. Research into population dynamics pointed to another anomaly. The 300+ elephants touted to be presidential could never have proliferated from a handful of 'original' estate elephants, as the public had been led to believe. Influx from the adjoining Hwange National Park was certain.

Various elephants were easily recognisable and, in my early years, did spend much of their time on the estate with their families. A few of these individuals were known by name to some resident safari guides. Most notable were two adult females, Inkosikasi and Skew Tusk, said to be two of the original presidential elephants. Yet I never encountered these two or any of their family members intermingling, so it was unlikely that they were related. Inkosikasi was a big, tuskless cow but, as I soon discovered, there were several such cows and all were mistakenly called Inkosikasi. When Skew Tusk broke her skewed tusk, as she did every few years, people with no knowledge of her ear patterns and/or family members could no longer identify her. Nor, therefore, would they know if she suddenly disappeared from her family. In fact, I soon realised that relatively little detail was known about this clan of elephants, apart from their being calm, friendly and used to the close presence of game-drive vehicles, something considered unusual at the time.

There was still more to contemplate. Though they enjoyed nominal protection under the head of state, they could hardly be considered a flagship herd when they were off-limits to all but

'With familiarity came trust, and there's nothing like being trusted absolutely by an enormous wild elephant.'

a privileged few with money to stay at a couple of private lodges. And no special security measures had been implemented to protect them. What happened, I wondered, when they wandered into nearby sport-hunting areas or the unfenced Hwange National Park (where ration-hunting regularly took place)? The park had wildlife rangers, but I never encountered them patrolling the estate, which didn't even have an anti-poaching team of its own. Equally concerning, trophy hunters had begun operating between two of the estate's photographic lodges without a murmur of public concern.

So I inherited a tangled web of inaccuracies and confusion, with nothing implemented on the ground to give weight to the decree. Dedicated, long-term monitoring, intimate knowledge and special protection measures had, in fact, been lacking during the 11 years since President Robert

Mugabe had issued his decree. Many, even within Zimbabwe, were not aware that these elephants existed, and there had been no notable government interest in them since their naming.

My timing was, in the eyes of friends and family, bizarre. I was working alone in a volatile country, not my own. People were fleeing violent land take-overs and here I was on a preapproved, but self-appointed mission to raise increased awareness of these particular elephants. What they urgently needed was an independent person with an open mind, a great deal of patience and tenacity, and a willingness to learn; someone who could be their voice among the political and economic madness that was Zimbabwe at this time. I was, it seemed to *me*, in the right place at the right time.

The elephants on the Hwange Estate were breathtaking. They were easy-going and glorious to behold. It was an opportunity and a privilege to get to know them intimately and I set out to learn everything I could about them as individuals and families. As a non-scientist I obtained input from members of Kenya's long-term Amboseli Elephant Project (run by pioneer Cynthia Moss), who were all happy to share their invaluable knowledge with me. Right from the start, I vowed to continually remind the Zimbabwean authorities about the unique status of these elephants and to raise their flag high in a bid to secure their wellbeing.

My first task was to introduce structure to the naming of individual elephants. As had been done in Kenya, I assigned a letter of the alphabet to each family and gave all the elephants within that family group names beginning with it. There was the extended A family, the B family, C family, and so on. It made no sense to me to use numbers, as science prefers.

I classed as presidential elephants those families which, in 2001 and 2002, appeared to be spending most of their time on the estate. How frequently I encountered each of the families during my daily 8 hours in the field, day after day, year after year, made it possible to gauge which ones didn't wander too far. Studies had found that older, independent males wandered over great distances, something my own sightings supported. So, given this fact, and also how regularly the adult males disappeared for good after wandering into nearby sport-hunting concessions, I concentrated on the family groups, taking thousands of identification photographs (right ear, left ear, front-on). Slowly and carefully, I began piecing together family trees.

With familiarity came trust and there's nothing quite like being trusted absolutely by an enormous wild elephant. Ultimately, I became a part of their families and a reassuring presence during troubled times. Often they would greet me with a rumble, as they would their own kind. Remarkably, they began coming to me from afar when I called them by name. It was a grand privilege when mothers chose to bring their newborns to meet me, regularly relaxing right beside my vehicle door for 30 minutes or more. Sometimes they would follow my slow-moving 4x4, as if I were their matriarch, taking time to browse whenever I stopped along the sandy roads.

I revelled in these new-found elephant friendships, which were to become, over more than a decade in the field, among the most remarkable relationships with wild, free-roaming elephants ever documented. During frequently challenging times in the remote Hwange bush, it was these special relationships that kept me motivated, smiling and sane. These exceptional creatures managed constantly to restore my spirit and keep me pressing on.

Snaring was widespread on the estate during my early years, and I would spend weeks searching for affected families, co-ordinating snare removals and monitoring progress of de-snared animals. This brought me even closer to them. Just as worrying were land grabs during Zimbabwe's 'land reform' programme, and trophy hunting in areas where it should never have been allowed. I chose to spend time calming families after bouts of gunfire. There were years when they ran from land claimants' vehicles and from some suspicious game-drive vehicles. With urgency, I helped kit out and deploy a dedicated anti-poaching team. Later, elephant families watched, seemingly in hope, as I oversaw scooping and clearing of neglected waterholes. These were highly intelligent animals. They knew I was their friend, an honorary elephant of sorts, someone who was wholly on their side.

Politics soon raised its head, but not in the way I expected. I came across dubious people, often in respected positions, who frequently and deliberately confused information regarding elephants in general. Why did they seek to mislead? Was it ignorance, ego, greed or apathy? Their uninformed message – that all's well and under control – was then, and still is, one of the very real dangers elephants face. Simply seeing lots of grey moving through the bush certainly doesn't mean that all is well within each family.

As with human friends, a caring touch was perhaps inevitable – although I never set out to touch a wild, free-roaming elephant. I knew the families, not just individuals, and I understood the various rankings within these families from years of monitoring. I was also familiar with a family unit's relationship to nearby families. I had learnt which individuals within each family were likely to cause trouble (like humans, elephants experience jealousy). I'd learnt to read their moods and their body language. And in time I could recognise them from a distance, just by the way they walked, the way they held their head or who was with them. I never pushed their level of tolerance, nor did I ever force myself upon them. I let them come to me.

It was an elephant I had named Lady – the matriarch of the L family – who was the first truly to accept me into her world and respond with excitement to my voice and presence. One memorable day, when she was right beside my 4x4, I leaned out of the window and placed my hand on her tusk. It was an instinctive moment, the same way I might greet a human being. She didn't flinch or react in any obvious way so I left my hand there for several minutes. This brief connection left my spirit soaring.

From then on, Lady always went out of her way to come and stand beside my vehicle, her trunk swinging like a pendulum as she hurried towards me. Then she would emit a deep, contented rumble. I always looked up into her amber-coloured eyes, talked to her and placed my hand on her tusk. I'd breathe deeply, inhaling all of the magic surrounding me: the beauty, the peace, the companionship, the learning and the extraordinary African light. I would close my eyes to heighten my sense of hearing and listen intently to the multitude of rumbles, trumpets and roars, the footfalls, the slurping of water, the crack of a branch or of a giant ear and the scraping together of leather hides.

One day, through the window of my vehicle, I placed my hand very gently on Lady's trunk. It felt as warm as Kalahari sand, much rougher than I thought it would be, and deeply grooved. Lady tensed a little, not knowing what this strange human appendage was against her skin, but did not try to evade my touch. In that moment, time stood still. A rush of adrenalin shot through me as we two creatures – such unlikely friends – momentarily became one. I felt as if I was dreaming: such trust from a fully grown wild elephant. Holding her gaze, I talked, and then I did the only thing that seemed appropriate: I sang 'Amazing Grace' to her. Tears sprang to my eyes as I struggled to comprehend the enormity of the privilege.

Later, I rubbed my hand up and down Lady's trunk, applying as much pressure as my own strength allowed. She appeared to revel in it, as did I. Sometimes when I did this, she concertinaed her trunk like an accordion and I always got the feeling that she was about to sneeze. Other times, when I crooned to her and looked up into her long-lashed eyes, her temporal glands erupted with liquid. It wasn't a trickle, but more like a bubble of liquid that sprang from within, before streaming down both sides of her face. This was a sign of excitement. It was obvious to me that both Lady and I thoroughly enjoyed our encounters. They are moments I will never forget.

I watched Lady's daughter, Lesley, grow from playful youngster to new mother. Her first-born, a boy, came into this world with some drama. Umbilical cord still attached, he was temporarily 'abducted' by three older teenagers from another family, who then bullied Lesley. In a submissive gesture typical of lower-ranked elephants, she backed into them continuously, trying desperately to get back her newborn. Then she suddenly raced towards my 4x4 and, within touching distance of me, turned abruptly and raced back towards her baby. She did this twice. I was both distressed and honoured that she was asking me for help. Thankfully, everything turned out okay and a few days later I leaned out of my window and gently touched her tiny baby's trunk, christening him Lancelot. Six months later, Lancelot became the third snare victim in the L family. A worrying 25% of Lady's family were ultimately caught and injured in wire snares, but thanks to dedicated monitoring and access to skilled darters, all of them were saved.

I had favourites in each of the 17 extended presidential families, elephants that I was particularly drawn to for various reasons. There were Misty and Mertle from the M family, who were about as different in character as two elephants could be. Misty was quiet and gentle, Mertle bossy and boisterous. I adored Misty's close presence right beside my door, where she regularly napped with Masakhe and other of her offspring.

When I was amongst the extended W family there were countless unforgettable moments with Whole (with a very distinctive hole in her left ear, but requiring a name beginning with W), Whosit, Wilma, Wonderful, Wish, Wanda and many others. But it was my first kiss that I remember as hauntingly special.

One day Willa unexpectedly lay down beside a mineral lick, under the blazing Hwange sun. I didn't often see adults lying down unless they were unwell, although I was aware that other elephant populations were observed asleep on their side quite frequently. She got back on her feet and wandered over to the shade of a sprawling teak tree, only to lie down on her side once more. I talked to her from afar. She eventually arose and wandered towards me, coming to a halt just centimetres from my door, as she always did. I continued talking to her gently as she rested her trunk in an L-shape on the ground and crossed her back legs – signs that she was particularly relaxed – and put what felt like her full weight against my door. I felt my 4x4 shift and realised she could toss it on its side if she wanted to. But I knew this wasn't her intention. She wanted company. She wanted comfort. Clearly feeling unwell, she wanted me to reassure her that everything would be okay.

While talking to her and tenderly touching her trunk with the back of my hand, I put my face against the long leathery nose of this wild giant and kissed her gently. This was not a hurried encounter. It was two beings, totally at peace with one another; a bond forged over many years with love, patience and understanding. I kissed her again and again. Willa stayed still, looking down at me with kind, wise eyes. This intelligent being, gifted with conscious thoughts and emotions, was clearly thinking. She may not have been able to speak my language, nor me hers, but she had chosen to commune with me nonetheless. We understood each other and she knew I was concerned for her. I recognised her own genuine warmth. This encounter with Willa (who gave birth to baby Wobble just a few months later) left me feeling euphoric. To have gained this level of trust from a wild elephant – one that had been through many difficult times over the previous decade – was just one of numerous encounters that made everything worthwhile.

But the good times always came with a considerable dose of bad. Politically connected land claimants and others, unethical trophy hunters and poachers continually tried to scare me off, have me expelled, and worse. Threats and increasing bouts of intimidation became common. I was accused of being a spy for the Australian Government. Later, my name remained for 12 long months on a publicly displayed 'Wanted' list at the local police station and an article in a government-mouthpiece newspaper stated that I needed to be dealt with 'once and for all'. I was verbally abused and once physically assaulted (resulting in a case I won in court) by accomplices of those who were grabbing presidential elephant land as their private property.

There seemed always to be someone trying to make trouble. My police file was as thick as a thesaurus, although I'd never been charged with anything, despite repeated attempts by antagonists. I had no reason to cause trouble myself, but I was not afraid to speak up on behalf of the elephants. I was also frequently having to flag lack of maintenance of key estate waterholes (forcing the elephants into other areas), increasing amounts of litter, noise, gunfire, speeding vehicles, off-road driving and other irresponsible behaviour. Egos and indifference continued to combine with other ludicrous carry-ons, while well-known elephants disappeared forever. One of them was Lady.

Eventually, by 2014, it was simply becoming too dangerous. After 13 years of exhilarating highs (including the release of several books, the reaffirmation of the presidential decree in 2011 and a world-acclaimed documentary), but unceasingly frustrated by crushing lows (young elephant captures, cyanide poaching, ongoing land grabs, corrupt and erratic officials and hunters), I had to find the courage to leave Zimbabwe and my elephant friends. It certainly felt like a life-or-death decision, and the most difficult I'd ever had to make.

Sometimes, as I was forced to rationalise, you must be out of a country to raise required levels of awareness. I knew the risks to me were not going to abate and I was completely burnt out from so many traumatic years full-time in the field. I had little choice but to leave, forlornly aware that problems would continue behind the scenes and, once again, that they'd likely go unnoticed, suppressed and lied about.

In 2017, nearly 3 years after returning to my homeland, Australia – having written the book *Elephant Dawn* and still missing my elephant friends deeply – I was further disturbed to learn that I have a rare, progressive and incurable autoimmune disease. Stress is likely to be one of its triggers and I've certainly had my share of that, and then some.

Today my special word – which I temporarily lost sight of in Zimbabwe – is hope. In 2001, I gave this uplifting name to a proud member of the H family: Hope. Now, with the world watching more attentively, I long for the day when Hope, her clan mates and all of her kind will finally be permitted to live out their lives, properly protected, and in peace and harmony.

Chilo Gorge
Safari Lodge on
the Save River.

25

Mahenye community – working with CAMPFIRE

Communities living alongside protected areas have a vital role to play – they're the frontline troops in the fight against poaching.

Clive Stockil

For centuries the wildlife-rich systems of the Save and Runde rivers, around what was to become Gonarezhou National Park, belonged to the Hlengwe tribe, whose occupation pre-dated the arrival of So-Shangaan in the early 19th century. They were skilled hunters and fishers – the presence of tsetse fly precluding domestic livestock farming. Although they engaged in subsistence agriculture along the river banks, the Hlengwe were essentially hunter-gatherers, developing an appreciation of the value of the natural fauna and flora in the area in which they lived. Elephants and humans in the area co-existed, sharing the shade of the ancient baobab trees, drinking from the same rivers, sharing the same space.

This was to change in the 20th century with the establishment of protected areas such as Gonarezhou. Villagers living within the gazetted park were relocated to surrounding communal lands. This resulted in restricted use of natural resources, the increase of crop destruction by elephants and livestock losses to predators. Human-wildlife conflict was inevitable.

The Hlengwe's exclusion created a clash between the park authorities and members of the Mahenye community, who had been located to the north-eastern edge of the Gonarezhou Park, close to the Save/Runde confluence. This resulted in mavericks such as Shadreck Muteruko and John Puzi, who challenged the park authorities and became professional poachers, enjoying the full support of the community. They were seen as modern-day Robin Hoods.

Muteruko and Puzi successfully operated in the park for 2 decades, targeting the legendary elephant tuskers and black rhinos, which were plentiful at the time. Following Muteruko's arrest in 1983, he admitted to removing 20 to 25 large tuskers from the park *per year*. Considering the number of years he successfully operated, this could have resulted in the removal of some 300 to 400 elephants and an unknown number of black rhinos. The ivory was sold to a Portuguese trader living in Beira, Mozambique.

Zimbabwe gained its independence in 1980, fostering expectations among the Hlengwe of a return to their ancestral village in the park. DP Chauke, the newly appointed councillor for the Mahenye ward, was mandated by the community to forward a request to central government that this permission be granted. After 2 years the Harare government refused. The return home of the Hlengwe was not an

The Chilojo Cliffs
and the Runde River,
Gonarezhou National
Park, Zimbabwe.

Scott Ramsay

option. The reason given was that the area was now a gazetted a National Park and the nation needed these parks to attract the international tourists who contributed to the economy. This resulted in the community upping poaching activities. Their logic was that less wildlife would result in fewer tourists, neutralising the government's rationale. This pressure created both an urgent need and an opportunity to engage in dialogue with the community leadership, including Chief Mahenye.

Rethinking community strategy

A historic meeting took place on 22 February 1982 under the shade of a pod mahogany tree (which still stands in Mahenye village), with over 70 community elders and the park warden attending. It was a highly charged and emotive meeting, held at a time when the sorghum and maize fields were reaching

maturity, and under siege from nightly raids by elephants. Villagers were sitting in dark fields and defending their year's supply of food against the hungry giants. An angered village elder, pointing at the park warden, lamented that 'his elephants cross the river and then destroy our crops, jeopardising our ability to feed our families. But when we then need to kill one of his elephants to feed our families, we are arrested as poachers and sent to jail.'

After hours of discussion, it was proposed that authority be requested from the Ministry of Environment to allow the community, through their traditional leadership, to legally use the natural resources in their area. This included the right to harvest, in a controlled and sustainable manner, elephants that exited the park. In May 1982 this authority was agreed and granted to the Chipinge Rural District Council, leading to the first Communal Area Management Programme for Indigenous Resources (CAMPFIRE).

This opportunity to empower the community was agreed on condition that the resource base was not used unsustainably and that there was to be a significant reduction of poaching within the Gonarezhou National Park. There were critics, mainly from within parks, who were sceptical, claiming the experiment would fail miserably.

In June 1982 the park authorities deployed Zephania Muketiwa, head of their law enforcement and investigation unit, to carry out anti-poaching patrols within the park, close to the Mahenye community. During this exercise, which involved 14 rangers and lasted for 3 weeks, 90 poachers were arrested, including commercial ivory and subsistence meat and fish poachers. These results supported the theory that communities were not willing to change and that co-existence was still a figment of the imagination.

However, that same year the community agreed to limit the elephant off-take to two mature males. The right to hunt them was sold to two American hunters in August and the meat was fairly shared amongst all the villagers. The community unanimously agreed that the revenue from the hunts would go to building the first classroom block for the Mahenye Primary School.

Because of the experimental nature of the programme, release of the funds needed to be channelled through the District Council 200 kilometres away. This proved to be challenging, as there are 30 wards in the Chipinge area, Mahenye being the only one sharing a boundary with Gonarezhou, and with potential income from wildlife. At a council meeting the chairman proposed that benefits accruing from natural resources should be equally shared by all wards. This posed a serious threat to the programme.

Mahenye councillor DP Chauke suggested that there should be no difference between elephants, buffaloes, impala, domestic cattle and goats. He suggested that, to be fair, a single fund be set up where all revenue from elephants and other wildlife, plus all domestic livestock, would contribute to this fund. This could be shared equally once a year and used at the discretion of the individual ward development committees.

Before this resolution was put to the vote, Chief Mutema, representing several wards with no wildlife, rose quietly at the rear of the room. Lifting his arms for silence, he requested an opportunity to contribute to the debate. He started his intervention by asking two questions:

- How many wards had suffered crop losses from elephants?
- How many wards had lost livestock to lions?

He asked those that could give examples to stand up. Only one councillor took to his feet – DP Chauke. Exercising his traditional authority, Chief Mutema ended the debate by stating that those who had not suffered any loss could not expect to benefit from the wildlife revenue, as these benefits belonged to those who voluntarily shared their space with wildlife and who bore the resultant costs. A few weeks later a ceremony was held at Mahenye, where the money was handed to the community for the construction of the first classroom block at the Mahenye Primary School.

This success encouraged the community to continue in its efforts to reduce poaching. To

Children of Mahenye Primary School,
Mahenye Village, Zimbabwe.

this end, the inhabitants of seven villages on Ngwachumeni Island in the Save River were asked to move back to the mainland, as this would create a community wildlife zone. They had been suspected of using the island as a stepping-stone to enter the park in pursuit of illegal activities. As they moved, the area quickly filled with several herds of elephants. With intelligence coming from informers within the Mahenye community, seasoned poacher Shadreck Muteruko was arrested with ivory and rhino horn. Zephania Muketiwa returned with his team and after 3 weeks of patrols made only nine arrests, most of these being fish poachers. This confirmed a massive reduction in poaching pressure compared to the previous year.

After 6 years, a joint review of this ground-breaking experiment was conducted by National Parks, the Centre for Applied Social Studies (CASS) from the Zimbabwe University, Worldwide Fund

for Nature (WWF) and the Zimbabwe Trust. The Mahenye initiative was recognised as a valuable pilot project, and the lessons learnt would later form the basic principles of community engagement and equitable benefit sharing. This programme was later expanded at national level and promoted in over 40 other districts within the country.

CAMPFIRE today

The bold and progressive decision made back in 1982 by the Mahenye traditional leadership and their elected CAMPFIRE committee needs to be recognised and lauded. They remain committed to pursuing an integrated and diverse economy for the sustainable development of their ward. Their initiative created new opportunities to improve livelihoods, while maintaining the biodiversity of the area.

They understood the need to diversify, which created an opportunity to develop the potential provided by their proximity to the Gonarezhou Park. Ten years after the start of the programme, the Mahenye community called for tenders to establish an upmarket tourist lodge on the banks of the Save River overlooking the Gonarezhou Park. This resulted in the Chilo Gorge Safari Lodge through a joint venture with the private sector. The lodge has significantly increased revenue for the community and has created employment for 40 members.

Recognising these benefits, the community then decided to set aside more land for wildlife. They approached a private partner for assistance in planning and funding the project. This initiative paved the way for the development of the Jamanda Conservancy, the first of its kind in Zimbabwe. Consisting of 7 000 hectares of community land that had no settlement on it, and which was normally used as relief grazing in the case of a drought, the area now encompasses a 10-kilometre frontage along the Save River, which separates it from the Gonarezhou Park. This creates a community wildlife conservancy that will be stocked with species traditionally occurring in the area before the tsetse eradication programme of the 1950s.

Following the community's request, a joint project proposal was developed between the private sector partner and the CAMPFIRE committee for the implementation of this ambitious project, and submitted to an international development agency. They agreed to fund the first phase of the project, focused on infrastructure development, including the construction of ranger accommodation, 25 kilometres of electrified game fence, entry and exit control booms on the main road, and the establishment of an operational headquarters. It will also provide water within the reserve for wildlife and extend the pipeline for community cattle.

The community requested a fence on three sides of the reserve to minimise potential human-wildlife conflict, leaving the boundary with the park, along the great Save River, open. At time of writing, 25 kilometres of electrified fence had been completed and three ranger bases constructed. Restocking the reserve with plains game such as eland, zebra, giraffe, wildebeest and possibly buffalo has been scheduled for 2019/2020, subject to funding. There is no need to relocate elephants as they have already taken up residence in sufficient numbers.

Reflecting on the events that have taken place over the past 35 years, it is encouraging to note some of the positive achievements in natural resource conservation. Initially, the hostility between the community and parks meant elephants were unwelcome on the communal side of the river. Today, they can be seen crossing the Save River both day and night, and resident breeding herds have taken up occupation of the Jamanda Conservancy and the Ngwachumeni Island.

During the 2014 Pan-African elephant survey of the Gonarezhou Park, it was decided to include the Mahenye ward, and the findings were amazing – 400 elephants were estimated to be resident within the Mahenye ward. The creation of the Jamanda Conservancy has re-established a corridor between the Gonarezhou Park and Mozambique, allowing elephant movement along the ancient migration route linking the lower Chimanimani montane forests and the elephant population from the south-east of Zimbabwe. Recently, more than 300 elephants have been observed moving through this corridor into their traditional range land.

In addition to the conservation successes, benefits resulting from both consumptive and non-consumptive tourism have contributed to the sustainable development of the community infrastructure, such as better education and improved health facilities, while empowering the community and allowing them to take ownership of, and responsibility for ,the environment in which they live.

However, the long-term survival of the African elephant in its natural habitat will continue to be threatened by the ever-increasing demand for ivory. Communities living alongside protected areas have a vital role to play – they are the front-line troops in the fight against poaching. The long-term survival of national parks, private conservancies and other conservation areas with local communities as neighbours will depend on their recognition of the need to develop and promote co-existence through equitable benefit-sharing of natural resources. Having hostile and negative neighbours will result in conservation failure.

Only when we are prepared to share our space, to co-exist, to drink from the same river and share the shade of that stately old baobab tree, will the elephant's future be secured for generations to come.

Villagers removing the meat from a trophy elephant in the Chitsa CAMPFIRE area, adjacent to the Gonarezhou National Park. CAMPFIRE is a Zimbabwean community-based natural resource management programme that was one of the first to consider wildlife as a renewable natural resource, while addressing the allocation of its ownership to indigenous peoples.

© David Chancellor

26

Gonarezhou: a place for elephants

How to think through an effective protection plan in the corner of a turbulent country

Hugo van der Westhuizen

Feeling the control stick of the Super Cub in my hand as I take off from Chipinda Pools in Gonarezhou National Park always puts a smile on my face. I clear the mopane trees and turn eastwards towards a spot where the sun should appear at any moment.

A few minutes into the flight, a movement from an elephant catches my eye and I bank the plane towards it. Big bull elephants naturally turn their backs towards an intrusion and this one seems particularly keen to hide something. I steer the Cub towards his front in the low light, but he keeps turning away.

Eventually I see what initially caught my eye: a left tusk hanging very close to the ground, and the right one broken. This is one of the biggest elephants I have seen in Gonarezhou. Where did he suddenly come from? He's an old bull and there is something peculiar about his reaction to the plane I can't pinpoint right away. I'll return on foot for a closer look, once I've finished with my morning patrol flight.

Gonarezhou is a 5 000-square kilometre park in the south-east of Zimbabwe, and has one of the highest densities of elephants in Africa – around two per square kilometre. In the 1980s, scientists put the park's carrying capacity at 3 000 elephants, though the concept and figure can be debated. The present population estimated in the 2016 survey is around 11 000.

Sections of the park were once fenced off, with elephants able to use only the space within the park boundaries, but fences have since disappeared and elephants can roam over a wider landscape. We are part of the Great Limpopo Transfrontier Park and Great Limpopo Transfrontier Conservation Area, and, undoubtedly, are favoured by elephants.

Although in the past few years two collared elephants have walked in from the Kruger National Park (KNP), none of the 20 Gonarezhou elephants we have collared since 2008 has returned the favour. To the east, Banine and Zinave national parks – which, combined, are one-and-a-half times the size of Gonarezhou – have precious few or no elephants at all. One of the Gonarezhou's collared elephants twice walked halfway to Banine, then turned back. The areas linking these parks to Gonarezhou are communal and hunting areas with low human impact and settlement.

On the Zimbabwe side, the park shares boundaries with a safari area, communal areas where the CAMPFIRE programme conducts hunts,

and a well-managed conservancy. Apart from the conservancy, elephants hardly remain for long outside the park. Any excursions are mainly at night for water or to raid crops, but they always return to the safety of the park. Elephants in Gonarezhou have had their fair share of persecution – wars, tsetse control, culling and poaching – but the population just keeps bouncing back.

With the sun now clearing the horizon, I can see several family groups moving about their daily business. I circle around their latest 'victim', a centuries-old baobab. The colossal tree has collapsed in a heap of fibrous material after being systematically disembowelled by elephant bulls. This is a sight I see too often these days. The bulls stand around these trees like men at a pub counter, idling away time in the shade during the midday heat. It's not clear why they're hitting trees near the perennial rivers – whether this is driven by thirst or hunger, or something else. Other evidence of elephant impact is canopy loss. The woodland in the north of Gonarezhou has lost more than 30% of its tree canopy in the last 60 years, with drought, fire and elephants cited as being the main reasons.

Gonarezhou stopped providing artificial water sources about 15 years ago, first from lack of resources, then as an ecologically motivated decision based on studies showing the detrimental effect of artificial water points on habitat and population dynamics in the KNP. The initial thinking was that waterholes away from the rivers would protect riverine forest and sustain populations of rarer species, such as sable and roan antelope. But more water just provides wider access to resources throughout the year for species such as elephants, wildebeest and zebras, causing their populations to grow without check.

In the absence of artificial water during the dry season, elephants in Gonarezhou remain within walking distance of the three main rivers and perennial pans. After the first rains they disperse into the back country, which has been under-utilised during the dry season, creating a landscape mosaic of variable use and impact.

Without artificial water and other human interventions, animal population growth should level off, with higher elephant calf mortality and longer inter-calving periods. Among Gonarezhou's elephants that's just not happening. Even if elephant

numbers tapered off, the lag effect and high impact of their browsing and trampling will have had a knock-on effect on other animal species.

How do we define 'acceptable' impact, and when should we decide to intervene? Even in large protected areas we have, among other things, blocked migration corridors, thanks to encroaching human settlement, which has changed the flow regimes and quality of water in the major rivers. What other means are there to control or change the impact of elephant numbers?

I mull over these questions as I reach the Mozambique border, along which I'll do my morning patrol. There are fewer elephants here, where there's been an increase in poaching. A heretical thought occurs to me: could poaching be a potential solution to high elephant numbers? Elephant losses to poaching in Gonarezhou have been kept within acceptable levels through a well-planned security system and a dedicated team of rangers on the ground. Why are we investing in law enforcement when poachers could solve our elephant problem?

Then I come to my senses. In 2012, poachers in Gonarezhou killed an elephant, cut the carcass open and laced it with the nematode poison Temick. This caused the death of at least 230 vultures, including lappet-faced and white-backs: Endangered and Critically Endangered, respectively. The loss of that one elephant was negligible in the bigger picture of elephant conservation, but the loss of those vultures was an absolute disaster. Poaching is wasteful, indiscriminate and often leaves animals wounded or entangled in snares. Removing a snare that has become embedded to the bone in an elephant's foot and dragged along for weeks, will make anyone realise that this form of elephant control is not the answer.

What about hunting? Gonarezhou is surrounded by hunting areas and is well known in the hunting fraternity as one of the last places hunters can bag a 100-pounder. In 2015, a hunter on the southern boundary of the park shot a 122-pounder in the adjoining safari area, apparently the biggest elephant legally shot in Africa in the last 30 years. The websites that advertise hunting around Gonarezhou showcase galleries of hunters, each standing proudly next to their very own big tusker. 'Around Gonarezhou', one hunter explains, 'you don't hunt in the traditional sense. You wait for a tusker on the boundary to leave the park at last light or return at first light.'

Shooting a collared elephant is illegal, but two have been shot in low light when hunters said they couldn't see the collar. Hunting doesn't necessarily reduce elephant numbers because hunters target only certain specimens – the biggest bulls. However, if well managed, hunting can play an important role in elephant conservation: marginal land outside parks could be designated as hunting areas, for the benefit of local communities who would otherwise farm the lands; and, by so doing, could provide a first line of security for inhabitants.

This is a controversial subject and often misunderstood. The revolutionary CAMPFIRE programme, which started in the 1980s, convinced communities to set land aside for wildlife and to continue living with wildlife in their midst, on the assumption that they would be able to derive benefits from this and offset costs from human-wildlife conflict. The potential for tourism in these areas is generally marginal, so hunting was an obvious alternative.

From the hunting of one elephant, a community can earn up to $25 000; much-needed money for things like schools, clinics and reinvestment in wildlife management. The alternatives are mostly farming with cattle and marginal crops, both of which can have a negative impact on this type of environment. But the hunting purse has not always been equitably distributed: with money being held by community officials and not filtering down to grassroots levels, some communities have turned their back on CAMPFIRE, increased their livestock – which they can own and trade – and cleared more fields closer to park boundaries.

With diminishing wildlife areas outside Gonarezhou, and few elephants moving out of the park, hunters are taking aim closer to the park boundaries. So the park is paying for added security but is still getting heavily impacted, while the hunters, sitting on the virtual fence, are picking off some of the biggest elephants in Africa. Am I against hunting? No, because the alternative could be even worse. We just need to find a way where hunting is transparent and sustainable, with proceeds contributing to elephant conservation and local communities, which is currently not always the case.

So what about contraception and translocation? Contraception is impossible due to the sheer size of the Gonarezhou population; it is just not practical. Translocation could create short-term breathing space, but to where do we translocate elephants? The 2016 Great Elephant Census report showed that, over the past 10 years, Zimbabwe's Sebungwe population had been reduced by 75% and the Zambezi one by 48%, while Hwange remained relatively stable, with a 10% increase. During this time, Gonarezhou's population increased by a staggering 134%.

Last year I visited the Chizirira National Park in the Sebungwe area (Chizirira is estimated to carry fewer than 2 000 elephants) to explore the possibility of moving elephants there. But, given that park's crash in numbers, it would be irresponsible to make such a move before the cause of the reduction is eliminated. As I flew over Sebungwe and viewed the extent of human population encroachment on the park boundary, it was clear that a large-scale relocation would not be a viable solution.

What about culling? Between the 1980s and mid-1990s, a great number of elephants were taken out through a systematic culling programme in Gonarezhou. This is obviously hugely controversial and widely condemned by many, including local and international conservation organisations. There was also some dispute about whether culling would actually help reverse the loss of big trees in the KNP.

Could I support culling in Gonarezhou? I tried to rethink the question while circling another baobab with most of its trunk shredded, and on the point of collapse. Faced with a large elephant population starving to death and creating a long-term negative impact on the environment, a cull could offer immediate, if temporary, reprieve. But there would be no win-win scenario. A positive aspect of a cull could be to provide communities with jobs, as well as cheap and subsidised protein. It would be an immensely difficult decision and, fortunately, is not mine alone to make. But I'm not yet prepared to remove the option of culling from the extremely limited available solutions.

The most important thing, really, is that we should create space for elephants. Poaching may be the main reason for diminishing elephant numbers, but in many areas, elephants are simply running out of space. Africa's human population is growing exponentially, and elephant refuges will be on the losing side. If we cannot create that space and find a way for rural people to live in harmony with these creatures that can eat an entire year's harvest in one night, the only option will be to contain elephants, physically, in fenced-off protected areas.

How do we create space? For a start, hunters must stop their peripheral hunting and allow elephants to move out into buffer areas, which should not become killing fields. Hunting clients must learn more about the areas where they hunt and the ethics of the operators they use to ensure that their activities end up working *for*, not *against*, conservation and local communities. Inevitably, we'll have to fence strategically to keep elephants out of sensitive areas, especially from human settlements close to park boundaries. In Gonarezhou we also need to establish safe corridors so elephants can move to Zinave and Banine national parks, which are within elephant walking distance.

A conservation organisation recently translocated a few elephants from South Africa to Zinave at great expense. Would it not have made more sense, both ecologically and financially, to have moved a few family groups from Gonarezhou, with the potential to resurrect ancient linkages? It would not take much to get elephants to move there and, one hopes, establish viable corridors. But this would need consensus and the resolve of people sharing a common vision. We need to think both big and strategically.

It is already late morning when I return to Chipinda and try to unwind my cramped body from the confines of the little plane. I grab my binoculars and set off on foot with one of our rangers to find that bull.

He's not far from the airstrip, and within 20 minutes we pick up his tracks. A few minutes later, we hear branches break about 100 metres in front of us. Checking the wind, we move closer. He's a really old bull, judging by the way his spine is protruding and the sunken patch above his eyes. I watch him closely and I get the uneasy feeling that all is not well. A puss-like liquid is coming from a wound on the side of his face, definitely not from his temporal

gland. His right tusk is broken crudely in the middle, maybe from a fight with another male? His left tusk, though, is incredibly long and symmetrical and reaches to within a few centimetres of the ground. It is one of the largest tusks I've seen on a live elephant.

But what I see next is difficult to believe. Right in the middle of the tusk is a perfectly round hole, and there can be no doubt that it was caused by a bullet. Only this type of projectile would have had enough velocity and strength to cause a mark of this kind. It could not have been from a heavy calibre weapon, because that would have shattered the tusk, so I speculate that it was from an AK-47. This could have been fired randomly, set on automatic, and there's a good chance that some of those bullets are still embedded in him, explaining the wound on his face. He starts to become aware of us, but there is simply no aggression. His right eye, with its long lashes, keeps turning towards us. Spending time with an animal this size and age, at such close quarters and with so much inherent trust, brings on emotions and a range of feelings that are simply impossible to explain. My eyes mist up against the binocular glasses. It's difficult to comprehend what this animal must have experienced in his lifetime. If only he could talk, maybe he could tell us where we have gone wrong as humans.

Over the next few days he keeps hanging around the airstrip, and whenever I have time I simply go to sit and watch him. Eventually he vanishes in the same way he appeared. I see him once again from the air about 20 kilometres away, but that's the last time. We consider fitting a collar to track his movements, but he's too old and, with the weight of that long tusk, he would struggle to get up again. I think of giving him a name, but then realise it would do him an injustice. He already carries one man-made curse in his tusk and possibly some in his body.

In the end, in his honour, we decide to use him for our logo. It's an emblem of true resilience, and a testimony that Gonarezhou is still a wilderness where it's possible for a bull this size to roam (almost) without anyone's knowledge. But the bullet hole in his tusk is evidence that, even in this vastness, he's not out of reach of man's greed and influence. Without mankind's understanding and protection, these animals are simply doomed.

The name Gonarezhou means 'a place for elephants'. It's fitting.

Scott Ramsay, Chilojo Cliffs, Zimbabwe

© Tami Walker

'The question is, are we happy to suppose that our grandchildren may never be able to see an elephant except in a picture book?'

David Attenborough

Beneath the the towering cathedral of *Faidherbia albida* thorn trees at Mana Pools in Zimbabwe, elephants look like mere miniatures. It's a place where they seem to belong – *their* ancient place.

Lugenda Wilderness Camp, Basili, Mozambique

27

Niassa's elephants

When the poaching slaughter hit northern Mozambique after 2009, the Niassa National Reserve was caught off guard. It desperately needs to make up lost ground.

Greg Reis

The Niassa National Reserve (NNR) in northern Mozambique, with its looming granite inselbergs, feels like a Jurassic wilderness. It's the third-largest reserve in Africa, spanning 42 300 square kilometres across parts of Cabo Delgado Province and nearly one-third of Niassa Province. It is unfenced and bordered by four rivers: the Rovuma to the north (the Tanzanian border), the Lugenda to the east and south, the Luatize to the south-west and the Lussanhando to the west. It's perfect elephant country.

Being on the 12th parallel, the reserve is hot and tropical, with temperatures averaging 30 to 40°C in October and November and 10 to 20°C in winter. Rainfall begins in November and ends in late April or early May. During this period, precipitation averages 250 to 350 millimetres a month.

Most of the Niassa Reserve is miombo woodland. In fact, it's one of the largest protected miombo forest ecosystems in the world. The area is interspersed with seasonal wetlands or *dambo* networks, and drier areas of wooded savanna in the larger, low-lying river valleys. Riverine forests line many large rivers and smaller streams. The reserve drains into the Rovuma and Lugenda rivers – large, braided,

sand-bed watercourses with strong perennial flows. A central watershed between these two rivers feeds many seasonal rivers as well as an extensive *dambo* network.

This pristine wilderness is home to the highest concentration of wildlife in Mozambique. In 2002, it supported more than 13 000 elephants – including many big tuskers – 9 000 sable antelope and several thousand each of Cape buffalo, Lichtenstein's hartebeest, eland, grey duiker, yellow baboon and zebra. An estimated 350 to 400 African wild dogs live in the reserve, making it one of the last refuges for this species.

It is one of six lion strongholds in Africa (more than 1 000) and home to leopards, spotted hyaenas, side-striped jackals, servals, honey badgers, civets, and small and large spotted genets. There are healthy populations of kudu, bushbuck, impala, wildebeest, waterbuck, reedbuck, bush pigs, warthogs, crocodiles and hippos, as well as three endemic subspecies: Niassa wildebeest (*Connochaetes taurinus johnstoni*), Boehm's zebras (*Equus burchelli boehmi*) and Johnston's impala (*Aepyceros melampus johnstoni*). Added to this are seven species of mongoose and over 400 species of bird. The reserve

'The people involved in protecting those elephants, like rangers on the ground, are so under-resourced. They have very few vehicles, they have very poor weapons – if any at all – and they are treated as the bottom of the tree when it comes to law enforcement priority.'

Allen Crawford
TRAFFIC

The Lugenda River weaves its silver thread beyond the Ngolonge Inselberg range, with Mecula Mountain rising to the right.

The vast Niassa National Reserve encompasses mountainous terrain, seasonally impassable rivers and a low-density road network, making conservation and anti-poaching efforts extremely challenging.

hosts remnant elements of East Africa's eastern arc forests, which are hotspots for endemism.

Despite this being excellent range country, the elephants are in trouble. A 2016 survey counted 3 675 elephants but, by late 2017, the 'gut feel' from operators on the ground put the number at just above 2 000. If accurate, this means an 85% drop in numbers over the past 15 years. In mid-October 2016, the estimate of mature bull elephants was 160, but these giants are rarely sighted in the reserve today. A decade ago it would have been unthinkable that the relocation and introduction of mature bulls might be needed to resuscitate the reserve's species.

During a trip to the reserve in November 2017, it was clear from the behaviour of breeding herds that they were terrified of humans. They displayed unusual and seemingly chaotic, leaderless behaviour, probably because of the absence of mature matriarchs and older female family members, who had fallen to poachers' bullets. The loss of mature bulls and the sorry state of breeding herds points to ongoing deterioration of the gene pool, and an unravelling of elephant social structures.

The impact of the illegal wildlife trade in ivory in the NNR can be seen from the numbers in the table below.

Formal reserve management was initiated at the turn of the century and an agreement signed in 2002 to form a public-private partnership known as the Sociedade para a Gestão e Desenvolvimento da Reserva do Niassa (SGDRN). This body managed the reserve for the next 10 years. In 2002, SGDRN awarded the first concession in the Niassa reserve and, by 2005, a protected buffer zone of six independently managed hunting areas had been created on the eastern, southern and western boundaries. Then, in 2007, through two tender processes, the SGDRN made available 10 operating

2016 Aerial Surveys of Wildlife in the Niassa Reserve Mozambique. * The 2009 elephant numbers are from the 2011 survey and the 2014 elephant numbers are from the 2016 report of the Great Elephant Census.

Source: SGDRN (Oct 2011) & NNR Management (Oct 2016) Aerial Surveys of Wildlife in the Niassa Reserve, Mozambique

Year	1998	2000	2002	2004	2006	2009*	2011	2014*	2016
Elephants	18 708	11 828	13 061	12 478	11 833	20 364	12 029	4 441	3 675
Carcasses	336	644	645	461	588	896	2 627	3 183	3 379

areas. They were mostly for photographic tourism, although two of the original 2005 hunting areas were also retendered. In September 2012, when its management agreement ended, SGDRN had secured long-term contracts with 11 operators, ranging from 15 to 25 years.

From October 2012, the Wildlife Conservation Society (WCS) and National Administration of Conservation Areas (ANAC), a newly formed Mozambican government department, entered into a co-management agreement to provide formal reserve management oversight, an arrangement that is still in place.

From the estimated increase in elephant numbers from 1998 to 2009, and particularly the large increase since 2006, it's clear that there was a definite benefit in having formal reserve management oversight, a buffer zone and the additional protected areas under long-term concession agreements.

Poaching tsunami hits Niassa

From 2009 onwards – as a repercussion of the 2008 sale of 108 tonnes of ivory by southern African states – poachers hit the reserve with a vengeance. The CITES-sanctioned sale had created a thriving legal ivory-carving industry in China. It provided a perfect opportunity for criminal syndicates to launder illegal, poached ivory into legal markets. Poaching surged throughout Africa, from minimal levels to catastrophic proportions.

The problem of running law-enforcement and anti-poaching operations over such a huge area, and of effectively combatting this sudden spike,

After the 2008 sale of 108 tonnes of ivory by southern African states, poachers hit the reserve with a vengeance. Here, a female carcass tells the all-too-often tragic tale of Niassa's elephants. Once bulls were hunted out after around 2012, poachers turned their attention to breeding herds. It is estimated that just over 2 000 elephants remain in Niassa today, a drop in the population of 85% over the past 15 years.

would soon become evident. In addition, of the 11 operators in place by the end of 2012, only a couple, at most, had been running formal law-enforcement operations for any meaningful period. This meant elephant protection across the reserve was low when it was needed most.

Close neighbours Tanzania and Kenya took the initial brunt of the killing epidemic in the region. But, as elephant populations dwindled further north, syndicates turned their attention south to an easier target across the Rovuma River. Neither the NNR nor Mozambique in general were prepared for the sudden poaching onslaught. The 2011 results, when compared to those of 2009, show a sharp decline in elephant numbers and a correspondingly sharp increase in elephant carcasses.

Bull elephants were the first to be targeted for their ivory. From 1998 onwards, there was a slight decrease in numbers, from around 1 300 in 1998 to around 1 050 to 1 150 by 2003. Under SGDRN's reserve management, together with the presence of research scientists and concession operators, by 2009 the bull population was at 2 400. This demonstrated the potential for species recovery under good protection and with low demand for ivory.

What followed from 2009 was a huge decline in the reserve's elephant bulls. By October 2011, the count was about 773 to 821 animals. Even after factoring in observer bias during the aerial surveys, this is still a massive drop. In the 5 years from 2011, the numbers dropped to 400 animals in 2014 and to 160 by mid-October 2016. Mature males were being hammered by poachers.

From 1998 to 2000, recorded elephant carcasses increased from 340 to 645, then declined and plateaued at around 460 in 2004 before starting an upward trend. In 2009, there were 900 elephant carcasses counted; by 2011, there were 2 627. Perhaps 200 of these were from natural mortality, but most were probably the result of poachers crossing into the area from Tanzania.

As bull elephant numbers dwindled, poachers turned to breeding herds. The subsequent years were heartbreaking. In 2014, some 3 183 carcasses were recorded; in 2016, it was 3 379.

The 2016 report of the Great Elephant Census showed how hard the Mozambican reserve had been hit. By far the greatest decline in the country

was in the NNR and adjacent areas to the south – a respective 63% and 58.6% reduction in recorded elephant numbers. Of the 4 460 recorded carcasses across all of Mozambique, 3 422 (76.7%) were in the Niassa area. Of the 9 752 elephants (48.6%) lost across Mozambique between 2011 and 2014, a worrying 9 074 (93%) were in the Niassa area, mostly in the reserve itself.

In want of protection

The NNR was created in 1954, while Mozambique was still Portuguese East Africa. Then it was abandoned and without effective protection from 1975 to 1992, during both the 2-year war of independence and the civil war.

A large area outside of the reserve between the Mozambican coast and Lake Niassa/Malawi is still mostly devoid of human population, and animal movement remains unrestricted. Going back in time, the area would have been teeming with wildlife, with elephants moving freely along ancient elephant migration corridors, going from east to west between the lake and what is now the Quirimbas National Park, as well as north and south across the Rovuma River.

Historically, hunting of big game in and around the reserve included elephants. When formal reserve management began after the civil war, hunting was seen as a solution to both the local economy and conservation. Northern Mozambique was placed firmly on the map and marketed as a safari-hunting destination. But in the face of a dramatic rise in poaching, the legal killing of elephants through hunting safaris in the NNR was stopped to help safeguard the animals. In the surrounding hunting blocks, however, there was limited to no protection. There, any remaining elephants were either poached or sought sanctuary within the reserve.

The NNR is made up of 19 operating areas or concessions. Three of these (two mountainous special-protection blocks and one zoned for public tourism) are under direct reserve management supervision. Of the remaining 16 operating areas, all but two are contracted to private operators under some form of agreement (eight zoned for hunting and six for photographic tourism). In terms of law enforcement and anti-poaching perspectives, the contribution of each varies greatly.

A formal reserve management plan was put in place during the first few years of SGDRN's management, in parallel with agreements with individual concession holders. Because these agreements were all signed at different times over a 10-year period, it led to different contractual and operating commitments, which unintentionally created silos and worked against collaboration.

Some areas where collaboration did take place included the Niassa Carnivore Project (NCP) team, which undertook trophy monitoring for the reserve to ensure ethical and sustainable hunting. Both during and after the initial 3-year co-management agreement between ANAC and WCS was signed in October 2012, several key strategic improvements and additions were identified in the areas of zonation, law enforcement, communities, mining, stakeholder communication and governance.

With the imminent signing of a 10-year co-management agreement between ANAC and WCS, a new and improved general management plan will use these insights and enhanced, reserve-wide collaboration to drive the required results over the next decade. While this is encouraging, the fact that it's starting from the current less-than-optimal baseline means that pragmatic support from government and prioritising of hard work will be essential.

Law-enforcement operations are mainly controlled from the reserve headquarters in Mbatamila under the overall authority of the park administrator (appointed by ANAC). A law-enforcement officer from WCS is responsible for the co-ordination across the reserve of all anti-poaching and law-enforcement operations – although individual operators manage their own teams and 'contacts' with illegal incursions. Reserve management deploys reserve scouts at outposts scattered across the reserve, including in several of the operator-controlled concession areas, and also provides the central infrastructure for the radio communication system, which operates with varying levels of reliability, mostly due to environmental challenges.

Until 2009, safari operations (mostly hunting), research-based activities by the NCP team and photographic tourism (Lugenda Wilderness Camp from 2006) provided enough movement on the ground to act as a poaching deterrent. This is no longer sufficient. Mozambique's weak, opaque application of the law – especially regarding environmental crime – has had a negative impact on the reserve's own law enforcement. Problems occur along most steps of the enforcement chain, from capture and evidence collection to prosecution. Many of the issues stem from insufficient regulatory officials, political unrest and war, lack of investment in infrastructure or equipment and proper training, and from poverty in the region.

The reserve was therefore largely caught off guard when the slaughter of elephants began in earnest. However, some redemption was found in the presence of an 8-year-old, strong, intact security force from one of the hunting operators (covering 11% of the reserve) based in the south-eastern section. In 2012, reserve management signed long-term agreements with two operators committed to making conservation a priority. Between then and now, protection and security have been driven mostly by reserve management together with these three operators who later formed the Niassa Conservation Alliance (NCA). This alliance now manages around 26% of the total area within the reserve.

With law enforcement being a key component of the NNR general management plan, a capable scout force is indispensable. Candidates for both reserve managers and operators come from the men within the local village communities, most of whom are illiterate and have little to no previous experience. Employment in a scout team should require completion of a basic training course but, given the demand for scouts, this is unfortunately not always the case. In 2015, SMART technology for anti-poaching data was introduced by one of the NCA operators, providing the potential for more centralised sharing of relevant data at a reserve management level.

There is formal law-enforcement support from operators covering one-third of the reserve. Together with central co-ordination and placement of scouts in key hotspot areas, the reserve now has a 'deterrent' presence in roughly 50% of its area. Reserve management law-enforcement personnel also co-ordinate with the mainstream Mozambique security agencies, including with local and national police and the newly established environmental police, the latter now based in the park and amalgamated with reserve scouts.

Despite the increased presence of law enforcement, the reserve continues to be hard hit by poaching. Law enforcement is further impacted by the ebb and flow of politics. An example is the withdrawal to district capitals of the newly established environmental police during tensions with Renamo in 2016 and into early 2017. There was little warning of this or consideration for reserve-enforcement requirements.

The handful of concession holders who have implemented scout teams or conducted organised security operations do so at their own cost. Many hunting operators ignore law enforcement, believing it to be a government responsibility. Official protection forces cannot match the well-orchestrated, -funded and -resourced poachers, and around 90% of the arrests and convictions remain the work of a small group of individuals.

An additional problem is population growth in the area. Records indicate that people became resident in the area only from around 1910, although some population groups have a long association, going back to around 1850, sustaining themselves largely through agriculture or fishing. The area was sparsely inhabited, with localised concentrations in settlements of varying sizes. When SGDRN initially formulated its management plan, there were about 20 000 people living inside the reserve, many as a consequence of the war.

By 2012, now with nine districts, three towns and more than 40 villages, the reserve was supporting over 35 000 people and, at the time of writing, there are over 40 000. These include Tanzanian nationals who have settled in the area within the past 8 years.

Going forward

Considering the current challenges (elephant poaching, the hunting of bushmeat, illegal artisanal mining, over-fishing), the present level of law enforcement and operation is inadequate. While some operators have basic law enforcement in place

Lugenda River, Mozambique.

at the ground level, this needs to be spread and owned more widely across the reserve. Improved use of data and collaboration across concessions will also be important. Optimising limited resources is critical.

Most importantly, sincere, dedicated and meaningful political support is required before it's too late. The Mozambican government needs to take urgent action to shut down heavily armed poaching groups. Concession holders cannot continue to

have their scouts face poaching gangs who carry high-calibre hunting rifles and AK-47 assault rifles. Concession-operator scouts are mostly unarmed or carry a few pump-action shotguns and the odd hunting rifle. Real protection has been left to a handful of key individuals working tirelessly for years to protect the reserve. Unless government assistance happens quickly, the reserve may lose some or all of this crucial independent support.

To quote the business-performance guru Stephen Covey: 'We all need to start with the end in mind'. The value of Niassa is that, despite the challenges, it remains what wild Africa once was across much of the continent: a landscape of tranquillity and serenity, only lightly touched by human footprints. The Niassa National Reserve is a sacred place and elephants are its ecosystem engineers. They – and their habitat – need our protection.

28

Desert-dwelling elephants of north-west Namibia

The extraordinary adaptability of elephants is tested to its limits in the oldest desert on Earth.

Dr Keith Leggett

There is a mystique about the so-called desert elephants that live in the arid and semi-arid north-west regions of Namibia, and range over great distances. They were once believed to be a subspecies of savanna elephants, in part because of certain perceived morphological differences, but this has been disproved through movement data and genetic analysis.[1]

Before 1900, the elephant population in the region was estimated to be from 2 500 to 3 500.[2] Groups would probably have moved from the wetter areas of northern central Namibia into the drier north-western areas to take advantage of seasonal and annual resource abundances. Once the resources were depleted, they would have left the area. This population was hunted extensively in the latter part of the 19th century without evidence of a decrease in numbers.[3]

By the late 1960s, however, increasing human population and settlements had interrupted these elephants' traditional migration routes. By then, settlements in the north-west and associated intensive hunting and poaching had caused the elephant population to decline to around 600 to 800[4] and, by 1983, war, drought and poaching had reduced this number to about 360.[5]

By the 1980s, the north-western elephant population appeared to have split into three distinct populations, with no apparent contact between two of them – the eastern and western groups.[6] Possible genetic exchange could have occurred via the third group, a transitional one that moved between the eastern and western groups, with both of which it was believed to make infrequent contact.[7] By contrast, Lindeque & Lindeque suggested that the distribution of elephants in north-western Namibia showed no definite interaction between the eastern and western population groups through the limited transitional population.[8] A low calving percentage was recorded for all groups, which was attributed to poaching and human disturbance. However, the overall population of desert-dwelling elephants has recovered, or at least remained stable, at about 317 since the 1980s.[9]

To understand how desert-dwelling elephants survive in these arid areas, it's necessary to understand their movement, social structure, activity and behavioural modifications. It is clear, however, that without the continuing conservation efforts of government and non-governmental organisations, these elephants would be unlikely to exist in these areas at all.

Elephant movement in north-west Namibia

The extent of the seasonal travels and home ranges of Namibia's desert-dwelling elephants is probably the greatest ever recorded for African elephants. GPS collaring of individuals has also solved one of the pertinent questions surrounding this population: we now know that they're not genetically isolated. Most collars were placed on males as they were thought to undertake longer-distance journeys, and collaring them is relatively simple. Collaring females is problematic, as the herd protects the drugged female, making it dangerous for the researchers and stressful for the elephant groups involved.

An initial GPS collaring exercise was undertaken in September 2002, with the collars collecting data every 24 hours. This was repeated three times, with a maximum of eight elephants in north-west Namibia collared by 2008. Some of the elephants were continuously monitored for 6 years. The collars collected data every 24 hours and transmitted it to a server in South Africa, which then forwarded it to scientists in Namibia.

The home range of adult males in the Omusati Region (north of Etosha and the easternmost area covered by the study) varied from 720 to 8 952 square kilometres; in the western Kunene Region, from 2 881 to 14 310 square kilometres; and in the eastern Kunene Region, from 2 168 to 12 150 square kilometres. Their home ranges were much larger than those of adult females and immature males (871 to 5 600 square kilometres, respectively) in the same area. A similarly extensive home range for free-ranging adult males has been reported by researchers in other parts of Africa. The larger home range of adult males has generally been attributed to their being in musth and searching for females in oestrus far outside their non-musth ranges.

The movement of most large adult males was from Etosha National Park (central northern Namibia) into the western and more northern areas. This means there has been constant movement in the gene pool from a relatively large population (Etosha has 2 000 to 2 500 elephants) to a relatively low population. The older males appear to have fixed seasonal migration patterns, moving through approximately the same areas annually. With young males, the movement appears to be random, or at least opportunistic; they have no set pattern and appear to respond more to seasonal availability of food and water.

Adult female movements are much smaller and more seasonally pronounced. Home ranges are usually smaller as the younger elephants are unable to make long seasonal treks, although family units have been recorded moving up to 35 kilometres a day. Seasonally, they move up to 90 kilometres (over a 48- to 72-hour period) between the Hoanib and Hoarusib rivers.

The timing of seasonal movements and the differential use of habitats has long been linked to rainfall, forage preference and availability.[10] It has been suggested that seasonal use of vegetation reduces the ecological impact of elephants.[11] This is particularly important in an arid environment where rainfall is variable, as is the subsequent availability of palatable vegetation.[12]

The most widespread large riparian tree in the western ephemeral rivers of north-western Namibia is *Faidherbia albida*.[13] These trees are particularly abundant in the western section of the Hoanib River and are unique in that they bear fruit from the end of the hot dry season (September to December) through to the end of the wet season (January to April). This represents a major food source for elephants, especially during the dry season, when precious little other vegetation is available.[14]

The best example of seasonal movement recorded in the area is that of the elephant identified as WKM-10. He was a large male, believed to be 40 to 50 years old, and regarded as likely to have been the dominant elephant of the Kunene Region. He was collared on four consecutive occasions and had one of the largest recorded home and seasonal ranges of any African elephant. From 2002 to 2008, his home range was 14 210 square kilometres.

While his movements varied annually, depending on rainfall and available vegetation, his cold, dry-season home range (June to August) was usually in the Hoarusib River, where he moved a total of 45 kilometres upstream and downstream of the village of Purros. In the early hot dry season (late October or early November), he moved south to the Hoanib River, in common with many elephants of the region, feeding off the ripening fruit of *Faidherbia albida*. He would sometimes spend the whole wet season in this area, but usually moved into the Etendeka Mountains south of Sesfontein early in the wet season (late January or early February), and then east into the eastern Hoanib River.

During the early wet season of 2008, he undertook an unusual trek and crossed back into Etosha National Park, proving that the gene flow was from Etosha to the west. Late in the wet season (late April or early May), he went back to the Hoarusib River, briefly stopping at the western Hoanib River, before returning the way he came. This was a total passage of around 625 to 650 kilometres and took nearly 6 months to complete.

Social structure

Male society

According to Poole, a family unit is defined as the basic unit of elephant social organisation, and consists of one or several females and their offspring.[15] Related families may form defensive units and kin-based allegiances which, in turn, may have a positive effect on calf-survival rate. Bond or kin groups are made up of several closely related family groups – up to five families – and form when family groups become too large and split.

When bond groups meet, elaborate greeting behaviour is often exhibited. Families and bond groups that have the same seasonal ranges are known as clans, a term used to define a level of association regarding habitat use, although it is not clear whether this is a functioning elephant social unit.

It has been suggested that the distribution of elephants in north-western Namibia reflects a skeletal elephant social organisation, and that the social structure of the elephants in the western Kunene Region exhibits similar features to that reported by Poole.[16]

During 2002–2008, there were from 12 to 16 males in the Hoanib and Hoarusib river area at any one time, although this was seasonally variable. Most observations were of single males, although a significant number were also observed in pairs. A surprisingly large number of adult males, mostly non-musthing, were with their bond groups.

It appears that adult males entered the bond groups for social reasons but, after a period ranging from hours to 2 days, they moved off by themselves. Adult females appeared to be remarkably tolerant of them and made no attempt to chase them from the bond groups, as has been reported from other areas of Africa. Longer-term associations (from 2 days to a week) between adult males and the bond groups occurred when the males were in musth and searching for females in oestrus. The lack of bachelor herds, a social grouping identified in other African populations, could be simply due to a lack of numbers.[17]

Female society

The more research was carried out on the north-west elephants, the less their social structure resembled that described by researchers in other parts of Africa. For example, over 500 hours of observation found that herd structure was very loose, with bond groups regularly meeting up, then joining or further splitting, without elaborate greeting behaviours and interactions. Bond groups were fluid, with family units coming and going, and only the central family unit remaining stable. There appeared to be an absence of leadership within bond groups that lacked a dominant female. It was clear that some other explanation was needed to describe their social structure.

From 2002 to 2008, there were 38 adult females, sub-adults and juvenile elephants in seven family units (comprising three to 10 individuals) in the western section of the research area. Each of these bond groups had different associations. One adult female (WKF-4) did not readily associate with other adult females, although was occasionally observed in the company of three of them.

Two adult females were observed spending the greater part of their time alone or with their family units, but rarely interacting with other adult females, although the two of them spent some time together. Yet three more females formed their own bond group, although they interacted more often with other females than the previously mentioned duo.

Four others formed their own bond group, but mixed freely with other females, as well as occasionally being seen alone. Two others were rarely separated from each other. While WKF-11 and WKF-3 spent long periods in each other's company, it was not unusual to find them separated.

It appeared that the traditional herd structure described by Douglas-Hamilton and Moss & Poole did not apply in the desert-dwelling elephants of north-west Namibia.[18] One of the reasons could be their history: in the late 1970s and early 1980s, they were severely depleted by poaching and disturbance.[19] Owen-Smith reported that the Hoarusib elephants, which had numbered about 80 individuals in the late 1960s, were shot out by poachers, leaving only three.[20] These three moved south to the Hoanib River and joined up with the remnant herds that had been similarly decimated. There were few or no family links between any of the units in the area. The group associations were more for social interaction and offspring nurturing than because of direct family relationships.

One bond group became the focus of the observational studies when it was noted that no female took responsibility for the leadership, control being determined by the individual making a first move, and who would then co-ordinate activities. It was difficult to assess who was leading whom, with leadership changing from situation to situation.

This bond group split apart often and it was not uncommon to find the various members separated by several kilometres (out of communication range), although they usually reformed after a period. All other bond groups in the north-west area appeared to consist of closely related family units, although even these groups would fragment into family units and spend long periods apart, rejoining periodically.

Genetic links

Mitochondrial DNA analysis confirms that there were three genetically distinct groups of females in the research area.[21] Two of the distinct genetic lines formed their own family units and bond grouping, rarely interacting with females from other bond groups; the third bond group enjoyed limited interaction with the other two groups.

During the cold, dry season, the desert-dwelling elephants reduce their daily resting to allow sufficient time for other activities, such as feeding.

Activity		Wet season avg. %	Cold dry season avg. %	Hot dry season avg. %
Feeding		48.1	45.9	45.3
Water		11.7	10.5	10.0
Resting		13.7	12.8	19.2
Social		1.8	2.6	1.8
Walking		24.7	28.2	23.7
Average defecations per hour	Male	0.53	0.30	0.46
	Female	0.21	0.24	0.21

TABLE 1

Average activities (as a percentage of time) of desert-dwelling elephants, 2002–2008

A DAY WITH DESERT ELEPHANTS

Elephants living in Namibia's dry areas, with little food and water, have adjusted their behaviour accordingly. The research looked at how much time they spent on daily activities in these extremes, and how it compared to elephants in other areas of Africa.

TABLE 2

Seasonal and diurnal activities (as a percentage of time) of desert-dwelling elephants in north-west Namibia, 2002–2008

Activity	Wet season			Cold dry season			Hot dry season		
	07h00-11h00	11h00-15h00	15h00-19h00	07h00-11h00	11h00-15h00	15h00-19h00	07h00-11h00	11h00-15h00	15h00-19h00
	Avg. %	Avg. %	Avg. %	Avg. %	Avg. %	Avg. %	Avg. %	Avg. %	Avg. %
Feeding									
(a) Grazing	19.1	7.6	6.4	6.5	5.4	4.2	4.5	1.7	7.4
(b) Browsing	30.6	23.4	59.9	54.9	39.4	50.5	48.7	29.0	46.5
(c) Debarking	0.0	0.0	0.0	0.0	0.0	0.0	0.3	0.0	0.1
Water									
(a) Drinking	5.0	7.6	2.6	3.4	6.8	5.9	3.6	8.2	1.8
(b) Wallowing	0.4	1.8	0.7	0.6	0.6	0.2	0.0	3.7	0.1
(c) Dust bathing	1.7	0.6	1.1	0.8	2.9	2.5	1.7	3.7	1.6
Resting									
(a) Standing in shade	10.9	42.7	2.6	6.2	20.0	6.5	15.8	33.0	10.8
(b) Standing in full sun	0.2	1.2	0.0	0.4	0.8	0.0	2.2	9.0	3.8
Social	0.4	2.3	1.5	1.1	2.2	2.6	1.4	0.3	0.6
Walking	31.7	12.9	25.1	26.1	21.1	27.6	21.9	11.5	27.2

Activity budgets

The research looked at the daily activities of these elephants between 2002 and 2008 to assess how much time they spent on daily activities in these environmental extremes; whether this varied seasonally; and how this compared to elephants in other areas of Africa.

The activity of each of the individuals within the observed group was recorded at 2-minute intervals for 30 minutes, with data documented in respect of feeding, water, resting, social and walking activities. A second method focused on a single individual, with data being logged at 5-minute intervals over at least 30 minutes, and sometimes up to 4 hours, and at a more detailed level. Defecation rates were also monitored, as this gave an indirect measure of vegetation consumption. Over 750 hours of observation details were obtained on elephants west of the 100-millimetre rainfall gradient in the Hoanib and Hoarusib rivers during the 2002–2008 period.

Feeding activities and defecation rates

The differences observed in types of feeding activity were probably attributable to variations in rainfall. In years of above-average rainfall, annual grasses and forbs were more abundant, and increased levels of grazing activity were observed. During average or below-average rainfall years, browsing dominated. This was completely different from what had been reported in other parts of Africa, where seasonally available grazing comprised a higher percentage of the diet.[22]

Levels of feeding activity decreased slightly during the cold and hot dry seasons, with browsing being the dominant feeding activity during these seasons. Elephants sought out the fruiting *Acacia erioloba* and *Faidherbia albida* trees during the cold and hot dry seasons, respectively.[23] This undoubtedly accounts for the higher levels of feeding activity observed in the dry seasons than those reported in Zimbabwe.[24] These levels were, however, lower than those reported in Tanzania.[25] Debarking was observed only during the hot dry season, and then rarely, in contrast to what was reported in Zimbabwe, where debarking comprised a substantial percentage of feeding activity during the hot dry season.[26]

Defecation rates of male desert-dwelling elephants during the wet and hot dry seasons were similar, with a decrease observed during the cold dry season. The decrease could have been due to the adult males being in musth, and spending greater amounts of time in socialising and walking in search of receptive females. A loss in body condition in musthing males was reported by Poole in Kenya.[27]

Female desert-dwelling elephants showed similar defecation rates during the wet and cold dry seasons, with a slight decrease during the hot dry season when vegetation was scarcer. Much the same defecation rates were reported in Zimbabwe, with a cold-season low, but the Zimbabwe rates were higher than those of the desert-dwelling elephants in all seasons.[28] Higher rates of seasonal defecation were found for male elephants in Tanzania.[29] This would imply that desert-dwelling elephants, while feeding for a similar proportion of their time to other elephants in Africa, take in a smaller quantity of vegetation.

Water

Annual and seasonal variations in activities associated with water were probably attributable to variations in ambient temperature and water availability. High daily temperatures in the wet and hot dry seasons resulted in a higher percentage of activities associated with water than during the cold dry season, and higher than those observed in the rest of Africa.[30] This was probably to reduce heat stress. Adult females and juveniles spent longer periods engaged in water activities than either adult males or sub-adult elephants.

Resting

During the cold dry season, the desert-dwelling elephants reduced their diurnal resting times to allow sufficient time for other activities. There was a slight increase in the rest period during the diurnal hours in the wet season. There was a large increase in the amount of time spent resting under shade trees during the 11h00 to 15h00 period of the hot dry season by all age/sex groups – undoubtedly due to the high midday temperatures. Juvenile elephants rested more than any other age group during this

The well-known Okaukuejo waterhole in
Etosha National Park, Namibia.

season. In response, adult females also increased
their rest periods during the hot dry season to
protect and assist the juveniles.

Social

Desert-dwelling elephants spent the least amount
of their time socialising, and this activity appeared
to be opportunistic. During the wet and hot dry
seasons, most of the social activity was observed
during the 11h00 to 15h00 period, when elephants
aggregated under available large shade trees to
avoid the heat. Social activity was higher during
the cold dry season, when adult males came into
musth and increased their sexual contact with

females. Courtship behaviour usually took the
form of a male smelling a female's genitalia and
laying his trunk along her back; in rare instances,
mating occurred.

This level of activity varied from year to year,
depending on annual rainfall and the amount and
quality of vegetation available. Increased play and
sparring were also observed between sexes and age
groups during the wet season.

Walking

Most walking during the wet and hot dry seasons
was during the cooler periods of the day. The
distribution of water does not generally increase

greatly in the wet season. Although the ephemeral rivers do flood, providing more water in the riverbeds, palatable vegetation may still be quite some distance away, requiring elephants to walk long distances.[31] In addition, with increased sexual activity observed during the cold dry season, adult males in musth tended to undertake almost constant movement searching for receptive females.[32]

Behavioural modifications

Coprophagy

In 2002, a dominant adult female was observed to excrete dung that had a loose texture. A young female calf scraped it together with her foot, then lowered herself onto her front knees to eat some of it. She twice repeated this action before moving off with the herd.

In 2003, a similar behaviour was observed between three adult males. One older male (about 45 years) excreted loose dung, which was consumed by a young adult male (about 20 years). After a few minutes, another older male (about 40 years) approached and took small amounts of the same dung and ate it. Similar actions by juvenile elephants were reported by Guy from Sengwa Research Station in Zimbabwe.[33] In that instance, it was the dominant adult female that excreted the dung and the juvenile animals (but no adults) that ate it.

While coprophagy has been observed more often in captive situations, it is regarded as abnormal behaviour, and termed faeces manipulation.[34] It is possible that, in the wild, it's an attempt by young elephants to obtain the necessary gut enzymes to start digesting vegetation. In very young elephants, such as in the Namibian study, the behaviour observed could also be playing, with youngsters simply learning about their environment by sampling anything in their surroundings.

The reason for ingestion of dung by adult elephants is unknown, although both observed incidences occurred during the time of year when vegetation was limited. It probably is an attempt by the elephants to obtain the gut enzymes required to digest the available vegetation.

Thermoregulatory behaviour

On seven separate occasions, adult male and female elephants were observed to place their trunk into their mouth and extract water from their stomach, which they sprayed over their back and behind their ears. This occurred on particularly hot days.

Another interesting thermoregulatory mechanism was observed when adult female elephants urinated on sand, and juveniles and sub-adults were seen to scoop up the wet sand with their trunk and throw it over their back and behind their ears. The desert-dwelling elephants appeared to use this thermoregulatory mechanism routinely when under no apparent stress other than high ambient temperatures.

Conclusion

Under the current conservation initiatives, government conservation policies and law enforcement, the desert-dwelling elephants of north-west Namibia are relatively secure. However, human-elephant conflict has been increasing and the Kunene Region has the second-highest number of conflicts reported in Namibia.[35] There is pressure on the elephant population from the increased number of people living in the area, who, ironically, are drawn by the employment opportunities in tourism, underpinned by the presence of these elephants.

An unfortunate outcome has been the death of several tourists and local Namibians, with the solution generally being to shoot the offending elephant as a 'problem animal'. These animals are invariably young males known for their aggression. Professional hunting of problem elephants has been controversial. Although the number of elephants shot on the professional hunting quota (problem elephants being included in the quota) are low, so too is the number of adult male elephants in north-west Namibia.

From the movement studies, we know that adult males are moving from Etosha National Park to the arid western regions. However, should increasing human populations block these migration paths and the western elephant population become isolated, then a re-evaluation of the conservation policies of the conservancies and government will be necessary to maintain elephant populations.

Elephants in Damaraland, in the Kunene Region of north-west Namibia, on a dawn trek in search of water.

29

Selous Game Reserve: paradise lost?

The grip the hunting industry has on the reserve and its control over the people who manage it must be broken.

Colin Bell

This is the chapter nobody wanted to write. We contacted many people who have been involved with Selous to tell their stories and comment on the alarming elephant-poaching situation and other significant threats. But no-one was prepared to put pen to paper. We do understand that the Tanzanian government and many within the hunting industries in Selous Game Reserve are renowned for 'disciplining' troublemaking whistle-blowers. The assassination of Wayne Lotter in the streets of Dar es Salaam in August 2017 is a clear example of how far the ivory kingpins will go to ensure that their turf is not disturbed. Between 2014 and 2017, Wayne's anti-poaching and counter-intelligence organisation PAMS, working alongside Tanzania's newly formed anti-poaching unit, arrested over 1 300 poachers and illegal ivory traders, two of the most notorious being Boniface Mariango and Yang Feng Glan.

Mariango ran 14 trafficking gangs across East Africa, and Glan, known as the Queen of Ivory, is probably the most notorious ivory trafficker to be arrested in over a decade. Krissie and their two children now have to face life without Wayne.

The Selous Game Reserve in south-eastern Tanzania stretches over 350 kilometres from north-east to south-west. At over 50 000 square kilometres, it's more than four times the size of the Serengeti National Park, more than twice the size of the Kruger National Park, and larger than Belgium, Switzerland, Denmark or the Netherlands, or Vermont and New Hampshire combined. The Selous is one of the most beautiful reserves in all of Africa and is also one of the most interesting and diverse. It's one of the last remaining large tracts of uninhabited wilderness left in Africa. The Selous is a special place indeed. Old Arab slaving routes to the coast passed through parts of what makes up today's reserve. Slaves were made to carry elephant tusks on their long trek to the sea. Livingstone ventured to just south of Selous, and early European explorers like Burton and Speke walked its plains, river beds and mountains in search of the source of the Nile.

With its formal origins dating back to 1896, Selous claims to be the oldest wildlife reserve in Africa, although it is probably only the sixth oldest; five were proclaimed a year earlier in Zululand to help protect its dwindling wildlife resources.

The German colonial governor of the time issued a protection decree on behalf of Kaiser Wilhelm II that created the Rufiji Reserve. Over time, more land was incorporated, creating the

'Funds alone have little guarantee of (conservation) success. More important is strong legislation, open and transparent governance and the institutional capacity to implement plans. This goes well beyond the expressions of political will that come so cheaply during multinational gatherings.'

Brian Huntley
Wildlife at War in Angola

© Robert J. Ross, Selous Game Reserve, Tanzania

The consequences of short-term thinking by governments

These maps are from a recent logging tender issued by Tanzania's Ministry of Natural Resources and Tourism, the very department that is tasked with protecting Selous Game Reserve and developing its tourism industry. The outcome from this tender will be that 2.6 million indigenous trees will be stripped bare from the heart of Selous Game Reserve solely to make way for a hydropower dam that the Tanzanian government is proposing to build on the Rufiji River (as outlined in red on the map below). Both the logging and the building of the dam are without regard for agreed-upon UNESCO World Heritage protocols and regulations. The logging is destined to go ahead unless enough pressure can be brought to bear on the Tanzanian government and the dam's likely financiers, the World Bank. Yet the dam has not been given approval, nor have the billions of US dollars been found that are needed to build it.

This logging tender will result in an enormous, barren, treeless desert being created in the middle of Selous. If Tanzania had no other power creation options, a case could possibly be made for building such a dam. But Tanzania is a country blessed with ample natural gas resources and plenty of solar options. Private-sector electricity companies are queuing up all around Africa to build solar-powered electricity plants at their cost, with no burden on involved countries' tax payers. The combination of natural gas and these new, large-scale, highly efficient and cost-effective solar plants (or even microgrids) can provide the electricity that the country is requiring to help drive its industrialisation plans. If either the logging or the dam goes ahead, Tanzania and the world will lose one of its last remaining true wildernesses. If it goes ahead, this initiative will become a massive blot on Tanzania's reputation. In the words of Wangari Maathai 'The generation that destroys the environment is not the generation that pays the price'.

'For Tanzania, the benefits of investing in centralised and microgrid alternatives to large hydro appear to be overwhelming. It would avoid the risk of damaging the Selous World Heritage Site and downstream livelihoods. At the same time, electrification for remote communities is likely to be delivered more rapidly through microgrids than through expensive centralised grid transmission upgrading.'

Ross Harvey

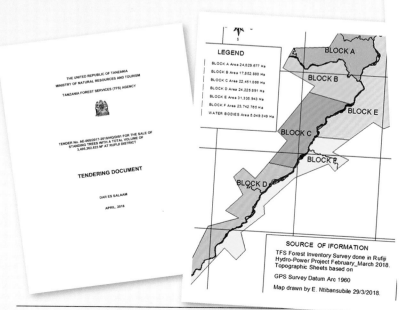

SOURCE OF IFORMATION

TFS Forest Inventory Survey done in Rufiji Hydro-Power Project February_March 2018. Topographic Sheets based on

GPS Survey Datum Arc 1960

Map drawn by E. Ntibansubile 29/3/2018.

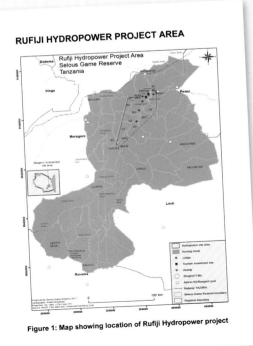

RUFIJI HYDROPOWER PROJECT AREA

Figure 1: Map showing location of Rufiji Hydropower project

5 million hectares that make up the Selous of today. The Maji Maji Rebellion of 1905 to 1907 was sparked by local opposition to the German colonial government's harsh agricultural and labour policies, culminating in an open rebellion that cost the lives of hundreds of thousands of villagers and tribesmen. Much of the fighting was within and around the current boundaries of the Selous. And the troubles continued. After the outbreak of World War I, British and German soldiers, led by Generals Smuts and Von Lettow-Vorbeck respectively, fought each other in the East African Campaign, using guerrilla tactics and the cover provided by the Selous's vast wilderness. In that campaign, the man who was to have the park named after him, Frederick Courtney Selous – one of Africa's more colourful early hunters and explorers – was shot in the head in the northern sector of the reserve near Beho Beho. Three years later, his name was given to the reserve in which he had hunted, fought and died. Tsetse fly and 'sleeping sickness' ensured that much of Selous remained uninhabited. As the reserve was expanded, some villages became incorporated, but by 1948 the park was devoid of people apart from a handful of hardy managers, rangers and workers. Iconic characters like CJP Ionides, and later, Brian Nicholson, ran the reserve successfully for decades.

In Nicholson's book *The Last of Old Africa*, he describes how, when he first moved into the Selous, it took him 14 days to get to his base camp, travelling by Dakota DC3 aircraft, train, bus and the last 5 days on foot – Selous was that remote and inaccessible. He describes how they had to shoot in excess of 1 000 elephants each year in the community lands around the reserve to stop crop destruction – elephants were that plentiful.

Because of its size, diversity and special wilderness qualities, in 1982 UNESCO, working with the Tanzanian government, declared Selous Game Reserve a World Heritage Site. At that point, there were around 100 000 elephants (probably the largest contiguous population anywhere in Africa) and close on 30 000 black rhinos, with lions and wild dogs in abundance.

That was as good as it got. Ever since, the threats to the integrity of Selous Game Reserve have been relentless:

- In 2012, a chunk of the reserve was carved out and de-proclaimed for a uranium mine.
- Plans are under way to build a massive hydro-electricity dam at Stiegler's Gorge in the heart of the reserve, which could cause devastating damage.
- Tenders were issued in April 2018 to logging companies to allow them to strip over 300 000 acres, at the very heart of the Selous, of all their hardwoods (around 2.6 million trees) to make way for the dam.
- Another dam is planned to the north of Selous, which will flood parts of the reserve, eliminating valuable grazing land.
- Prospecting licences for gemstones have been issued within the reserve.
- Hunting-block concessions have been further divided up, and some of these smaller blocks have retained the same quota as the original larger block, effectively multiplying its quota many times over.
- A lack of adequate funding and training for effective management has been an ongoing constraint.

All this comes under the watch of a government that has proposed building a new highway through the middle of the Serengeti (when there are other, more viable and rational options) and has moved an entire group of tribesmen off their land abutting the Serengeti to make way for Arab hunters. In May 2018, the deputy minister responsible for the Environment, Kangi Lugola, went one step further when he said in the Tanzanian parliament that 'those resisting the [Stiegler's Gorge hydropower] project [which will denude of the heart of the Selous Game Reserve of all its hardwoods] will be jailed'.

In the Selous of today, rhinos are close to being extinct. The 2013 aerial elephant census counted only 576 live elephants (and 314 carcasses) within the reserve. When the statistical extrapolation was applied, the estimated total number of elephants within the Selous was set at just 10 000, a more than 90% drop since Iain Douglas-Hamilton's aerial counts in 1977. In the aerial census, the Kilombero Region to the west of Selous is recorded as having zero elephants (down from around 5 000 in 1998). In acknowledgment of the dire state of affairs on the ground, UNESCO has formally placed Selous's World Heritage Status 'In Danger'.

Two contrasting scenes in the
Selous Game Reserve, Tanzania.

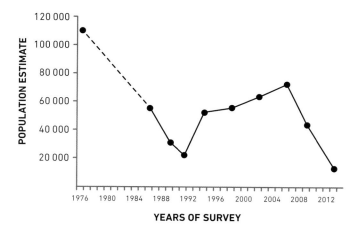

What has gone wrong in what should be one of the planet's finest wildernesses? Everyone has different theories; these are mine. Let's start with elephant numbers.

Many a tourist vehicle and safari camp around Africa was trashed by angry, harassed elephants during those years of intense ivory poaching in the 1970s and 1980s. I had been guiding safaris in Botswana for a decade and vividly remember my first safari to the Serengeti in 1987 where, to my immense surprise, my guests and I did not see a single elephant. Poaching had taken that much of a toll. In those traumatic years, elephants would hide deep within forests and dash to water in fading light to quench their thirst before returning swiftly to the security the forests provided. The heavy poaching throughout Africa during those years ended after the Nairobi ivory burn of July 1989, when all elephants around Africa were moved to CITES's Appendix I protection. At that burn, the Kenyan president, Daniel arap Moi, was ahead of his time when he said:

To stop the poacher, the trader must also be stopped. And to stop the trader, the final buyer must be convinced not to buy ivory. I appeal to people all over the world to stop buying ivory.

The concerted international demand-reduction campaign that followed the ivory burn resulted in demand for ivory plummeting. Poaching was reduced to insignificant levels, leading to the start of the elephants' 'golden years'. With little or no international market for ivory, elephant numbers throughout Africa started climbing again. From the early 1990s, elephants were slowly becoming more relaxed and many more youngsters were being born.

Around the same time as the 1989 ivory burn, the newly created and well-funded Selous Conservation Programme started to gain traction and effectiveness. Rangers were trained; equipment, uniforms and vehicles were bought; and a renewed sense of energy and commitment was created by individuals dedicated to Selous. Better policies and policing on the ground, in tandem with collapsed demand for ivory internationally, ensured that poaching within Selous was significantly reduced. Elephants in Selous started recovering from a low of just 30 000 in 1989 to around 70 000 by 2005. The good times were returning!

Those special conditions lasted for around 15 years, but ended abruptly when a few southern African states, led by South Africa, pleaded for CITES to grant permission for a one-off sale of 108 tonnes of ivory. The application was made on the undertaking that the ivory would be from stocks of natural mortalities, and the proceeds from the sale would be applied to elephant conservation. It was a convincing story, and CITES and the world bought it.

The sale finally went ahead in 2008. China and Japan colluded to buy the entire stock, and a legal market for ivory was stimulated by the Chinese government dribbling 2 tonnes into their carving and ornament market each year. Criminal syndicates seized the opportunity to use the cover provided by the legal Chinese ivory market to launder their illegal, poached ivory. The result is today's ivory crisis, where around 30 000 elephants are poached annually throughout Africa – an elephant dies every 15 to 20 minutes. To make matters worse, not one cent of the proceeds from the ivory sale was ploughed back directly into conservation. That sale was a dreadful self-inflicted 'own-goal'. CITES, the scientists at IUCN, and a host of conservation organisations have made sure that legal ivory sales have not been repeated.

By the late 2000s, the Selous Conservation Programme's good work started to wear off after initial funding from the German government had run its course. The systems and discipline implemented by the programme waned, while the demand for ivory was reignited by South Africa's one-off ivory sale. The deadly combination of lack of effective policing and renewed demand for ivory meant that elephants were again in poachers' crosshairs at even higher levels than in the 1980s. The booming Chinese middle class had developed an insatiable appetite for ivory, and as much as 400 000 kilograms of poached ivory was leaving Africa's shores each year.[1]

The 2013 aerial census and the Great Elephant Census report of 2016 confirmed everyone's worst fears: Selous's elephants had been decimated, with a mere 10 000 animals estimated within the reserve's boundaries. The heart of the Selous Game Reserve's problems can be summed up this way:

- endemic, entrenched corruption at both the highest governmental and reserve management levels;
- the historic non-transparent, often closed-shop tendering system for hunting blocks/concessions;
- an entrenched and sometimes unethical hunting brotherhood;
- unsustainably high, non-scientifically derived hunting quotas;
- minimal effective policing and monitoring of those quotas;
- the lack of a gradual morphing of prime concession areas over time from hunting to photo safaris, which would create more jobs, skills and money, and benefit local communities; (to this day, hunting is still conducted over 90% of the Selous's total land mass);
- the lack of meaningful, effective community involvement in the wildlife and tourism industries; and
- the lack of adequate finances for management to do its job and pay its wages. Insufficient earnings from Selous's tourism revenues are being returned to the reserve by central government. Paying staff on time and providing necessary equipment are basic, essential ingredients for any well-managed reserve.

Contrast this with the Kruger National Park, which at just 40% of the size of the Selous, earns around US$100 million a year from tourism receipts, concession fees, etc. Kruger's profits help to fund its management costs as well as those at other smaller, lesser-known, but equally important loss-making national parks around the country. At Selous, the benefits that an expanding photographic safari sector (capable of creating much more revenue and many more jobs) have not been capitalised on, relative to potential. Photo safaris within Selous are physically constrained to just the northern corner of the reserve, offering guests little of the privacy and space that such a wonderful reserve should have in abundance, and which, in turn, would help to grow the tourism arrivals, revenues, jobs and eyes on the ground to deter poachers.

Many of these issues are, of course, interrelated. Corrupt hunters operate in cahoots with corrupt park and government officials who, in turn, look after the interests and cover the backs of the corruptors. Sadly, this is not unique to Selous. This collusion occurs in many parts of Africa and is one of the main reasons hunting gets such a bad reputation. In many ways, Selous is a microcosm of the trouble that some parts of Africa and their elephant populations now face through poaching, human encroachment, inadequate finances, bad management, a lack of training and skills, and little integration of the local people into the tourism and wildlife industries at scale.

In Zimbabwe's Hwange National Park, many elephant bull hunting licences were quietly sold to the Zimbabwe hunting fraternity for less than market value. The dollars earned were said to be earmarked for park management and conservation but, some argue, were merely a profitable collusion between park officials and certain hunters. Similar hunter-reserve collusion in Selous ensures that retendering and reallocation of hunting blocks is not necessarily awarded to the highest bidding, most ethical hunting company. It has also ensured that leases have sometimes been awarded below market rates (although in recent times the arrival of quota speculators has shaken up the system, forcing some hunters to close shop). This collusion has ensured that money has neither reached neighbouring communities, nor made its way to fund the management of the reserve. This sleaze goes right to the top. I remember years ago listening to a hunter bragging about arriving at a meeting with an official in the office of the Tanzanian president, armed with a briefcase full of bank notes to help negotiate the leasing of a hunting block.

The hunting/photographic safari conundrum

In the early 2000s, *National Geographic* signed a contract with acclaimed documentary filmmakers Dereck and Beverly Joubert to produce two blue-chip wildlife films in Selous that would have been viewed by over a billion people. Those documentaries would have put the Selous Game Reserve and the rest of Tanzania's travel industry firmly on the world's tourism map.

The Jouberts are well known for their principled stance against corrupt hunters. Once the hunting fraternity heard they were about to make Selous their home, the hunters' connections within government ensured that their film permits were perpetually delayed. As a result, the Jouberts switched their contracts from Selous to the Okavango Delta, and Botswana benefited from the extraordinary exposure and positive publicity given by several documentary films they went on to produce, among others, *The Eye of the Leopard* and *Relentless Enemies*. Safari tourism in Botswana has never looked back, but it was an immense opportunity lost by the Selous and by Tanzania.

Don't get me wrong. I'm not railing against the ethical hunting industry, but against corrupt government officials, unethical hunters and unscrupulous safari operators who line their pockets for the short term instead of doing what is right for the greater good of the reserve and the country.

For the hunting industry to have a future, there needs to be a radical clean-up from within, led by ethical hunters (as has now started in South Africa). Equally important, the competition with photo-safari tourism for the rights to operate in prime areas needs to end. Each industry should operate only on land where it offers the most profitable macroeconomic outcome for the country and its rural neighbours.

When that happens, both industries can maximise their contribution to their national economies. But for this to happen, wildlife land needs to be better zoned and divided into the following areas:

- core areas – with no tourism infrastructure or activity except on foot, horse or *mokoro* (traditional canoe);
- prime areas – only low-volume, low-impact, high-revenue, high-job-creating photographic safaris;
- semi-prime areas – medium-density/medium-revenue photo tourism;
- buffer areas – no tourism (even if this zone is merely a few kilometres in width); and
- marginal areas – for hunting.

If all wild lands in Africa were zoned along these lines, there would be little animosity between hunting and photographic safaris, and tragedies like the Cecil and Skye lion 'hunting'/poaching fiascos would not happen. Clear zoning with well-enforced buffer zones would enable all types of land use to prosper.

There are, of course, some prime areas where few photo tourists will visit, areas such as those that dominated the Selous in the 1960s and 1970s when hunting first started. In such areas strictly controlled, low-volume, sustainable, ethical hunting can pay its fair share to the game reserve, its neighbouring communities and the country. But as the photo-safari tourism industry starts to grow and mature, and to provide better finances and more, long-term jobs for all stakeholders, the hunting industry in these areas needs to evolve and morph into photo-safari opportunities.

My single biggest concern about the Selous Game Reserve is that this evolution from hunting to photographic safaris has not happened, and photo tourism still occupies only the same small portion of the northern section of the reserve as it did 40 years ago. While large parts of Selous may not be suitable for photo tourism, there are many prime areas that could successfully be converted into productive low-volume, high-tariff, low-impact photographic-tourism concessions, and thus generate more for the economy and provide more, and more long-term jobs for local communities.

The southern sector of the Selous Game Reserve in the Lukula Concession, Tanzania.

© Dana Allen

What pays best?

For decades there's been an often acrimonious debate between hunters and photographic-safari operators about who contributes most from the same piece of land. The debate is about to be settled by a study on an ex-hunting concession in Botswana, which has converted from hunting to photographic safaris. Early indications are that the government of Botswana, local communities and Maun businesses will earn something like 1000% more each year from photo safaris than they earned from hunting safaris from the same concession. Moreover, the local communities will be working an additional 40 000 'people-working-days' a year in service of photo safaris than they worked when the concession was hunted.

What is clear is that, in marginal lands, ethical hunting can outperform photo-safari tourism. And limited, ethical hunting in these lands is often preferable to livestock grazing when the communities gain jobs and revenues from what would otherwise be relatively unproductive land. What is important is that safari disciplines operate in areas where they are most productive. Selous will be a better place when that happens, and some of the better concessions and blocks morph from hunting into photographic tourism.

Can Selous regain its place as one of the finest wilderness areas on Earth? Right now, the reserve may be close to the point of no return, especially if the logging within its core goes ahead.

But nature is remarkably resilient and can recover if afforded the space and opportunity to do so. What is needed is prudent, sensible leadership at the highest level that focuses on what is best for the reserve and the surrounding people in the medium and long term, backed up by effective, adequately funded and well-motivated management on the ground. The book *The Selous in Africa: A Long Way from Anywhere* by Robert J Ross illustrates the beauty and the potential of the Selous, and clearly portrays what is at stake and what we cannot afford to lose. Selous needs new, transparent, open tender processes. Each concession requires its own individual, pragmatic management plan, with longer leases and agreed tourism densities that encourage significant investment along with the checks and balances that limit serious transgressions. And

areas that are designated for hunting must have scientifically-determined quotas that are sustainable. The grip that the hunting industry has on the reserve and its control over the people who manage the reserve must be broken. The new policies must bring the neighbouring rural communities into the mainstream of the revenues generated by its tourism industry – both hunting and photographic – so that they become significant beneficiaries.

Then maybe, just maybe, Selous can develop to its full potential and avoid being downgraded by UNESCO. There are only two World Heritage Sites that have ever lost their status. Stakeholders need to make sure Selous is not number three, for the sake of its elephants and for all the people of Tanzania. If Tanzania's current leadership could align itself with what was recognised by their still much-revered first President, Julius K Nyerere, there would be hope:

'The survival of our wildlife is a matter of grave concern to all of us in Africa. These wild creatures amid the wild places they inhabit are not only important as a source of wonder and inspiration, but they are an integral part of our natural resources and our future livelihood and wellbeing. In accepting the trusteeship of our wildlife we solemnly declare that we will do everything in our power to make sure that our children's grandchildren will be able to enjoy this rich and precious inheritance.'

Julius K Nyerere

Digging for water during the dry season in
Selous Game Reserve, Tanzania.

© Colin Bell

Conserving baobabs

In many parts of Africa, elephants feed off the nutritious wood pulp of baobab trees, especially towards the end of the dry season when food, nutrients and water are scarce. In areas of high elephant densities, some baobabs take a battering and fall. They can be protected from over-utilisation by placing large logs or sharp rocks around their base or wrapping the trunk in chicken wire. This young elephant in Ruaha National Park, Tanzania, was killed by a falling, over-browsed baobab, and the double tragedy soon attracted predators.

Effective anti-poaching efforts have resulted in a fourfold increase in the elephant population at Singita Grumeti.

© Singita

30

The success of Singita Grumeti

Building an integrated wildlife system
along the Serengeti's western edge

Dr Neil Midlane

**It's early afternoon when a watchkeeper in the
ops room of Singita Grumeti's Joint Operations
Centre in the western Serengeti takes a call on the
community hotline. It's a desperate villager from
a nearby community: elephants are approaching
his homestead and he's convinced they're about to
destroy his maize crop. He's relying on it to feed
his family and pay his children's school fees. The
watchkeeper assures him help is on the way, hangs
up and gets on the radio to dispatch the human-
wildlife conflict mitigation unit.**

The team of armed, well-trained former game
scouts jump into their Land Cruiser and head for
the area from where the report was received. Upon
arrival, they divert the elephants from the crops
and back towards the protected area from where
they came. The crops are saved and, through the
team's training and experience, the villagers are
not exposed to the risk of chasing off elephants
themselves. And the elephants avoid a potentially
lethal encounter with a group of frightened people
ill-equipped to deal with such a volatile situation.

Human-wildlife conflict (HWC) is ever
present across much of Africa, and remains
the greatest challenge faced by rural Tanzanian
communities living near protected areas like the
Serengeti ecosystem. While jackals killing sheep on
commercial farms or baboons raiding city homes
are costly and inconvenient, the potential loss of an
entire year's income in a single elephant incident
means the stakes are much higher for both people
and wildlife.

The village here is one of more than 20 located
along the northern edge of Singita Grumeti's
concessions: a privately managed wildlife area
adjacent to the north-western boundary of the
iconic Serengeti National Park. Here, high-density
human habitation, agriculture and livestock are
separated from a thriving wilderness area by
nothing more than a dry river bed, easily crossed by
elephants, lions and other wildlife.

Although increasing human and livestock
populations (cattle numbers alone have doubled
in the last 10 years) are a major driver of escalating
conflict, the recovery of wildlife numbers within
Singita's concessions has also played a significant role.
With around 1 500 elephants now using the reserve –
up from 350 just 15 years ago – it's not surprising that
levels of conflict between people and wild animals in
this zone have been increasing steadily. It's a situation
that led to the establishment of the specialist human-
wildlife conflict mitigation team.

The dusty grasslands of Singita Grumeti were virtually devoid of large mammals 20 years ago. Today the 350 000-acre Western Corridor is home to an abundance of wildlife.

At Singita Grumeti, an elite special operations group of 18 high-performing anti-poaching scouts are provided with ongoing advanced training.

The Singita
conservation model

Singita is a conservation company active in five African countries. It employs a high-value, low-volume photographic tourism model to partially fund the conservation management of almost a million acres of wilderness under its stewardship.

The Singita Grumeti Fund is a Tanzanian non-profit organisation dedicated to the conservation of the Western Corridor of the Serengeti ecosystem. It's responsible for the management of some 141 640 hectares comprising the Grumeti and Ikorongo game reserves, Ikona Wildlife Management Area, Sasakwa concession and associated communal lands, jointly referred to as Singita Grumeti.

Although it's widely believed that the Serengeti has always been teeming with game, in the system's Western Corridor at the turn of the 21st century, poorly managed hunting and extensive poaching had decimated wild populations. With uncontrolled wildfires and widespread alien plant infestations compounding the problem, this crucial block of the Serengeti-Mara system was on the verge of ecological collapse.

This desperate situation caught the attention of American philanthropist and conservationist Paul Tudor Jones. Following an agreement with the Tanzanian government, he established an in-country, non-profit organisation to take on the task of rehabilitating the severely degraded habitat and wildlife populations. A meeting of minds with Luke Bailes, CEO of the conservation tourism company Singita, resulted in a long-term partnership and the genesis of the Singita Grumeti Fund.

The complexity of the Fund's task is reflected in the structure of its departments: Conservation, Anti-Poaching and Law Enforcement, Community

Outreach, Research and Monitoring, Relationships and Special Projects. Each team has a clearly defined role to play, but close collaboration between them is essential as they attempt to solve the problems at hand.

The Fund's first priority was to reduce the haemorrhaging of animals to illegal poaching and poorly managed 'licensed' hunting. The team recognised, however, that for many people living in the area, the illicit wildlife trade was their only livelihood. An independent study around the time found that more than a third of traders in the region relied solely on bushmeat for income. More than 75% of poachers arrested were planning to sell or trade the spoils of their hunt; just one in four kept the meat for personal or family consumption. The Fund realised that, although enforcement of the laws was critical, it had to be done with an element of compassion.

Recognising the need for an alternative source of income, the Singita Grumeti Fund offered these men, predominantly local community members driven by poverty and desperation, the opportunity of training and employment as game scouts. Their extensive knowledge of the area and its wildlife, as well as the habits and tactics of the poachers, meant they quickly became effective at protecting the animals that they used to hunt. More than 100 ex-poachers have since been converted, trained and recruited into the reserve's anti-poaching unit, ensuring a stable income for them and their families.

At the same time, the Fund launched a Community Outreach Programme (COP) to support villagers neighbouring the reserve, and to share the benefits of a sustainable wildlife-based economy. The programme has evolved over the years, with monitoring and evaluation of projects a critical element as the team looks to continuously improve the level of positive, sustainable impact in communities. In 2016, a needs assessment in the 21 neighbouring villages led to a new strategy: UPLIFT, an acronym for Unlocking Prosperous Livelihoods for Tomorrow.

The three pillars of UPLIFT are education, enterprise development and environmental awareness. Targeting these pillars provides benefits to communities as a result of the presence of wildlife, and empowers them with knowledge to sustainably manage the natural resources of their own environment. Programmes within each pillar include scholarships for secondary and tertiary education, a community agricultural co-operative that supplies the staff and lodge kitchens, enterprise development training and an environmental education centre.

Since the Fund's inception, more than 700 students from local communities have received bursaries, an investment of over $450 000. Between staff meals and food requirements of the tourism business, Singita Grumeti buys around $250 000 worth of fresh produce a year from the agriculture co-op. Close on 2 000 students and over 250 teachers have attended courses at the environmental education centre. The Fund also employs 163 staff, while Singita's tourism operation employs almost four times as many (640). Some 60% of those employed are from neighbouring communities, and a further 31% are from elsewhere in Tanzania. In a country and region with high levels of poverty and unemployment, the salaries of these staff are a lifeline for families and the local economy.

The Singita Grumeti Fund also helps communities to understand the link between these benefits and the wildlife-rich environment. In that way, they become its ambassadors and protectors, a voice for wildlife in communities where it is seldom heard.

While the community outreach team is mostly busy outside the reserve, a dedicated conservation management team works within its boundaries to restore and maintain the system's ecological functions. Controlling fires and combatting invasive alien plants need constant attention.

Fire is a natural component and important part of the Serengeti ecosystem and is used as a management tool for the benefit of the habitat. Burns at the right time reduce rank standing grass, allowing it to put more energy into its roots. This way, grasses become more palatable for wildlife. Fire also improves nutrient recycling in the soil, leading to more nutritious forage. If not managed appropriately, however, fires that burn too often or too hot hinder ecological function. The conservation team's fire management protocols have resulted in measurable improvements in quality and quantity of grass available to wildlife in the reserve.

Twelve permanent anti-poaching scout patrol camps and a network of high-lying observation posts are manned 24 hours a day at Singita Grumeti.

Alien plant species are harmful to both animals and indigenous plants, and dealing with this threat is a constant battle. Singita's alien plant control programme works both inside the concessions and in neighbouring villages identified as being overrun by the primary invaders in the system, *Chromolaena, Opuntia, Parthenium* and *Tithonia*.

Mechanical and chemical clearing methods are the main tools but, in partnership with the Centre for Agriculture and Biosciences International, an innovative biological control is being pioneered against *Chromolaena*. These types of collaborations with conservation NGOs and academic institutions are often initiated to address a particular management challenge or question. Outputs of projects are then evaluated and management strategy is adapted accordingly. Alongside focused research projects, ongoing monitoring of mammal populations is carried out through biennial aerial wildlife censuses and a reserve-wide remote camera array to ensure that the team has its finger on the pulse. Satellite sensing and vegetation surveys provide additional data, giving a fuller picture of the health of the ecosystem.

The success of the Singita Grumeti Fund's conservation approach is best seen through the growth in its population of large mammals. Between 2003, when the Fund was established, and 2016, the biomass of resident large herbivores has nearly quadrupled. Although almost all species surveyed (12 in total) showed a positive growth, the biomass increase is driven mainly by buffaloes and elephants. Buffalo numbers have increased from around 600 to over 6 000 in this period. Elephants have increased fourfold, from 355 in 2003 to 1 499 in 2016. Aerial surveys suggest there has also been a significant increase in the lion population. And, whereas the annual zebra and wildebeest migration spent just 3 weeks in the Western Corridor in 2003, these animals are now present in high densities for up to 6 months of the year.

This incredible ecosystem recovery has enabled Singita's tourism business to develop its Grumeti concession into one of the most sought-after photographic-safari destinations on the planet, simultaneously adding to the stability of the employment from local communities.

Despite great strides, poaching and other illegal activities have remained persistent threats. For this reason, the anti-poaching and law enforcement team accounts for some 70% of the Fund's staff. From 12 strategically placed scout camps, each with a dedicated observation post, 100 game scouts, in conjunction with their colleagues from the Tanzania Wildlife Management Authority, conduct daily patrols and 24-hour surveillance. The importance of their work is borne out by statistics. In 15 years of operation, the anti-poaching team has dealt with almost 5 500 illegal activities. Of these, almost half involved poaching and about 1 600 were illegal grazing of livestock. More than 12 000 snares have been recovered and close to 25 000 illegal traditional weapons (used for poaching purposes) collected. Without this law enforcement effort, the spectacular resurgence of wildlife in the Western Corridor would not have been possible.

Most poaching in Singita Grumeti is for the bushmeat market, both local and further afield. But the continent-wide escalation in elephant poaching continues to threaten the Western Corridor. While elephant numbers plummeted in other areas of Tanzania, according to the Great Elephant Census, the Serengeti sector is one of the few that had a positive growth rate – more than 5% over the 10 preceding years. In spite of this particular triumph, the spectre of well-armed, well-organised criminal syndicates moving into the region is very real.

To increase capacity in response to this threat, a low-tech detection and tracking dog unit was deployed alongside the development and implementation of high-tech digital solutions. Working Dogs for Conservation, a US-based NGO, trained a team of four dogs and their handlers. The two Labrador mixes and two Belgian malinois, all rescue dogs from the US, are now highly proficient at sniffing out ivory and ammunition, as well as rhino horn, pangolin scales, wire snares and bushmeat. In collaboration with local law enforcement, this canine unit is deployed at roadblocks as well as in raids on suspected poachers.

On the high-tech side, Seattle-based Vulcan Inc. works closely with the Fund on the installation and deployment of a Domain Awareness System (DAS). This web-based tool aggregates the positions of radios, vehicles, aircraft and animal-borne sensors to provide users with a real-time dashboard showing the wildlife being protected, the people and resources protecting them, and potential illegal activity threatening them. Combining these data on a single screen in real time and using the system's analytical capabilities allows the Fund to efficiently identify threats and deploy law-enforcement resources for effective intervention.

A major focus of DAS is the development of drones specially designed to detect illegal wildlife activities. Success in this field will be a game changer for anti-poaching operations. Further data sources such as satellites, camera traps, animal sensors and weather monitors can all be integrated with the DAS system.

Ultimately, though, response speed and effectiveness will determine successes. That's where the Special Operations Group within the Fund's anti-poaching unit comes in. This comprises 16 scouts selected for their integrity and strong work ethic. They're given ongoing advanced training and have access to the latest equipment. With these systems in place, the Fund aims to apprehend would-be poachers before they are able to kill any wildlife.

Is the Singita Grumeti Fund's approach to conservation working? According to the elephants, it would seem that the answer is yes. These intelligent animals are known to favour areas where they feel safe and secure, and where resources are plentiful. Aerial surveys indicate that as many as 20% of the elephants in the Serengeti-Mara ecosystem can be found on the Grumeti concessions, which cover just 5% of the surveyed area.

Success, however, has come at a price. The ever-present threat of elephant poaching syndicates pushes up security costs, and more elephants means that human-elephant conflict continues to intensify. A study commissioned by the Fund found that, from 2012 to 2014, the annual number of elephant-related crop-damage incidents in nearby villages increased by 400%. Villages adjacent to the concessions were more likely to be affected (80 to 85% of incidents), but those further away were not immune. Surprisingly, elephant incidents also led to more cattle being injured or killed than by lions, leopards and hyaenas combined. Livestock incidents associated with elephants – in comparison with other wildlife species – were the most geographically scattered, most severe in terms of the number of animals impacted, and involved more high-value livestock types. Although elephants do not target cattle or sheep, the damage occurs when they are chased away from crops by frightened villagers, and they end up running over livestock in the ensuing chaos.

The reality is that, although the Fund has made a significant contribution to positive development in neighbouring communities, the sheer human density means that many people see no value in living close to the protected area. The damage caused by elephants to crops and livestock reduces tolerance of their presence.

With no fence or physical barrier to separate high-density human and livestock areas from the high-density wildlife on these concessions, it's likely that conflict between the two will continue to escalate. Singita Grumeti's HWC mitigation unit can expect busy times ahead.

The long-term ecological sustainability of the Western Corridor of the Serengeti, and the security of the elephants living there will, therefore, depend on the ability of the Singita Grumeti Fund, with support from communities, government and other partners, to adapt, innovate and finance its successful conservation model in an environment of constantly evolving threats and challenges to its existence.

Opposite
The goal of the Singita Grumeti Fund is to contribute to the conservation of the Serengeti ecosystem, its natural landscape and its wildlife.

Tim and Mount Kilimanjaro
– Africa's biggest elephant
and highest mountain.

© Johan Marais

31

Beneath Kilimanjaro: elephant conservation in Kenya

The most significant achievement of the combined efforts of the state, the private sector and civil society organisations is that elephants are no longer a minority concern in Kenya.

Dr Paula Kahumbu

In June 2014, people in Kenya and across the world were shocked by the news that Satao, Kenya's biggest elephant, had been killed by poachers. He was celebrated as one of the last surviving 'great tuskers', bearers of genes that produce bull elephants with huge tusks reaching down to the ground. This news followed hard on the heels of the slaughter of another legendary tusker, Mountain Bull, deep in the forests of Mount Kenya.

This was perhaps the lowest moment for defenders of elephants in Kenya, already reeling from an upsurge in poaching that had been gathering pace, and which the authorities seemed powerless to control. Official figures that recorded 384 elephants killed by poachers in 2012 and 302 in 2013[1] were certainly underestimates.

By 2016, the situation had changed dramatically for the better. Monitoring data shows that elephant poaching fell by 80% nationwide between 2012 and 2015, while the proportion of illegally killed elephants (PIKE) in Samburu-Laikipia fell from 72% in 2012 to 37% in the first quarter of 2015.[2] The most recent data shows that poaching fell by 84% in the key Tsavo Conservation Area from 2014 to 2016. According to a report released by CITES in March 2017, PIKE in East Africa dropped from 42% to 30% from 2015 to 2016, with the carcass numbers from Tsavo having the highest influence on this decline.[3]

The first section of this chapter examines the reasons why Kenya was able to turn around the poaching crisis in such a relatively short time. However, Kenya's elephant populations are still under pressure from threats that are even more intractable and complex than poaching and illegal wildlife crime. The concluding section of the chapter analyses these challenges and how they can be addressed.

'From spending so much time with elephants, I know they are not just another animal. They are beings with personalities; with families; with feelings. It is such a horrendous injustice to allow them to be slaughtered ... imagine if you went to a human family and shot the parents in front of the kids.'

Dr Paula Kahumbu
WildlifeDirect

Background

In 2017, Kenya's elephant population numbered between 25 000 and 35 000, according to the IUCN African Elephant database. In contrast to many other African countries, this population is currently stable or possibly increasing. Yet these numbers are a far cry from the late 1970s when the country's elephant population was estimated at around 275 000. Between 1970 and 1977, Kenya lost more than half of its elephants and, by 1989, fewer than 20 000 were left. They were being systematically culled to supply international ivory markets. Poachers were able to operate with impunity in a climate of endemic corruption at all levels of the institutions charged with protecting the nation's wildlife.

In April 1989, President Moi appointed Richard Leakey as head of the Kenya Wildlife Service (KWS). Within months, Leakey had cleaned out the corruption at KWS and created well-armed anti-poaching units that were authorised to shoot to kill. In July that year, Moi and Leakey set fire to an 18-foot pyramid of tusks, Kenya's entire stockpile of ivory worth millions of dollars, in a dramatic gesture intended to persuade the world to halt the ivory trade. This undoubtedly influenced the decision of the Convention on International Trade in Endangered Species (CITES) to declare a worldwide ban on commercial ivory trade in the following year. The ivory industry rapidly contracted to a fraction of what it had been. Poaching evaporated and illegal elephant deaths decreased to insignificant levels. Elephant populations in Kenya and across Africa stabilised and began to increase again.

Then, in 2008, CITES authorised a one-off sale of ivory from four southern African countries to China and Japan. This decision triggered an explosion of demand, and poaching once again erupted across the continent, with Kenya being swept up in this second wave of elephant killings. In many ways, the poaching crisis of 2012/2013 recalled the situation nearly 25 years earlier; but this time the scale of the trade was far greater, the prices of ivory were unprecedented and the demand, mainly from China, seemed insatiable. The criminal cartels involved in trafficking ivory were also far more sophisticated, with the supply chain between poachers to markets being facilitated by large numbers of Chinese people visiting and working in Africa.

Success factors

Ivory poaching affects all African countries with elephant populations, and different countries have adopted a range of strategies to attempt to control the scourge, with varying successes. In Kenya, although by 2013 the situation was dire, there were a number reasons for optimism:

- the existence of a large body of African ecologists and bold activists who were championing the cause of elephants;
- Kenya's vibrant civil society and freedom of the press;
- political support from the new president, Uhuru Kenyatta, who took office in April 2013; and
- the previous successes of the 1990s, which inspired confidence that it could be done again.

These conditions proved favourable for the development of an effective, country-wide response to the poaching crisis. Its success was based on a range of mutually reinforcing factors: Kenyan ownership, a holistic approach to law enforcement, effective information and intelligence gathering, and nationwide cross-sectoral collaboration.

Kenyan ownership
of the anti-poaching drive

As poaching spiralled out of control in 2012, the KWS and other government agencies at first seemed to be in denial. One conservationist was even arrested for undermining the state authority after making public data that exposed the seriousness of the crisis.

The turning point came in February 2013 when the head of Vision 2030, Mugo Kibati, persuaded the government to call a special session of the National Economic and Social Council (NESC) to discuss wildlife conservation. This body handles urgent emerging matters affecting the economy, and had previously discussed terrorism, coffee production and declining tourism. This landmark meeting to address poaching was attended by dozens of representatives of ministries, law enforcement agencies, the private sector, academia and civil society.

Leading Kenyan conservationists warned the government that thousands of elephants were being killed each year, and highlighted the threat this posed to tourism and the economy. They questioned the capacity and commitment of KWS and border agencies to control poaching and trafficking, and presented a 14-point response plan that included more boots on the ground, improvements in the criminal justice system and destruction of Kenya's ivory stockpile.

The NESC adopted most of the recommendations and instructed authorities to urgently adopt a 'whole government' response to the crisis. A task force was gazetted to examine the state of wildlife security in Kenya. Its report revealed that the conservationists were not only right, but had underestimated the scale of the challenge. It concluded that 'serious reforms are required if the current siege on Kenya's wildlife is to be reversed'.[4]

The fact that these initial efforts by conservationists were channelled through existing democratic institutions played a vital role in ensuring Kenyan ownership of subsequent efforts to defeat the poachers and traffickers and to safeguard elephant populations. Kenyan ownership of anti-poaching efforts is manifested at many levels: as political leadership, public support and endorsement of government agencies. In his inaugural speech in 2013, President Uhuru Kenyatta declared:

My fellow Kenyans, poaching and the destruction of our environment has no future in this country. The responsibility to protect our environment belongs not just to the government, but to each and every one of us.

Presidential support for elephants is a tradition in Kenya that dates back at least to the 1970s when, in response to a letter-writing campaign by Kenyan schoolchildren, the first president of Kenya, Mzee Jomo Kenyatta, declared the great tusker Ahmed a 'Living National Treasure' and assigned five armed rangers to protect him, day and night.

Throughout his first term of office, current President Uhuru Kenyatta has not only supported elephant conservation in Kenya, but has also actively promoted African leadership of efforts to combat the illegal wildlife trade – the first pillar of Kenyan ownership of anti-poaching efforts. He was one of the founder members of the Giants Club, a forum bringing together African heads of state, global business leaders and elephant-protection experts in order to 'secure Africa's remaining elephant populations and the landscapes they depend on'.[5] President Kenyatta has been in the forefront of efforts to secure a global ban on all trade in ivory, efforts that were dramatically highlighted by the destruction of Kenya's entire stockpiles of ivory and rhino horn in the historic ivory burn in Nairobi National Park in April 2016.

This leadership on the global stage has been complemented at the domestic level by the inspirational leadership of civil society initiatives by First Lady Margaret Kenyatta and by the tireless efforts of Professor Judi Wakhungu, cabinet secretary for the Environment, who has played a key role in facilitating collaboration among governmental and non-governmental organisations.

Public support has been the second pillar of Kenyan ownership of anti-poaching efforts. A key role has been played by the campaign Hands Off Our Elephants, launched in 2013 by the NGO WildlifeDirect.[6] The campaign, whose patron is the First Lady, has enlisted the support of businesses and celebrities, and partnered with other civil society organisations, schools, universities and youth organisations to get Kenyans on to the streets and on to social media in support of elephants. Hands Off Our Elephants has succeeded in mobilising ordinary Kenyan citizens to take action in support of wildlife conservation to an extent that is unprecedented on the African continent. The high level of media coverage of the campaign played a key role in raising public awareness of the alarming threats facing Kenya's elephants. So calls for action to save Kenya's elephants have come both from above and below.

The third pillar of Kenyan ownership has been the positive response by state agencies and institutions. As described in the following section, KWS and the judiciary, in particular, have made important progress in developing the institutional capacity required to fulfil their specific organisational responsibilities in the wider war on poaching and ivory trafficking.

When the savanna dries out in Amboseli National Park and the ground has been cropped bare, elephants use the park's swamps as their pantry.

Warrior tribesmen gather for an event in Kenya. The future of free-ranging elephants in the region is very much in the hands of rural tribesmen who have to live with these animals, constantly guarding their crops from being raided, sometimes putting their lives at risk in the process. The meaningful inclusion of communities into the wildlife and tourism industries at scale is crucial if elephants are to have a chance of surviving the population pressures that the next decades will unfurl.

A holistic approach to law enforcement

In 2013, the president prioritised the anti-poaching drive as one of a number of rapid-results initiatives that were required to adopt a whole-government approach. Law enforcement is a key area where this approach has paid off.

In accordance with NESC recommendations, additional funds were allocated to anti-poaching activities, allowing the recruitment of 577 more rangers at KWS. An elite, multi-agency anti-poaching squad was formed, trained and deployed in poaching hotspots. Training for gazetted scene-of-crime officers at KWS has significantly improved investigations, while additional support in cash and kind from diplomatic missions and funding organisations has further enhanced the mobility and rapid-response capacity of rangers.

As a result, poachers are more likely to be caught than ever before. But arrests have no deterrent effect unless they lead to serious consequences. Before 2013, poaching was treated as a petty offence. Maximum penalties were seldom enforced and even these were derisory compared to the vast profits made by organised wildlife crime. Following intensive lobbying by a wide range of organisations and citizen groups, the Wildlife Conservation and Management Act 2013 (WCMA) came into force in January 2014. The new law makes poaching and ivory trafficking a serious crime in Kenya, on a par with terrorism and drug trafficking. Penalties for wildlife crime in Kenya are now among the harshest in the world, including life imprisonment in some cases.

Following the introduction of the new act, a major effort was launched to improve the handling of wildlife crime cases by prosecutors and the courts. A specialised wildlife crime prosecution unit, established under the Office of the Director of Public Prosecutions (ODPP), has taken over the prosecution of wildlife crime across the country.

The ODPP and KWS have collaborated with the NGOs WildlifeDirect, Space for Giants and the UN Office on Drugs and Crime (UNODC) to produce a Rapid Reference Guide for prosecutors and investigators, setting out standard operating procedures and point-to-prove for different categories of wildlife crimes.[7]

These have been supplemented by a Wildlife Crimes Case Digest, giving prosecutors access to case summaries and full texts of judgements in important wildlife crime trials and a guide to the Wildlife Act to support training of rangers and other frontline staff.[8] These documents are being used as training materials in ongoing courses on best practice in wildlife crime trials that have so far benefited more than 500 legal officers (prosecutors and magistrates) working in courts across Kenya.

These developments have been described as a game changer by people monitoring development in the courts. The rate of convictions in wildlife crime trials has risen, and sentences have increased in line with the provisions of the new law. The jailing of convicted poachers is up from 4% from 2008 to 2013, to 44% in 2016. Several poachers have been jailed for up to 20 years and many have been fined hundreds of thousands of US dollars. Suspected traffickers have had their assets seized and bank accounts frozen as the law on proceeds of organised crime can now be applied to wildlife crimes.

Perhaps most importantly, for the first time in the country's history, major ivory traffickers are being prosecuted in Kenyan courts. One of the most notorious suspected traffickers, Feisal Mohamed Ali, was arrested, with the support of Interpol, following the seizure of a huge haul of ivory in Mombasa in 2014. In July 2016, he was convicted of illegal possession of ivory and jailed for 20 years, with a fine of 20 million shillings (about US$200 000).

When the plains offer little food in Kenya's dry season, elephants move to the Amboseli Marshes. The park protects two of the country's five main swamps and includes a dried-up Pleistocene lake. Amboseli is famous for being one of the best places in the world to get close to free-ranging elephants.

Effective data and intelligence gathering

All the progress made by law-enforcement agencies, as well as improved evidence-based management in the field, depends on effective data collection and intelligence gathering.

WildlifeDirect's courtroom monitoring programme, 'Eyes in the Courtroom', has played a key role, both in exposing systemic failings in the prosecution of wildlife crimes before 2013 and, since then, in monitoring progress and ensuring transparency and due process in ongoing trials. The first report was published in 2013 and its conclusions, based on 5 years of courtroom monitoring, were stark.[9] Most suspects in wildlife-crime cases were walking free. Suspects were able to plead guilty and pay a paltry fine or, if they chose to contest the case, could be confident that it would be dismissed due to procedural irregularities. This report was a wake-up call and provided incontrovertible evidence to back up calls for reforms to the legal system.

The courtroom-monitoring programme is continuing, recording the process and outcomes of wildlife-crime trials at courts across the country. Work is now under way on the development of an Android web-based application that will provide law-enforcement personnel and other users with access to historical and real-time courtroom data.

Enforcement of anti-poaching laws also depends on the effectiveness of tracking seized ivory so that it doesn't re-enter the market through the actions of corrupt officials. Kenya now has an excellent ivory-seizure database. Between June and August 2015, the country completed the largest-ever inventory of national elephant ivory stocks. A total of 105 tonnes of ivory was inventoried by KWS, supported by Stop Ivory and Save the Elephants (STE) using bespoke software developed by Bityarn Consult. By April 2016, the country had completed its first electronic tracking of national ivory stocks, and all 105 tonnes of ivory were moved from stores around the country to the secure stores in the KWS headquarters in Nairobi National Park. They were then moved into the park by an army of volunteers for the ivory burn on 30 April.

Anti-poaching actions in the field have benefited enormously from high-quality data produced by

KWS in collaboration with scientists from many conservation organisations, including STE, the Mara Elephant Project and the Amboseli Trust for Elephants. Mortality data is collected for the CITES programme 'Monitoring of Illegal Killing of Elephants' (MIKE) in four key areas: Samburu-Laikipia, Tsavo, Meru and Mount Elgin. The Samburu-Laikipia MIKE site, co-ordinated by STE, is the most data-rich on the continent.[10] Conservation organisations also participate in national elephant counts and provide unique data on elephant movements through radio-tracking initiatives. All this information provides vital inputs for evidence-based decision making.

Investigation in the field is increasingly supported by state-of-the-art information technology. An innovative wildlife security framework, tenBoma, has been implemented by the International Fund for Animal Welfare (IFAW) with KWS and other partners at strategic locations across the country. The project supports enhanced collection and analysis of field data using smartphones, geospatial analytic software and mobile device forensic kits (for rapid extraction of data from phones and other devices) in order to deny operational space to wildlife criminal networks. Since the beginning of 2016, no poaching events have occurred in tenBoma focus areas, despite a heavy concentration of poaching in the same areas in the years leading up to 2015.

The KWS forensics laboratory, set up in 2015, and forensic training for KWS first-response staff have significantly improved the ability of prosecutors to identify the source of seized ivory and track links along the supply chain from poachers in the field to middlemen and traffickers. Amendments to the security laws introduced in 2014 allow prosecution-guided investigations into all reported seizures, and establish the admissibility of electronic evidence (photographs and mobile records), canine evidence and DNA evidence in wildlife-crime trials.

The increasingly sophisticated data-gathering and intelligence capabilities of Kenyan law-enforcement agencies and their partner organisations are safeguarding elephants in the field and making it more difficult than ever before for poachers and traffickers to escape justice.

Working together to protect elephants

The NESC meeting in 2013 set the pattern for a cross-sectoral approach to the poaching crisis, a strategy that has been continued and deepened to the present day. The country is blessed with having numerous conservation organisations that specialise in elephants, focusing on fields of interest that range from research and management to social justice, law enforcement, intelligence gathering, investigations and prosecutions. The creation of the Kenya Conservation Alliance, an umbrella for dozens of conservation organisations, has created an opportunity for a unified voice from the conservation sector to lobby government.

The government, for its part, has embraced the opportunity to draw on the expertise of non-government organisations, both to advise on policy and facilitate the achievement of targets and goals. Since 2013, civil society and government agencies have been conducting regular dialogue meetings initiated by the NGO Africa Network for Animal Welfare (ANAW), which have helped galvanise inter-agency collaboration, particularly in the field of law enforcement. Partnerships between non-state and state actors have provided the framework for experts in civil society to work in concert with government agencies. Prime examples of these are between WildlifeDirect and the Judiciary Training Institute in monitoring of wildlife-crime trials and training of judges and magistrates, between KWS and Stop Ivory in organising the inventory and burning of ivory, and the KWS partnership with various NGOs in the tenBoma initiative.

Equally important has been the support of leading figures in the government for awareness-raising campaigns such as Hands Off Our Elephants and, more recently, NTV Wild, a partnership between the Nation Media Group, WildlifeDirect and KWS which, for the first time in Africa, broadcasts regular wildlife programmes on primetime TV.

In the field, space and habitats for elephants have been enhanced through the collaboration of landowners and communities in securing large tracts of privately and community-owned land through conservancies, and managing them for conservation. This rapidly growing movement under the umbrella of the recently created Kenya Wildlife Conservancies Association, currently includes over 200 conservancies, with a total area nearly equal to the 10% of Kenya's national territory protected by national parks and other state-run protected areas.

Conservancies are registered by KWS and, in addition to conservation and restoration of habitats for wildlife, also provide benefits to communities. They are supported by international donors and Kenyan NGOs, including Big Life, the African Wildlife Foundation (AWF), African Conservation Centre (ACC), Space for Giants and the Northern Rangelands Trust (NRT), as well as private-sector organisations.

Private entities and conservation organisations have enhanced the government anti-poaching effort by providing aerial and other logistical support for rangers, and by hiring and training hundreds of local people as rangers or scouts. Examples of community-based wildlife-protection initiatives include those of Big Life, AWF and IFAW in Amboseli; the Tsavo Trust and the David Sheldrick Wildlife Trust in Tsavo; privately owned Lewa and Ol Pejeta conservancies in Laikipia; NRT in Samburu; and the Mara Triangle and the Mara Elephant Project in the Maasai Mara. By employing scouts from local communities, these organisations have played a vital role in fostering goodwill and the support of local people for elephant conservation.

Challenges

Despite the major achievements made in bringing poaching under control, the future of elephants in Kenya is still not secure. As Richard Leakey has often said, defeating poaching is not a difficult problem *per se*: all that is needed is political will and sufficient resources. Elephants in Kenya, like those elsewhere in Africa, face a range of other, more intractable threats intertwined in complex ways with other continent-wide trends and wider socioeconomic, cultural and moral issues. These include human encroachment into elephant ranges and human-elephant conflict where farmers, understandably, attempt to protect crops, the loss of which could wipe out a family's annual food supply in a single night.

Ivory trafficking through Kenyan ports

Kenya – and specifically the port of Mombasa – continues to be the most important transit point for illegal ivory shipped out of Africa on its way to Asian markets. Kenya's strategic position on the supply chain gives the country a special role and responsibility in the global effort to crack down on the transnational crime cartels that control the ivory trade. Halting the trafficking of ivory is not only our duty as a 'good neighbour' to other countries in the region afflicted by wildlife crime. We know that, as long as we allow the traffickers to maintain a base on our soil, they will be ready to take advantage of any opportunity that arises to revitalise poaching operations in Kenya.

Halting ivory trafficking is becoming increasingly difficult because it involves many different agencies and ivory is difficult to detect inside locked containers, where it is usually hidden among other products. Criminal cartels use increasingly ingenious ways to hide their cargo. Recent seizures have found ivory inside containers of vegetables, in coffee, timber shipments, mixed with waste materials and submerged in oil drums – places where it can be detected by scanners. Unfortunately, not all containers can be scanned, some of the reasons being out-of-order scanners, and corruption. Some of the largest hauls have been made in tea containers, which are exempt from scanning and searching due to the perishable nature of the product. Kenya has made efforts to strengthen enforcement at the ports, including the introduction of detection dogs. This is co-ordinated by the new multi-agency Joint Port Control Unit in Mombasa, which has been set up as part of the UNODC-World Customs Organisation's (WCO) Container Control Programme. The unit brings together KWS, Kenya Forest Service (KFS), Kenya Police Service (KPS), Kenya Revenue Authority (KRA), and Kenya Ports Authority (KPA), whose officers work together and exchange information in real time.

However, despite these efforts, there was only one major seizure of ivory in Mombasa in 2016, while several other large consignments known to have passed through the port were seized on arrival by authorities in South-east Asia. We can only guess at the amount that continues to flow through the port undetected. Even when traffickers are caught and apprehended, the courts seem unable to enforce the law. Despite the widely publicised success in convicting the ivory trafficker Feisal Ali Mohammed, currently serving a 20-year jail sentence in Mombasa, other cases relating to major seizures remain stuck in the courts, in one case since 2012.

Corruption

The limited success in detecting ivory in transit and bringing known ivory traffickers to justice is certainly due, in part, to corruption among port officials and law-enforcement agencies. Ongoing efforts are being made in Kenya to reduce corruption among agencies responsible for the implementation of wildlife laws. A Corruption Prevention Committee set up at KWS has led to the identification of risk areas for corruption and the implementation of risk-mitigating measures.

The permanent presence of courtroom monitors at serious case trials now makes it much more difficult for corruption (such as tampering with evidence and misplacement of files) to remain undetected. Nevertheless, corruption remains endemic at all levels in Kenya, as in many other African and ivory-consuming countries, and is one of the greatest obstacles to success in combating elephant poaching and ivory trafficking around the world.

Human-elephant conflict

Even if poaching could be eliminated, elephants in Kenya and across Africa would face many other threats. These are linked to population increase and the rapid pace of socioeconomic development. Elephants, in particular, are affected by:

- the encroachment of urban and agricultural development onto traditional elephant range areas;
- illegal grazing of livestock in national parks and protected areas;
- overgrazing by livestock (inside and outside protected areas); and

- large infrastructure developments that cut through national parks and other elephant habitats.

These pressures are felt by elephants as loss, fragmentation and degradation of their habitats. Increasingly hemmed in on all sides by these developments, they are brought into ever closer contact with humans in surrounding areas. As a result, human-elephant conflict (HEC), is emerging as the greatest threat to elephant populations in many parts of Kenya. Elephants not only damage crops and property, they sometimes kill people who get in their way, leading to often deadly retaliation attacks. In the Amboseli ecosystem alone, more than 10 people were killed by elephants in 2016 and, though the data has not been released, many more elephants were killed by irate communities. According to the World Wide Fund for Nature (WWF), about 30 people are killed by elephants each year in Kenya, as a result of which, the authorities kill from 50 to 120 elephants each year.[11]

A number of initiatives are under way in Kenya to mitigate HEC, including the use of tracking collars to provide early warning when elephants approach human settlements, and deterrent methods such as broadcasting the sound of bees to keep them away.[12] However, the only lasting solution to HEC is for humans to adopt sustainable patterns of development and resource use that leave elephants with the space to get on with their lives. Advances in elephant tracking and monitoring are providing vital inputs for the identification of elephant-dispersal areas and corridors so that these can be incorporated into spatial planning and regional-development programmes.

At the same time, Kenya urgently needs innovative ways of effectively rewarding communities and people living in elephant range areas. At present, local people bear disproportionate costs of protecting elephants, which are a national (and global) resource.

Community ranger programmes are helping to redress this imbalance by generating employment for local people in elephant protection. In northern Kenya, NRT Trading is a for-profit social enterprise that is exploring innovative ways to link wildlife conservation to better pasture, higher income and the growth of sustainable businesses within the Northern Range Trust (NRT) conservancies. The aim is to move from aid dependency to financial self-sufficiency as a sustainable model for the management of the 33 conservancies, covering more than 32 000 square kilometres under the NRT umbrella.[13]

Outlook

Kenya is an innovator in elephant conservation. Ivory burns, campaigns to mobilise public support, courtroom monitoring and capacity building for law-enforcement officers are all, to some extent, Kenyan initiatives that are being replicated in neighbouring countries.

The most significant achievement of the combined efforts of the state, the private sector and civil society organisations is that elephants are no longer a minority concern in Kenya. Success in combating poaching has been achieved by transforming elephant conservation into a mainstream political issue.

Nevertheless, Kenya will remain complicit, albeit unwillingly, in the ongoing slaughter of African elephants until it can secure its borders, bring the ivory traffickers to justice and stop the transit of illegal ivory through its ports and airports.

Moreover, elephants will never be really safe until sustainable solutions to the problem of human-elephant conflict are developed that bring benefits to both elephants and people. We need to devise ways of creating wealth for our people that leave space for elephants as a continuing source of wonder and inspiration for future generations.

'When a country burns ivory it means that they put behind them any chance of underhanded dealings. It's not like pretending to have a ban and then selling the ivory under the carpet, which often happens. I think it was a very courageous decision and would suggest that donors buy up ivory and give the money to governments. Then I'd like to see that ivory burned.'

Iain Douglas-Hamilton CBE

Beneath Mount Kenya on the
Borana Conservancy.

© Elsen Karstad

32

Elephants of the north

This is the story of how a few thousand animals became the nucleus and catalyst for change. Now communities previously in conflict meet to plan a future for their people and their elephants.

Ian Craig OBE

A hundred years ago northern Kenya's elephants were a massive interconnected population of many thousands of animals ranging across today's Ethiopia, South Sudan, Uganda, Kenya and Somalia. They moved freely, unrestricted by humans, as the seasons dictated, living with the privilege of dying of old age.

The ecological connection between the vast untouched mountain forests, massive swamps, grassland and deserts were seamless, with elephants using the entire diversity of habitats with which northern Kenya is blessed. They were part of a rich and diverse ecosystem with no borders.

A story told recently by an elder from the Pokot community on the Kenya/Uganda border, where a small remnant population of 600 elephants remains, captures this movement perfectly: 'Elephant bring rain and grass; they move with the seasons bringing grass seeds from Sudan, ensuring our cattle always have grass and are fat. Since the elephants were killed our grazing has gone.' His story captures perfectly the threats facing northern Kenya's elephants: ivory value, human overpopulation, climate change, the challenge of livestock and declining rangelands.

Over just three human or elephant generations, this tranquil seasonal movement has changed. In the past, Arab, African and European ivory traders pursued elephants for their ivory. In response, elephants learnt to avoid man. The number of elephants killed for ivory or meat by the Il Ngwesi hunter-gatherers was small and, in the overall regional population, the offtake was well within sustainable limits.

In the 1950s and 1960s, low ivory values and strong colonial law enforcement kept poaching at bay and allowed elephants to flourish across northern Kenya. The area became internationally renowned for its massive tuskers. It was the focus of big-game hunters seeking a bull elephant with ivory over 100 pounds (45 kilograms). These hunts established the notion of 'safari' in its purest form.

Famous names like Roosevelt, Eastman and Johnson flocked to northern Kenya, their exploits recorded on film and in writing, providing evidence of a time in history when man and animals coexisted. One professional hunter, Andrew Holmberg, killed 34 bulls with ivory exceeding 100 pounds over a 15-year period.

© Michael Poliza

'If we are serious about conservation and about saving wild animals, we have to accept that African leadership must drive it.'

Dr Paula Kahumbu
WildlifeDirect

A sky island, surrounded by plains that form part of the Mathews Mountain Range in northern Kenya. These are known as the Lenkiyio Hills, and are home to the Samburu people.

The Ewaso Nyiro River in Isiolo County flows down from the slopes of Mount Kenya to water the dry plains that stretch east from Kenya's Great Rift Valley.

'Earlier, 100 000 elephants lived in Kenya and we didn't have a noteworthy problem with it. The problem that we have today is not that there are more elephants.'

Dr Richard Leakey

At this time, Kruger National Park was celebrating its Big Seven 100-pounder bulls. The Mathews Mountains in Kenya hid comparable groups of big bulls, including old elephants that had traversed the mountains all of their lives. The genetics of big ivory was strong and the habitat perfectly supported these wandering giants.

The devastation begins

Following Kenya's independence in 1964, a brutal war raged between Somalia and Kenya. Somalia claimed parts of northern Kenya, but the newly independent Kenyan government saw its country's boundaries as sacrosanct and protected them forcefully. The war was gradually won by Kenya. But 'shifta' (armed Somali bandits) turned their guns on elephants for their commercial value. The devastation that followed was almost beyond comprehension and elephant numbers plummeted. Politically protected cartels flourished openly, moving ivory across the country to the international airport under police escort.

Northern Kenya's elephants fled from their historical Eden to the refuge of a small number of national parks such as Meru, Samburu, Buffalo Springs and the safety of private ranches within Laikipia. Lands that had seldom, if ever, seen elephants, suddenly became crowded with animals seeking safety among people. Most of the large bulls were killed and family units were disrupted and divided, leaving disjointed and un-associated animals joining together in large, defensive but unstructured concentrations.

The Mathews Mountains, where author Robert Ruark reported seeing 2 000 elephants in one day, were emptied. Prime habitat with water and food in abundance, which had once echoed to the trumpet of happy, satisfied elephants, fell silent, with not even the tracks or signs of a single animal. By 1980, the Mathews Mountains forests and valleys were virtually lifeless, the vegetation having become a matted barrier, impenetrable in the absence of the elephants' 'gardening'.

The Boni Forest on the north Kenya coast adjoining Somalia – Kenya's most remote, intact and untouched forest complex, once home to over 20 000 elephants – was not spared. Within a short time numbers were reduced to fewer than 200 individuals, fugitives hidden in the deepest parts of the forest and venturing out to feed only at the dead of night.

The poachers were tough ex-soldiers who had fought and lost a bush war. In the elimination of elephants, they were thorough. But what must have begun as a frenzy of killing, feeding tonnes of ivory caches, must have ended as days of searching for the last remaining elephants for the sake of a few kilos of ivory. Yet their determination was relentless.

By 1989, northern Kenya's elephants were a sad and disparate population, devastated through poaching, crowded into small areas, where the presence of humans afforded a certain safety. Being so compacted, they destroyed their small habitat. They also made enemies, raiding agricultural land adjoining the protected areas. This hastened the spiral of destruction of this remnant population, along with any hope for a stable future.

Elephants that had never met a fence now met barriers to their movement. After the freedom of hundreds of thousands of square kilometres for centuries and across generations, they were now contained. They responded by discovering their tusks were non-conductors of electricity, destroying the electric fences restricting them.

A new beginning

It was this small remnant population of a few thousand animals that would be the nucleus and catalyst for change. The country's Ivory Burn in 1989 was the first turning point and an innovative community-led model started shaping the future of not only northern Kenya's elephants but the principle of conservation across the area.

As a result of this burn, the world became aware of the plight of Africa's elephants and ceased to buy ivory. Financial returns to poachers and their brokers for killing an elephant fell from $150 per kilogram to $20 per kilogram, reaching a point where it was no longer worth the effort and risk of poaching. It bottomed out the market. Slowly, over two decades, the deeply rooted norms and historical movement of elephants started to return. Among the herds were a handful of matriarchs who knew the routes trodden by their ancestors. They knew the last water sources that remained by the end of the dry season, and where the most nutritious vegetation was when the rains arrived, filling waterholes originally established by the herds of a previous era.

The bones of their forefathers, poached by the shifta, had crumbled to dust. The bark of the massive *Newtonia* trees so favoured by elephants, scarred and peeled for centuries, had now grown back undamaged after 20 years without them. Slowly the impenetrable valleys of the Mathews Range were opened up with contoured elephant paths – interconnected highways through the forest – that had been untrodden for years.

These elephants were understandably cautious. Families had been disrupted, social bonds broken. Groups of unrelated individuals built bonds that would grow into families. Matriarchs emerged as leaders among these newly formed groups. Males bonded together as cheeky, petulant youngsters without the wisdom or guidance of the 40- and 50-year-old bulls that could pass on the traditions learnt from their forefathers.

Gradually, northern Kenya's elephants became colonisers and opportunists. They tested the edge of their range and found safety as they expanded, pushing into areas where they'd not been recorded since the early 1970s. Human communities that had never seen an elephant found they had new large, grey neighbours.

By 1990, technology was coming into play, initially as a research tool among scientists eager to learn more about where elephants moved and why. VHF radio collars monitored from the air started to reveal secrets previously known only to matriarchs and bulls; inherited knowledge evolved over time to keep the herds safe and well fed.

As technology progressed and access to satellites became available, it was discovered when elephants watered; and how, when and why they would 'streak' at night from one safe area to another. Scientists could see how elephants came together in massive concentrations during wet periods. They could trace where the well-trodden paths were aligned, all contoured perfectly to suit the movement of these animals trekking thousands of kilometres every year in search of the best habitat and social gatherings. Within a few years, these secrets enabled conservationists and local communities to provide added security to elephants throughout their range.

A new way of thinking

Kenya is inherently a capitalist country. By 2002, the concept of conservation capitalism through communities was emerging as a new tool, an opportunity for communities and for peace in northern Kenya. It grew roots among the diverse pastoralist communities of the area. Elephants were to become the greatest beneficiaries of this evolved thinking.

In Kenya, conservation has historically been the preserve of government, with small but effective private sector involvement. With 70% of Kenya's wildlife outside of protected areas, communities owning the land on which this wildlife depended were the key to the future of elephants. For this reason, the survival of northern Kenya's elephants now rests firmly in the hands of the local communities.

In 2004, these communities came together to establish the Northern Rangelands Trust (NRT). It was community owned and led, multi-ethnic in composition, development orientated in its objective. It directly links conservation to the everyday lives of communities, is transparent, and managed to world-class standards.

This was the first time in East Africa's history that conservation and community development had merged and demonstrated success. It showed that stopping elephants being killed could improve lives and stimulate employment in areas that traditionally had none. It proved that conservation could generate new economies for livestock, facilitate further revenues from tourism, bring clean water to local people and enable additional education opportunities for communities aspiring to move from traditional pastoralist-based livelihoods to modern lifestyles. It opened a door to professional career paths.

Kenya is a country plagued by ethnic tensions. Adding to this, northern Kenya has an abundance of illegal firearms flowing from South Sudan and Somalia. This is a toxic mix for wildlife, people and development. But, against expectation and against all odds, NRT has harnessed this situation to build a force for good.

Highly trained and well-resourced community policing teams are now protecting people and wildlife. Communities previously in conflict are meeting to plan a future for their people and their

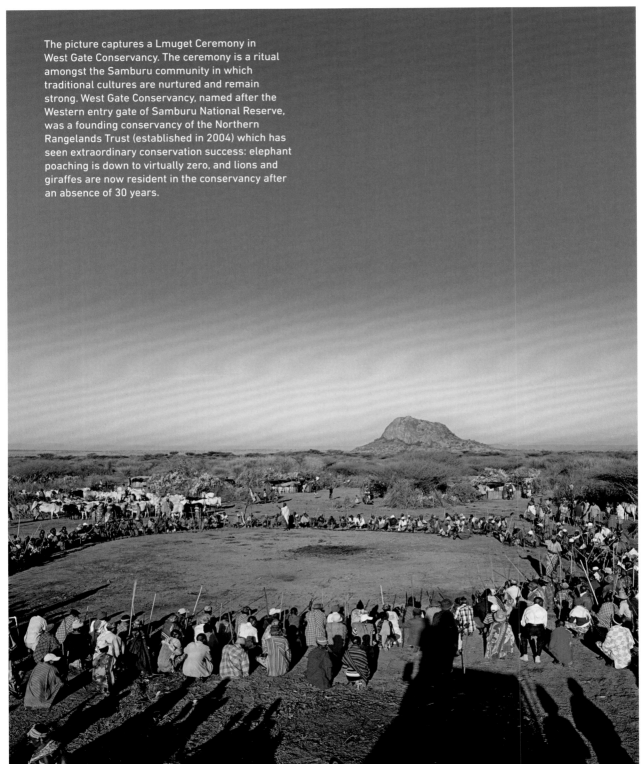

The picture captures a Lmuget Ceremony in West Gate Conservancy. The ceremony is a ritual amongst the Samburu community in which traditional cultures are nurtured and remain strong. West Gate Conservancy, named after the Western entry gate of Samburu National Reserve, was a founding conservancy of the Northern Rangelands Trust (established in 2004) which has seen extraordinary conservation success: elephant poaching is down to virtually zero, and lions and giraffes are now resident in the conservancy after an absence of 30 years.

© David Chancellor

elephants. Partnerships with national and county government agencies are evolving and having a lasting impact. Novel economies based on wildlife are developing, providing new job opportunities for young pastoralists.

Where elephants were previously harassed and shot, they are now accepted as part of the landscape. Elephants have rallied to the opportunity, moving east and north out of crowded parks and private ranches of Laikipia, back into the mountains and deserts in which their ancestors once thrived.

As Kenya's human population expands, land pressure will continue to increase, water will become scarcer and, inevitably, elephants will be pushed to the peripheral land with no agricultural potential. While national parks are sacrosanct, private land is under severe threat. Northern Kenya, however, still has huge areas of prime elephant habitat, adequate water, and is unmarked by fences on land owned by pastoralist communities.

The landscape may in time be spoilt by ribbon development, oil pipelines, roads and high-speed railways, but I believe that under the present custodians, there will still be adequate connected habitat between protected areas and communities to ensure the expanding elephant population has a stable home for the next 50 years.

The 12-kilometre corridor joining Mount Kenya National Park, Lewa Wildlife Conservancy and the Samburu/Buffalo Springs/Shaba complex demonstrates the adaptive nature of the elephant. With careful planning, it is possible to maintain connectivity. In the case of Mount Kenya, this corridor reconnects 2 000 elephants, isolated by agriculture for 40 years in the Mount Kenya National Park, to an estimated 7 000 elephants in the Samburu/Buffalo Springs/Shaba complex.

What has been preserved are the genes of northern Kenya's famous tuskers: Ahmed, Mohammed and the many others whose ivory was shipped to make trinkets for affluent and poorly informed markets. Bull elephants with ivory of over 100 pounds are emerging from the Mathews Forest – individuals completely unknown to the scientists and researchers studying northern Kenya's elephants.

On Marsabit, an area that once had up to 500 elephants, the 150 remaining animals are thriving. Poaching for ivory there is at an all-time low, with only small numbers being killed for cultural wedding ceremonies in Ethiopia. The connectivity and movement between the Mathews Mountains, Marsabit and Mount Kenya are now restored. Elephants are again in the South Horr Valley and Ndoto Mountains where they were once reduced to zero.

As northern Kenya's elephants move into a fresh phase of expansion, they take pressure off the vegetation within the protected areas and private reserves. While protected areas will remain at the core of their range and go-to places in times of drought and trouble, the bulk of the population will adapt to populate the vast areas of rich elephant habitat of north-eastern Kenya. Pastoralists and elephants there will learn to co-exist, share water and grazing, although, undoubtedly, there will be areas of high human-wildlife conflict, specifically within Laikipia and the south-western edges of their range.

There is a good chance that breeding herds will re-establish their social and migratory patterns, knowing where to go in which season, joining up during the rains into concentrations of over a thousand in small but rich environments, enabling social interaction that ensures the diversity of the population. In the dry season, they will split up and move back to their dry-season ranges in the parks or now-safe forested mountains.

Groups of old bulls are re-establishing their home ranges, living in small areas with little need to move until their time of musth, when they choose to find the breeding herds. These bull groups are growing in number as the population ages. Groups of 20 or more big, mature bulls – something not observed across much of northern Kenya for decades – are becoming a regular occurrence.

An important requirement for this stability, however, is that ivory must become valueless. It is vital to ensure that the abundance of firearms and the greed of cartels do not once again focus on northern Kenya's elephants. There is hope: China and the United States have banned ivory trading and, through the Elephant Protection Initiative, there's growing recognition among African range states that the continent needs a cohesive and common stand on the sale of ivory – that a sale by one country stimulates demand and catalyses poaching across the continent. Right now, under the custodianship and care of the pastoral communities, the future of northern Kenya's elephants looks bright.

'I've focused on elephants all my working life. But you also have to get deeply interested in politics, community issues, human welfare – because man is the most important determinant of the elephant's future.'

Iain Douglas-Hamilton CBE

The Kauro River in the Sera Conservancy, where annual elephant sightings have risen from 3 000 during pre-conservancy days to 11 000. Grévy's zebra sightings have increased from fewer than 250 to over 2 000.

© Anthony Njuguna

The way it was in the 1960s and 1970s: a legally hunted 142-pound (64-kilogram) tusker in Selous Game Reserve, Tanzania. It had the 8th-longest tusks ever recorded – 3.1 and 2.98 metres. Luca Belpietro's interest in nature began as a young boy spending time in the bush on hunts with his father (pictured here). But the world is now a very different place, and today Luca has dedicated his life to non-consumptive, non-hunting conservation projects in Kenya.

© Luca Belpietro Archives

33

The elephant and the kid

In the Chyulu Hills people and wildlife have learnt to work together. Tribesmen are aware that elephants are worth more alive than dead.

Luca Belpietro

'Coki, stay!' my dad tells my beloved black-and-white pointer. I'm frightened. Everything's dark outside. I can see stars through the mosquito net of the tent window. Now that I'm about to be by myself, I am no longer so sure this was such a great idea.

I still remember the smell of the night. The fresh air.

I still remember the crickets' serenade …

This is my first night in a tent by myself. My dad says good night. I crawl near Coki, excited, but scared … and finally fall asleep. I wake up only once, hearing some distant barking. Not of zebra, but of dogs. I'm not in Africa, but in the vineyards of our country house in Italy. It's 1968 and I'm 4 years old.

This was my first adventure. It took another 7 years before my dad took me with him on a real African safari. His safaris, in the 1960s and 1970s, were hunting ones. In those days that's how things were. There were no small tourism camps and lodges, just a few big lodges in very few locations; and there was no room in a hunting safari for a young kid.

My dad finally took me to Tsavo and Amboseli on a photographic safari in 1975. I remember it

vividly. A stray memory comes to mind, of being in trouble because the driver couldn't brake at a river ditch. Though I'd been told countless times not to, I had climbed onto the roof of our Land Rover, eager to see everything around me: the zebras, the giraffes, the baobab trees; and so, when the driver tried to brake, I was abruptly catapulted from the roof … years later, serendipity brought me to live exactly where that incident happened, and today I cross the same river ditch with my safari guests, telling them the story of a naughty boy getting reprimanded by his dad.

I remember first visiting the Maasai village, the same one I now visit with my guests. I still have a photo of that visit: my mom, my brother and me. In the background is Longido and the Chyulu Hills, which I now call home, and have done so for the last 22 years.

Those hunting days with my dad seeded my passion for Africa. Without them, I would not have been living and working with the Maasai. And I wouldn't have renounced my Italian citizenship to proudly become a Kenyan. This is not unconnected to the legendary elephant my dad shot in Selous in October 1970. His hunting forged my dreams of Africa.

The rolling Chyulu Hills were formed by lava flows (most recently in 1856) and are some of the youngest mountains anywhere. From their base on the dry savannas, the Chyulus rise 2 188 metres to their mist-shrouded peaks.

The huge elephant you see in the photo on page 348 was in the dusk of his life; probably in his late 60s or perhaps even 70 years old. His tusks were enormous, one more than 3 metres long, the other just short of that: the 8th-longest ever. This story is not about my dad's trophy, but what that elephant and those tusks have meant to me.

Through my dad's hunting, I learnt the complex dynamics of African conservation. Things are never easy or easily understandable in Africa. For example, the national park that borders Kuku, the Maasai Group Ranch where I live, once had more than 30 000 elephants[1] and 8 000 rhinos. The 1960s and 1970s saw an upswing in poaching to meet a corresponding increase in the global demand for ivory. As a result, one day in 1977, hunting was abruptly banned.

The expected protection of wildlife was not achieved. Quite the opposite happened. The safety network provided by professional hunters, who were present in the hunting blocks, was not substituted by similarly effective policing and anti-poaching. Illegal hunting skyrocketed. In a decade, the elephant population plummeted to 6 000 and the rhinos were practically annihilated.

Does this mean we should re-open hunting in Kenya? Absolutely not! Kenya of today is not the Kenya of 40 years ago, nor is the world the same as it was. There were fewer than 20 million Kenyans at the end of the 1970s; the number is now approaching 50 million. People and wildlife are increasingly in competition for the same land and for the scarce resources of the African savanna: grasses and water.

How can we make local people understand that a foreigner with a hunting licence is allowed to come in for a couple of weeks and hunt antelope, buffaloes, lions and even elephants, while the people co-existing with that wildlife cannot – and would be arrested for doing so?

Sustainable tourism is the best conservation tool we have. Especially in the Maasailand where I live, between the iconic national parks of Tsavo and Amboseli, where big tuskers still roam free.

They're protected by the same people with whom they share the land, and it's a story worth telling. For me, it began with my dad's hunting, but it ended with protecting wildlife through a tourism and conservation partnership with the 15 000 Maasai of Kuku Group Ranch, on 283 000 acres of pristine African wilderness.

In my early twenties, I began spending all my summers in Kenya. With its teeming nature and its wonderful people, I realised I could live nowhere else. I was sure of that. On completing a thesis on sustainable development and environmental conservation in Kenya, I engaged the local Maasai community of Kuku and suggested a tourism and conservation partnership. For the first time, Maasai landlords in the area could earn a direct income from their wilderness and wildlife. In these 'Green Hills of Africa', celebrated by Ernest Hemingway, I built my home: Campi ya Kanzi, a community boutique ecolodge.

There was not a road; we (my girlfriend Antonella and I) did everything from scratch. After two years in a small tent, and with lots of hard work, the lodge was built and our dream came true. A wedding and three kids later, we happily call this paradise home.

Africa tests you – constantly. But we're stubborn and, so far, haven't given up. There were some extremely tough moments, such as when a fire destroyed the lodge, coinciding with the birth of our daughter Lucrezia.

Here in the Chyulu Hills, we have demonstrated that the wilderness teeming with wildlife is worth protecting precisely because it pays economic dividends to the Maasai landlords in a sustainable way. To enhance the impact of our community lodge, we created the non-profit Maasai Wilderness Conservation Trust. It employs 265 Kenyans, providing education, health and conservation services for the local community.

The desire to make a difference is what got us where we are, and we're moving ahead by developing a carbon project based on the protection of the Chyulu Hills cloud forest. Payment for ecosystem services is simple and ensures that natural resources are economically productive for the people living around them. Tourists pay a US$116 (at the time of writing) conservation fee, per guest per day, to compensate the Maasai for livestock losses caused by wildlife, and we are working on a similar fee for water. We also have a carbon credit payment for ecosystem service. (The Chyulu Hills is a critical water catchment area. A pipeline at the foothills of the Chyulu takes millions of litres to Mombasa, Kenya's second-largest city.)

In two decades, through our tourism and conservation partnership, we have proved to the Maasai that wildlife is worth protecting. It sends their children to school (we employ 56 teachers and support 22 schools), it pays for the doctors and medical care, it creates employment, and it compensates for livestock killed by predators.

Elephants, lions, cheetahs, leopards, giraffes, zebras, antelope have become assets to the Maasai community. We don't have to invest in drones and we don't need to go after poachers – the community keeps them out by patrolling their land and defending *their* wildlife.

Instead, we invest in teachers, nurses and other such service providers. And we have found that one initiative leads to another. Three years ago a guest at Campi ya Kanzi asked how she could help support Maasai girls. We suggested a secondary-level scholarship. We knew of a brilliant student and were keen to assist her in attending a good high school. The guest agreed. The girl's father, we discovered, had just sold 50 goats to send his daughter to a boarding school near us. We offered her the scholarship and he was delighted: his daughter would get an education, and he could get his goats back. That same guest has now agreed to pay for the girl's university fees.

A month later a Maasai was killed by an elephant outside his boma at night. The community demanded a retaliatory killing of an elephant. The father of the girl we had helped immediately went to the village and addressed the emotional crowd. He said something I couldn't have brought myself to say: 'If you get drunk and walk at night in the wild, you might get killed. The problem here is not the elephant, the problem is drinking and then wandering out in the dark. My daughter is in school because of elephants and you have no right to kill any of them.'

More recently, a young man was killed by an elephant, his body torn apart. He was simply in the wrong place at the wrong time. A waterhole near cultivated areas attracted elephants, which raided the nearby crops, and the herd was set upon by the community. The young man found himself in the path of the elephants as they tried to escape, and he was attacked and killed. This left two young children without a father.

The community asked to kill the elephant. Our community rangers, along with Kenya Wildlife Service rangers, the police and our Group Ranch chairman, managed to appease the rightfully angry crowd and get them to agree that no elephant should be killed. In spite of the community's distress, they were able to appreciate that the elephants had not simply turned on the young man, but that the animals themselves had been under attack. Only by knowing that elephants are among their greatest assets were they able to view the tragedy in this way.

These two stories are among many, where sometimes painful community decisions have taken into account the value of wildlife, and shown their support for long-term ecological sustainability.

Here in Kuku in southern Kenya, we believe that humans and wildlife can, and should, co-exist, and we are doing our best to ensure that wildlife remains a productive resource for the local landlords. There's no need to fence off elephants from the few watering holes that support farming: resources such as water are scarce, and wildlife needs them as much as humans do. Where conflict of interest arises, alternatives can be pursued.

The people of the Chyulu Hills are aware that humans and wildlife are interdependent, that elephants and other iconic African species are worth more alive than dead, and that it benefits all to live and work together in this magical corner of Africa.

My story, which began with a boy and a dog in a small tent in Italy, has seen me nurture and fulfil a dream to protect Africa's diverse wildlife. Children instinctively know the importance of Earth's creatures – the elephants, lions, giraffes – and that this Earth is for sharing. In the Chyulu Hills, we have found a way.

'I had to measure every single tusk, which is a great source of information about the sex and age of the elephant. We gathered data from about 3000 tusks and drew a timeline, as every tusk had a marker of when and where it was seized. What the timelines showed was alarming. The tusks got smaller and smaller. This showed poachers were going for smaller elephants, calves essentially.'

Dr Paula Kahumbu
WildlifeDirect

© Campi Ya Kanzi, Chyulu Hills, Kenya

34

Urgent intervention needed to save forest elephants

Gabon is perhaps the only place in the world where you can see gorillas, elephants and whales on the same day. It's a treasure that needs protection.

Wynand Viljoen

When the conservationist Michael Fay discovered the Minkébé Archipelago, he thought he'd found Africa's ultimate primal wilderness. One evening in 1999, while on his MegaTransect fact-finding expedition, he left his team in camp and scrambled to the top of an isolated granite inselberg with a 360-degree view of the surrounding landscape. When he reached the summit, the setting sun was casting shafts of golden light into the steaming rainforest. 'I felt complete bliss', he wrote in his journal, 'like I was alone on a virgin planet'.

His ecological census and information-gathering expedition lasted for 455 days, during which he walked over 3 200 kilometres through some of the most inhospitable and least-explored tropical rainforests of Central Africa. It cost him bouts of malaria, tropical illnesses and multiple elephant charges, but for the first time the world was alerted to the rich diversity of wildlife in the region.

The country is on the equator, nearly 10% larger than Great Britain, but with a population of only around 2 million, most of whom live in the main towns and cities. Dense tropical rainforest covers more than 80% of its land area, where over 600 species of bird and 200 species of mammal live,

including iconic species such as elephants, sitatungas, chimpanzees, western lowland gorillas and mandrills.

As a result of Fay's MegaTransect discoveries, 13 national parks were established to protect the region's extraordinary natural heritage. Gabon's national parks are breathtaking and diverse, varying from primary and secondary forest, swamp forest and coastal forest to mangroves, each with rich biodiversity. It is there that 50% of Africa's forest elephants live.

Minkébé National Park, created with Fay's assistance, had the highest density of forest elephants in Central Africa at the turn of the century, and was considered a critical sanctuary for them because of its relatively large size and isolation. These secretive animals were once thought to be a subspecies alongside their savanna family, but genetic analysis recently found them to be a separate species, *Loxodonta cyclotis*.

Fay's paradise was not to last. A survey by John Poulsen, Fay and others comparing elephant numbers in the park and surrounds between 2004 and 2014 showed a dream on the way to becoming a nightmare. In 10 years, Minkébé had lost from 78% to 80% of its elephants – more than 25 000 –

Gabon is perhaps the only place in
the world where you can see gorillas,
elephants and whales on the same day.
It's a treasure that needs protection.

Paul Augustinus. Oil on canvas, 61 x 122 centimetres

Koungou Falls, Ivindo National Park, Gabon.

undoubtedly to poaching.[1] The population in 2004 was calculated to be 32 851 elephants; by 2014, that had dropped to 7 370. 'Their loss from the park', wrote the researchers, 'is a considerable setback for the preservation of the species. The documentation of significant declines in forest elephant populations is not new, but a 78 to 81% loss of elephants in a single decade from one of the largest, most remote protected areas in Central Africa is a startling warning that no place is safe from poaching.'

It appears that poaching from within Gabon reduced elephant numbers in the south of the park, while poachers from Cameroon emptied the northern and central sections. Cameroon's national road lies just 6 kilometres from Minkébé at its nearest point, making access to the park relatively easy. Cameroon plays a major role in the ivory trade, with Douala serving as an important exit point for ivory. In 2011, the National Parks Agency (ANPN) expelled over 6 000 illegal immigrants, mostly Cameroonians, from an illegal gold-mining camp within Minkébé. The site was a hub of criminal activities, including poaching, originating from the Cameroonian town of Djoum.

Protection a priority

The high level of elephant poaching is not an indication of government neglect, but of the sheer scale of the poaching operations. Gabon is a country serious about environmental protection. The ANPN was initially underresourced and understaffed. But following reports of increased poaching, the government raised the status of the forest elephant to 'fully protected', doubled ANPN's budget and created the National Park Police. In 2012, Gabon became the first Central African country to burn its ivory stock.

In 2002, the late President Omar Bongo Ondimba created 13 new national parks covering 25 000 square kilometres – around 10% of the country. At the time, this was considered a politically risky move. However, his audacious decision was rewarded with a US$53-million grant from the US government for the Congo Basin Forest Partnership, a regional alliance of governments, conservation groups and industries.

In 2017, President Ali Bongo Ondimba, Omar's son and successor, announced the creation of 20 new marine protected areas covering 26% of the country's territorial waters. Around 20 species of whale and dolphin are regularly seen in Gabonese waters. The marine proclamation has ensured the protection of one of the world's largest breeding grounds for leatherback and olive ridley turtles.

Most conservationists had believed that a combination of the remoteness of the country's parks and low human population in rural areas would be enough to protect Gabon's wildlife and, in particular, the forest elephants. But the Poulsen survey showed that inaccessibility and remoteness in Gabon has not been a deterrent to determined commercial poaching syndicates, especially along its borders with Cameroon and Congo.

There *have* been some notable successes on the part of the authorities, such as in Wonga Wongue National Park. Under Fay's guidance, the government evicted Sinopek, the Chinese oil company that was operating there without many checks and balances. After its eviction and the arrest of the region's poaching kingpin, elephant poaching inside the reserve was halted within a year.

A significant problem is the commercial and cultural demand for bushmeat. The geography and climate of Gabon are not suited to cattle, so bushmeat is the natural protein substitute, both in the cities and around logging, oil and mining concessions in remote areas. These concessions drive access roads into forests, which attract new human settlements and, with them, hunters and poachers.

For forest animals, especially elephants, the conflict with humans occurs on yet another front. Because of a drop in oil prices, local companies are laying off staff who are returning to their ancestral villages and to subsistence farming. These farms and small villages are generally clustered along the forest edge, where plantations are raided by elephants. In the conflicts that follow, elephants always pull the short straw. Emergency call-outs to villages to deal with snared elephants are becoming increasingly common. In many cases, the elephant can be released with minimal effort and side effects, but in others, the animal must be put down.

Gabon has experimented with different ways of solving this conflict, from peri-peri shotgun ammunition to chilli hedges, bees and noisy bells. It seems, though, that the only reliable solution is electric fences, which are now being tested at a few sites. Although this works, fences are slow to erect

and are labour-intensive, maintenance is ongoing and installation is expensive. A countrywide satellite-tracking and -collaring programme has been started to record which corridors elephants use between parks and protected areas. With this information, Gabon's park authorities and scientists hope to offer better protection for elephants or even create new protected areas.

The magic of Loango

Unlike Minkébé, Loango National Park along the Atlantic coastline in the south of the country has not experienced major poaching. With its open savannas, coastal forest, swamps and beaches, the park is perhaps the only place in the world where you can see gorillas, elephants and whales on the same day. It's one of the few places in Africa where you are able to watch and photograph elephants, buffaloes and sitatungas on the beach, while hippos surf the waves. It is for good reason the park is described as Africa's Last Eden.

Loango receives only a few hundred visitors each year, mainly hardier travellers wanting to view lowland gorillas or sport fishermen looking to lure the powerful tarpon on a catch-and-release basis. The park is also a bird haven and most species found in the Congo Basin can be viewed here.

Loango, however, is not without its poaching incidents. In June 2017, poachers killed two large male elephants on the banks of the Akaka River. In response, Loango intensified its patrols, manned roadblocks and did random checks along the rivers and lagoons. At the time of writing in mid-2018, no incidents of new carcasses or poaching had occurred in the previous 9 months. Part of this success stems from the use of informants who have been carefully nurtured over the years, and who help keep poaching down to manageable levels. They only get paid when the information they supply leads to an arrest, and the amount paid is linked to the value of their information.

A recent success story was when an informant knocked at my door around 02h00. Half asleep, I asked what he wanted. 'They're going to hunt tomorrow near Bon Terre and they will be there with a dugout canoe', he said. With two of my rangers on the patrol boat at first light, we searched the banks of the Iquela Lagoon and, as my informant had predicted, we found a dugout with an outboard, hidden in the mangroves. Following their tracks, we spotted four poachers, two of whom were armed. Being unarmed, we would lose that fight, so we backtracked to our boat and waited in ambush.

They returned, loaded their boat with their kill and headed into the lagoon. When they were midway, we suddenly rammed their dugout and all four, plus spoils and weapons ended up in the water. We hauled them aboard and into handcuffs, then dived for the weapons as evidence. All four landed in jail.

Not all information leads to arrests. We were approached by a driver who'd been fired from a nearby Chinese logging company. He told us the company smuggled ivory and iboga wood hidden on trucks transporting pallets of legal cut wood (iboga is a rare and endangered tree in high demand for its roots that have psychedelic qualities). The driver explained how loggers fitted hidden compartments inside wood pallets. We searched their logging trucks, but didn't find any secret compartment. However, we hope our thorough inspections made the loggers more circumspect about trying to smuggle forest contraband.

Looking ahead

Gabon has the potential for a thriving tourism industry. It's a 7-hour flight from most European centres and offers the sort of magic only a tropical rainforest can. What it now needs to do is attract investors and experienced safari professionals to target the high-end/higher-priced, low-volume/low-impact sector of the international wildlife and safari tourism industry. Tourism will mean more eyes and ears on the ground and can help keep the poachers at bay.

For this to happen, it's necessary – urgently and effectively – to conserve and protect the last remaining forest elephant strongholds. The Poulsen report, quite rightly, called for listing of the forest elephants to Appendix I, and to recognise them as Critically Endangered on the IUCN Red List:

'The international community must recognise the species [in order] to engender the multinational support necessary to prevent its extinction', it reads. 'To save elephants, nations must co-operate by designing multinational protected areas, co-ordinating law enforcement and prosecuting nationals who commit or encourage wildlife crimes in other countries.'

For the sake of the forest elephants, the time for forceful and decisive action is now. Not a moment should be lost.

© Wynand Viljoen

Apprehending an illegal Chinese fish-poaching (and sometimes ivory-smuggling) trawler offshore of the national park is not a normal park ranger's job.

Opposite, top

Lopé National Park is situated in the centre of Gabon and became the country's first protected area in 1946. The park covers over 4 000 square kilometres of mainly tropical rainforest, which merges into savanna grasslands in the north. Primates are well represented, including chimpanzees, lowland gorillas and large aggregations of mandrills. The park is a UNESCO World Heritage Site.

Opposite, bottom

Langoué Baï, located in Ivindo National Park, is the largest forest clearing in Gabon. Forest elephants congregate here in large numbers, especially during the dryer months, and their trampling helps to keep the baï open for other animals like sitatungas and lowland gorillas.

© Lee White

© Paul Augustinus

© Robert J Ross

© Lee White

© Robert J Ross

Conflict gold

In 2011 the Gabonese National Parks Agency and an elite paratrooper regiment closed an illegal gold camp that had grown from a few hundred to almost 7 000 illegal immigrants mining at an industrial scale within 2 kilometres of Minkébé National Park. They found organised criminals trafficking gold, ivory, bushmeat, drugs and arms, as well as women and children.

The Gabonese authorities treat poaching and trafficking as not just a wildlife issue but as a threat to national security and it is thanks to this that they are winning the battle.

'Elephants along with other endangered species could be driven to extinction unless western governments begin to take the illegal wildlife trade as seriously as terrorism or drug running. We cannot win this battle alone ...'

President Ali Bongo Ondimba of Gabon, calling for a concerted international intelligence and law-enforcement effort to break up the trans-national criminal groups who now dominate the trade in ivory

Gabon's Ivindo National Park was one of the 12 new national parks created by President Omar Bongo in 2002. The highlights of Ivindo (located in the central north-west of the country) are Langoué Baï and the spectacular waterfalls of Koungou, Mingouli and Djidji. This is Djidji, with a drop of over 60 metres, making it the highest in equatorial Africa, and of great spiritual value to local people.

Field rangers in
Garamba National Park.

© Marcus Westberg

35

Garamba National Park: conservation on the continental divide

What it takes to keep elephants alive in the northern DRC's war-torn tribal lands

Naftali Honig

The Garamba savanna in north-eastern Democratic Republic of the Congo (DRC) slopes down from the border with South Sudan. Before lines were drawn on maps, it was simply the boundary between two great African river basins. The park's northern edge is a continental divide: from hills to the north, water flows into Africa's longest river, the Nile; and through the savanna and woodlands to the south run the headwaters of the Congo, Africa's largest river by volume.

From outer space, the Garamba National Park, the heart of a vast complex of protected areas, appears as a light patch in Africa's deep equatorial shades of green. Natural maintenance occurs in the form of seasonal bushfires and the grazing and browsing of large mammals, especially elephants. Like water and fire, the elephant is one of the park's natural elements.

Elephants have never recognised artificially imposed borders, and neither have many of the human actors of the region's tumultuous past. The hilly border acknowledges streams and rivers but holds no consistently enforced law over its inhabitants. This subtle ridge bisects tribal lands, like those of the Kakwa and the Zande. Some say Garamba was originally a refuge for wildlife *because* it was itself a frontier between different societies, a no-man's-land

boundary that ignored the maps of colonial reverie. Whatever the reason, elephants populated Garamba in great numbers and helped shape the land.

The park's first systematic aerial survey in 1976 reported over 20 000 elephants. Today, the elephants still gather in immense herds and their stomach rumbling and trumpeting is an awe-inspiring sound. One can only imagine what they are thinking and communicating to each other at these grand gatherings. What lies within their cells is fascinating too: genetic evidence suggests that Garamba elephants are hybrids of the savanna species and their forest cousins.

Garamba is an ecological and cultural crossroads, where lion prides overlap with communities of chimpanzees. In a corner of the bush graced by the DRC's last giraffes, a researcher recently took the first known photo of a water chevrotain in the area. Nilotic peoples to the east converge with Bantu tribes of the forest. Over the last couple of decades, Falata pastoralists from further north have expanded southwards, now grazing their livestock even south of the DRC's Uele River. Nomadic raiders from faraway Darfur regularly visit the area. Their periodic forays are centuries old yet, no longer scouting for pasture, most seek only products and profit.

Garamba's natural beauty makes it a spectacular tourism destination, but the security situation needs to improve before it fulfils its potential.

© Marcus Westberg

ACAN PARKS

A/P

AN PARKS

AUF
OPEN

If poachers are spotted in the park, Garamba's rangers are usually deployed by helicopter, enabling them to strike quickly and without much warning.

'Wildlife trade is big business ... it needs big, innovative solutions to make sure that it does not threaten species and undermine the livelihoods of the poorest communities.'

Allen Crawford
TRAFFIC

Historic routes, fortified by the Ottoman Empire, established a firm taproot into Central Africa, with raiders penetrating deep into the heart of the continent for slaves and ivory. A single powerful trader, al-Zubayr Rahma Mansur, 'reigned over the country', in his own words, in the mid-19th century, exporting northwards everything from human beings and gold to elephant tusks across an empire reaching from Zandeland to Darfur.

Today, outside of the continent's great deserts, the continental divide is one of the emptiest areas of Africa. Its inhabitants were long ago dragged to Khartoum and sold on to the Middle East. Young women were taken from their families and sold as servants or concubines. Men were wrestled from their societies and forced to be soldiers or doorkeepers in societies not their own. Europe half-heartedly pushed for the end of slavery, for the sake of its conscience, but kept the parallel ivory market flowing.

With populations depleted, the demand for slaves wound down in the 20th century, but the demand for ivory did not. Across Africa, trade routes penetrated into the continent's rich interior and, as both technology and infrastructure improved, ivory from great tuskers that had walked the banks of the Garamba River left north-eastern DRC, first on the shoulders of men, then as cargo in aircraft.

Another, parallel story of the Garamba elephants unfolded: attempts at domestication. Lieutenant Pierre Offerman, who founded the Gangala-na-Bodio elephant training station in 1927, noted that 'it had taken centuries of patience and effort to make the Asian elephant render service to man; and people expect the African elephant to be as good as the combustion engine after only a few years'.[1]

Maintaining such a training centre proved challenging, with periodic conflict shifting priorities. (A nostalgic revival was attempted in the late 20th century, but these ambitious domestication efforts were incongruent with the challenges of protecting the park.) The unusual initiative did, however, result in some of the first black-and-white photos of Garamba; and the Gangala-na-Bodio training centre became part of the park's unique heritage, along with the presence of northern white rhinos and Kordofan giraffes. This fuelled growing momentum for Garamba's conservation and, on 17 March 1938, the Belgian Congo gazetted Garamba as its second national park.

The killing fields

Garamba National Park lies about 1 800 kilometres from the capital, Kinshasa, and clings to the north-eastern edge of the DRC. Along with a number of other protected areas to the north, it is situated within the sphere of influence of Sudan. For centuries, Sudanese horsemen moved deeper and more frequently southwards as elephants were extirpated to the north. In the old days, elephants were hunted by spear. However, ongoing civil wars in Sudan have resulted in a proliferation of weapons and ammunition. Now killers carry more deadly weapons. And while their camels don't fare well as far south as Garamba, their rugged horses manage.

Destabilising cultures within reach of Khartoum's long arm has been the hallmark of successive Sudanese regimes, generating massive loss of life and power vacuums deep in the interior of the continent. Amidst the chaos, organised and well-armed poachers have emerged, who have siphoned off untold tonnes of ivory. They have dramatically reduced the elephant populations that had survived the ivory trade into the second half of the 20th century, and eliminated the northern white rhino from many of its Sudanese strongholds.

The power vacuums saw combatants in these conflicts, such as the notorious Lord's Resistance Army (LRA), often playing more than one role. Originally a rebel group from Uganda, in South Sudan the LRA changed its focus. In both contexts, the LRA was a viable proxy for anyone wishing to destabilise first Uganda and, later, South Sudan (not yet independent at the time). Armed, equipped and adjacent to the park, they drifted into Garamba with its rich natural resources and chipped away at decades of conservation efforts. This included the trafficking of ivory, which was far more lucrative than pillaging rural communities.

In 2008, when Ugandan forces went on the offensive, launching Operation Lightning Thunder against LRA camps in the national park's savanna and adjacent forested reserves, most LRA fighters escaped and went on a rampage. They killed 150 people in Faradje and attacked Garamba National Park headquarters in Nagero on 2 January 2009. The leader of the LRA, Joseph Kony, established himself in nearby Kafia Kingi, where Sudan, South Sudan and Central African Republic meet. Here he traded ivory and sought an alliance with armed forces in Sudan.

The ongoing instability in South Sudan has proven complex and dangerous for humans and wildlife alike, both before and after its independence in 2011. As recently as late 2016, the former vice-president and hundreds of rebels holed up within Garamba National Park, having retreated from fighting in Juba and the Equatorias (a region of southern South Sudan). A complex operation was undertaken to get them out as swiftly as possible, and no elephants were lost. While the impact of civil wars in the DRC itself has decreased in the new millennium, Garamba, because of its position on the map, is forced to remain vigilant. It's a daunting challenge to establish a sense of stability in the park, knowing that threats may be brewing hundreds of kilometres away. On Garamba's savanna, the elephants are aware they're on a precarious island of stability.

Far too often, bullets have ended the lives of rangers dutifully fighting to protect the park's wildlife. Increased infrastructure has redistributed and diversified illegal flows of ivory, but it begins with a burst of gunfire or a well-placed round piercing the elephant's thick hide. Without a doubt, the Garamba savanna falls within one of the most anarchic corners of the planet.

Seeking solutions

In the early days of Garamba as a protected area, conservation was the work of naturalists. Legal protection generally focused on the commercial value of a species. But from early in the 20th century, the animal's natural history started to be a consideration. In the new millennium, and against this historical backdrop, a novel initiative in protected-area management was unveiled.

In 2005, the protected-area agency Institut Congolais pour la Conservation de la Nature (ICCN) began to work alongside protected-area management specialists of the non-profit African Parks in a public-private partnership to manage Garamba and its adjacent reserves. For the DRC, the partnership was a seminal act of improving governance, inspiring confidence and building local capacity.

Full management of a protected area precludes the relegation of conservationists to mere pieces of the puzzle, such as ecology. African Parks management can investigate the finest granularity of the challenges and seek solutions, knowing that we can act upon them. This includes a strong emphasis

For decades it was believed that only Asian elephants could be domesticated, with one exception: those at the African Elephant Domestication Centre at Gangala-na-Bodio, a small settlement just south of Garamba in the DRC. Most of the elephants in these 1947 photographs were born around 1930 at the domestication station.

© H. Goldstein, Congopress and African Park archives

'Despite all the human destruction on the planet, there is still a natural order and it is necessary for us to do everything we can to protect that.'

Kate Brooks

on capacity building. Local communities are empowered across a wide range of skills. To them, Garamba and the protection of its elephants is not just conservation, it's empowering local people to be a part of the stabilisation of north-eastern DRC. The problems and solutions in Garamba are the problems and solutions of the people.

Some have accused this type of conservation as 'militarisation' – an unfortunate misunderstanding. The professionalisation of rangers is their future and the future of the area. Their ambition is not to conduct counter-terrorism or counter-insurgency campaigns; but, for the sake of their communities and elephants, they often unintentionally find themselves caught up in such campaigns. While debate continues, so does the plunder: the northern white rhino has been lost, the illegal trade in minerals remains a significant source of funding for many armed groups on the continental divide, and elephants remain in their sights. Professionalising Garamba's personnel is a moral imperative.

Of the weapons needed to succeed, understanding is undoubtedly the most powerful. Rather than regarding poachers on our borders as faceless enemies, we must know them and understand their lives. Dialogue has proved critical in facing challenges coming from over the horizon. In mastering our understanding of Garamba and all of the diverse forces drawn to it, a more mature humanity will be able to restore its relationship with elephants to the equilibrium it deserves, and honour Garamba for the World Heritage Site that it is.

Studies confirm that there are clear genetic and physical differences between forest and savanna elephants. However, Garamba National Park in the DRC is one of the few areas in Africa where the elephants show characteristics of *both* species, yet they remain fertile and able to breed through generations. Scientists attribute this unusual hybridisation to poaching and pressures induced through habitat changes.

© Marcus Westberg

'What I have seen in a lifetime of roaming the tropics in quest of unspoiled nature has dispelled any complacency and replaced it with panic. What I see are inadequate parks, unstable societies and faltering institutions. Nature cannot long survive any of these debilities: all three in combination add up to a recipe of extinction.'

John Terborgh
Requiem for Nature

© Marcus Westberg

An African Parks ranger displays a pair of 70-pound (32-kilogram) tusks taken from the carcass of a magnificent bull killed by poachers in Garamba National Park in the DRC.

36

Odzala-Kokoua
National Park

The Odzala-Kokoua is another African Parks success story, achieved against considerable odds. Odzala is located in the north-west of the country and was proclaimed in 1925, making it one of the oldest national parks on the continent. The park forms an important component of the Congo Rainforest Basin (the second-largest rainforest basin after the Amazon) and is an important refuge for forest elephants.

This and the next four pages are a photo essay of Odzala, photographed by Marcus Westberg of Life Through A Lens Photography.

Pages 384–385

Odzala-Kokoua's elephant population has suffered heavily from targeted poaching, dropping from over 18 000 in 2000 to fewer than 10 000 by 2014. Under African Parks management the tide is beginning to turn and forest elephants are now among the park's star attractions, along with its population of around 22 000 western lowland gorillas.

Pages 386–387

At 13 500 square kilometres, Odzala-Kokoua is one of the continent's largest, most diverse protected areas – and one of the most difficult to manage. Bushmeat consumption in the surrounding communities is high, and patrolling is a challenge. Many parts of the park are accessible only by boat or on foot. At park HQ, confiscated weapons, ammunition and animal parts, including ivory, are kept in this secure lock-up.

The Last Place on Earth: Nouabale Ndoki National Park

Kyle de Nobrega

There are not too many places left on Earth today that have an unbroken link with an ancient world where we can still tap into pure, primeval energy. It makes sense, then, that to reach such a seemingly improbable place, the sturdiest travellers need to travel very far, across harsh terrain and by many modes of transport – planes, boats, hardy Landcruisers and eventually, and most importantly, using our legs, feet and backpacks to get there.

Incredibly, such a place does exist, where highly intelligent Great Apes live, unaffected by the contamination and traumas of the outside world. In this place, observations of Central African chimpanzees expressing honest curiosity and awe when confronted with humans have given rise to the term 'naive chimps'. Their calm reactions to rare human explorers tell best how remote this part of the world is.

Here, Earth's forests have lived in harmony with ancient rhythms, linking all the way back to the planet's origins, perhaps to when there was only a single landmass. This place is called Nouabale Ndoki, a World Heritage Site deep within the Republic of Congo at the heart of the giant Congo Basin. No adjectives adequately describe Nouabale Ndoki: a shimmer of the ancient natural world nestled under a grey sky, dissected by black rivers and covered by the thickest of deep green forests, inhabited by forest elephants, primates, parasites and sweat bees.

The fact that this unique place still exists is largely thanks to the work of a dedicated and patient team under the management of the Congolese government and in partnership with the Wildlife Conservation Society. These champions of conservation have, for 25 years, had the vision and determination to work in a tough and most challenging part of Africa, protecting what has been called 'The Last Place on Earth'.

Bongo antelope and forest elephants in Dzanga Baï, Central African Republic.

© Andrea K Turkalo

37

The other African elephant

Exploring the secretive world of the Congo Basin's elusive and charismatic forest elephants

Andrea K Turkalo

The historic range of forest elephants (*Loxodonta cyclotis*) was from the forests of Senegal in West Africa to the Congo Basin of Central Africa. The decline of both forest and savanna elephant numbers started with the ivory trade in the 15th century and was amplified by the West and Central African slave trade. Slaves transported tonnes of elephant tusks from the interior to the coast, from where both humans and ivory were sent to Europe and the Americas. In this exodus, both savanna and forest elephants were exploited, with no distinction made between the two species.

Following the abolition of slavery, forest-elephant range continued to decline as human populations and their expansion into wilderness areas increased. At one time, human settlements were islands surrounded by elephant range; this has steadily been reversed. In Central Africa, human expansion has been driven, in part, by extractive industries such as commercial logging and mining, enterprises that attract people and provide the possibility of economic development where there is high unemployment.

This has put tremendous pressure on all species of wild animal, which are not only a protein source where livestock and animal husbandry is minimal or absent, but also provide income for local populations as commercial bushmeat and ivory. Coupled with extensive civil unrest, this has led to an increased availability of arms, resulting in an increase in poaching across forest-elephant range.

In West Africa, the range of forest elephants is now limited to several small populations, with the last great bastion being the forests of the Congo Basin. The highest number occur in Gabon. Census data determined that between 2000 and 2011, their numbers declined by more that 62% over most of their range. They are presently at 10% of their potential, occupying a mere 25% of their prospective range.

The total number of forest elephants has been estimated to be one-quarter to one-third of the total number of elephants in Africa, with current estimates of around 100 000. As of 2011, when the data for forest elephants was analysed over all of its range, Gabon contained over 50% of the total, followed by the Democratic Republic of the Congo (DRC) with 19%, Republic of Congo with 20%, with Cameroon and the Central African Republic (CAR) at 7% and 2% respectively. Poaching continues and numbers are declining.

Dzanga Baï is probably the most spectacular and interesting clearing in any African forest. Sitting high in an elevated hide is a magical all-day affair, watching relays of elephants, forest buffaloes, giant forest hogs, sitatungas and rare bongo antelopes.

Dzanga Baï clearing in the CAR

My personal experience with forest elephants began almost 3 decades ago at a unique site, the Dzanga Baï clearing in south-western CAR. Dzanga is one of the few places in the region where it's possible to observe this species over extended periods of time. Located in the Dzanga Ndoki National Park, it is a naturally occurring forest clearing, which attracts hundreds of elephants as well as other species of mammal. It was known as the place to observe forest elephants in the open, but no one had attempted a full-time study because of the remoteness and difficulty of working in the area. With this window into the forest-elephant world, my study yielded the first long-term data set on a previously unstudied species.

Starting with basic drawings, my team composed identification cards – a technique used to study savanna elephants –which enabled us to follow the lives of individuals. The area was unlike anywhere I'd ever worked: a small patch of primary forest with high densities of both forest elephants and other mammals. A definite plus was the company of Bayaka pygmies, who not only knew the forest but enjoyed spending extended periods in an isolated area. Annually, they spend several months of the year in the forest hunting by traditional methods, and for the rest of the time, live in permanent villages along roadsides.

Working from a field camp in the forest 2 kilometres from the clearing (so as not to disturb the animals), we made daily observations. Though we tried to minimise our impact, the elephants were never far away, with several well-known individuals making forays into our camp.

At the beginning of the Dzanga study we knew nothing about forest-elephant numbers or behaviour. The big-picture study was to identify as many individuals as we could and track their life histories. The biggest obstacle was the setting. Unlike savanna studies, where you can drive to areas to find elephants, the clearing was a fixed setting. Although we observed from 40 to 100 elephants daily, there was no way to predict the presence of specific individuals. Our task was simply to wait. There was never a guarantee that the same elephant would reappear in the clearing, and some were more seasonal than others, although we saw about 80% of the 4 000 we

identified a second time or more, and many made the area around Dzanga their range.

As we augmented elephant identities, we trawled through ID cards daily until we were able to identify individuals without their aid. On the cards we recorded salient features, such as unique holes and tears in the ears, and other features such as the form of the tail, body scarring and gender. Aging them was difficult, but as time went on we began to read their relative body sizes and were able to use this information too. Once we became aware of newborns, we were able to follow their physical development and behaviour, enabling us to pinpoint their exact ages.

Knowing identities gave us a connection to this elephant population. By observing these amazing animals over several decades, we saw in them a commonality with human existence. These were wild animals, but with family attachments, extended relationships and a rich emotional life that not only brought them joy, but also trauma. Reading literature that scoffed at the notion of animals' emotional life made me wince. Surely the authors had never made long-term observations such as we had? I found their view ludicrous, that an animal such as an elephant, with its complex nervous system, prodigious memory and emotional attachments, would be incapable of feelings for another of its own species.

Perseverance began to pay off. After 5 years of intensive observations we had identified thousands of individual elephants, many more than I had anticipated. Patterns began to emerge. The elephants had a daily cycle: during the morning, numbers in the clearing decreased, but in the afternoon and throughout the night they entered the clearing in numbers sometimes exceeding 150 individuals. Daytime observations were therefore done in the afternoon while light was available.

Because we were seeing so many elephants, including a constant flow of new individuals, our impression was that all elephants in the area eventually visited the clearing. However, after visiting other areas in the vicinity of Dzanga, we began to realise this wasn't the case. During the mid-1990s another study in northern DRC, based on the Dzanga methodology, made it clear there were many elephants that had never been seen at Dzanga. We did, however, log 80 elephants that were travelling between the Dzanga site in the CAR and the adjacent Nouabale Ndoki National Park in the DRC, a distance of over 80 kilometres. One individual, Basil, the son of a well-known adult female seen at Dzanga, regularly made the trip between the two sites.

Dzanga Baï was not only the first study of forest elephants, but it was instrumental in providing data on other forest species such as bongo and giant forest hogs. Monitoring of clearings is invaluable to park management because these sites provide a window into areas where, in surrounding forest, monitoring is done on foot and visibility is difficult.

One finding of the study has been that forest elephants are one of the slowest-reproducing animals on Earth. It overturned the assumption that their reproduction paralleled that of the savanna species. Based on lengthy observations of known individuals, we found that forest females not only conceive less frequently, but start reproducing much later than savanna elephants. Compared to savanna reproduction, with a generational rate of 20 years, for forest elephants it's 60 years. This reproductive pace, coupled with continued poaching, does not bode well for their survival.

The forest canopy at first light at Dzanga Baï is alive with the deafening chirps of hundreds of African grey parrots. And then, with one swoop, they are on the floor for a quick drink before heading back off into the forests.

Colin Bell

Forest vs savanna elephants

Like savanna elephants, their forest cousins are highly social and form close family bonds based on a matriarchal model. In Dzanga there were extended family groups and, although poaching has disrupted social structures, we still found groups numbering up to 24 individuals. But the more typical group was an adult female and two offspring.

Adult males are solitary and have never been observed in bachelor herds. In the clearing they compete with each other over the best mineral sites, with the biggest males generally dominating. Young males leave maternal groups at as early as 5 years old, probably because of the absence of large predators in the forest. They greet their mothers even after permanently leaving their family groups.

Forest elephants are smaller than their savanna relatives, with adult females averaging 2 metres at the shoulder, males 2.4 metres and the tallest male 2.8 metres. The quality of ivory also differs: forest ivory is harder and preferred by artisans.

The future of forest elephants

The survival of forest elephants will depend on management of protected areas staffed by well-trained personnel. However, many areas are under-staffed, skills are deficient and parks not sufficiently funded. Many national parks in Central Africa exist on paper only, and there is little or no management or protection.

While rich in natural resources and human capacity, the region is rife with civil unrest, and poor governance has led to decades of instability. Natural resources are being exploited at an unprecedented rate, driven by government corruption.

Extractive industries such as mining and logging are short-term activities profiting a few and leaving most people in the same state as – or poorer than – they were before. It's a boom-and-bust economy. Well-managed resources would assure a sustainable future for even the poorest, and a chance for wildlife to survive.

With the precipitous decline in forest elephant numbers, their future does not look good. Without considerable improvement in the economic condition of local populations and a return to peace in the region, the conservation of forest elephants and other wildlife species poses a massive challenge for both national governments and NGOs.

© Andrea K Turkalo

38

Ivory and terrorism

Brent Stirton

Containers with 4 tonnes of illegal ivory were confiscated in January 2014 by the Togolese customs office from its new deep-water port, Lomé, Togo. This ivory has been directly linked through DNA evidence to the elephant massacre that occurred in Dzanga Baï, Central African Republic, in 2013.

That massacre was perpetrated by Seleka rebels who climbed the observation towers at the famous forest-elephant gathering place in Dzanga Baï and gunned down the elephants with automatic weapons. This is the same observation tower from which Andrea Turkalo spent decades observing and learning about Dzanga Baï's elephants (see pages 390 to 397).

The Seleka rebels would have used the proceeds from this ivory sale to fund some of the violence that plagued the Central African Republic for much of 2013 and 2014. Togo, with its new deep-water port, has been viewed by ivory smugglers as offering new trafficking opportunities.

However, customs officers with new container-scanning technology have made the efforts of these smugglers more difficult.

Peace has returned to Dzanga Baï and the elephants are back. With new investment into Sangha Lodge, travel to this fabulous part of the Congo Basin is possible again and should be on every adventurer's bucket list.

Some of the many guns confiscated by Yankari rangers. These traditional muzzle-loaders are locally made and are known as Dane guns. They are lethal and are responsible for the deaths of many elephants.

© Natalie Ingle

39

The elephants of Yankari

With boots on the ground and smart tracking, the park's guardian angels are keeping alive the largest surviving elephant population in Nigeria.

Nachamada Geoffrey & Andrew Dunn

The elephant, proud emblem of the Nigerian Police Force and also of the oldest bank in the country, has declined dramatically across Nigeria and faces an uncertain future. Formerly widespread throughout the country, elephants have declined by more than 50% in less than 20 years and it's likely that fewer than 300 survive across Nigeria today.

Although it was traditionally believed that both savanna and forest elephants occurred in West Africa, recent genetic studies suggest that a single form occurs here, whose taxonomic status is yet to be determined. This suggests that West African savanna elephants are very different from those of eastern and southern Africa and warrant special attention. Only small pockets remain, most notably in Cross River National Park, Omo Forest Reserve and Okomu National Park. The large herd that formerly roamed throughout Adamawa and Borno states has now disappeared. One bright spot where this trend has recently been reversed is Yankari Game Reserve in Bauchi State.

Yankari

Yankari Game Reserve is in north-eastern Nigeria, the country's richest wildlife oasis. It contains the largest surviving elephant population in the country, estimated at 100 to 150 individuals, and one of the largest in West Africa. In addition to elephants, Yankari supports many other endangered species, such as the West African lion, slender-snouted crocodile, hippopotamus, buffalo, leopard and several species of vulture. The reserve is dominated by the perennial Gaji River which bisects it and provides an important dry-season refuge for the park's large population of various antelope.

Originally created as a game reserve in 1956, Yankari was upgraded to a national park in 1991 and was managed by the National Parks Service until 2006, when responsibility was handed back to the Bauchi State government. It has long been a popular tourism destination and, although recent insecurity in northern Nigeria has affected tourism revenues, there are signs that the situation is gradually improving.

The park covers 2 244 square kilometres of the southern Sudan savanna zone. Its vegetation is characterised by *Burkea africana* and *Combretum glutinosum* trees and woodland savanna, with an open canopy and continuous layer of annual and perennial grasses. Tree and shrub density is high, while the grass is dense and tall in the wet season. Along the Gaji River Valley, there's a

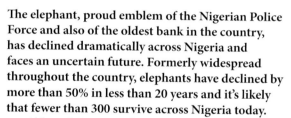

Natalie Ingle

With a WCS-donated vehicle, ranger teams can cover more ground, but most of the real work has to be done on foot.

swampy floodplain bordered by patches of swamp forest, gallery forest and riparian woodland. The topography is rolling, hilly country with altitudes from 200 to 640 metres above sea level.

Yankari forms a large basin and almost all streams in the area flow towards the Gaji River at its base. The river and its tributaries form the only watershed and the only source of permanent water during the dry season – late December to April. At this time of year, large herbivores are concentrated in the valley, a critical area for both wildlife and tourism.

More than 50 small communities – 100 000 to 150 000 people from around 15 different ethnic groups – live along the edges of the reserve. Most of the local population are Muslim; literacy levels are very low and they are generally extremely poor, living on less than US$1 a day. Agriculture is the mainstay, with the crops being largely sorghum, millet, maize, rice, groundnuts and cassava. The rearing of livestock is also important, with large numbers of cattle moved into the area each dry season by nomadic Fulani herdsmen.

Threats to Yankari

From 2006, when the Bauchi State government took over management of the reserve, it was neglected and poorly funded. Poaching increased and it's likely that large numbers of elephants were killed during the period up to 2014 to supply Nigeria's ivory trade (the domestic ivory trade being legal). In addition, community support for conservation diminished following crop damage caused by elephants to farms on the reserve's boundaries.

The main threats to the reserve are hunting of large antelope and warthogs to supply the bushmeat trade, elephant poaching for ivory, illegal livestock grazing and human-wildlife conflict. Bushmeat is mainly for commercial purposes and is sold to middlemen who transport it to larger cities, where it is highly prized. Ivory poaching by specialist hunters from outside the region has been a problem in the past, particularly from 2006 to 2014, when management and protection of Yankari were neglected.

The Wildlife Conservation Society

The goal of the Wildlife Conservation Society (WCS) is to conserve the world's largest wild places in 16 priority regions, home to more than 50% of the world's biodiversity. It has supported conservation and related research in Nigeria since 1996.[1] It initially focused on surveys of the critically endangered Cross River gorilla and biodiversity surveys of key taxa. It has expanded this programme and now works in five different areas in Nigeria: the Oban and Okwangwo divisions of Cross River National Park, Afi Mountain Wildlife Sanctuary and the Mbe Mountains (all within Cross River State) as well as Yankari Game Reserve in Bauchi State. It supported the creation of Afi Mountain Wildlife Sanctuary (2000) and the Mbe Mountains Community Wildlife Sanctuary (2007), and completed the first landscape-level survey of Cross River gorillas (2006–2008), and the first nationwide lion survey of Nigeria (2009).

Its work in Nigeria focuses on six species: the Cross River gorilla, Nigeria-Cameroon chimpanzee, elephant, lion, drill and Preuss's red colobus monkey. Its oldest and largest project in Nigeria is the Cross River Gorilla Landscape, contiguous with the Takamanda-Mone Landscape in Cameroon. The major focus of the Nigeria programme is to:

- strengthen law enforcement and monitoring within existing protected areas;
- support measures to tackle the illegal wildlife trade;
- promote conservation education and awareness in local communities;
- provide support for alternative livelihoods for local hunters;
- promote community conservation, including the creation of community-managed protected areas;
- encourage local participation in the management of protected areas;
- monitor elephants, lions and gorillas; and
- promote trans-boundary conservation between Nigeria and Cameroon.

WCS in Yankari

With funds from the US Fish and Wildlife Service, the Elephant Crisis Fund and others, WCS has worked in Yankari since 2009 and has established good working relationships with local communities. It has developed essential partnerships with key government agencies and gained a thorough understanding of the local ecology and political and legal landscape.

Their work in Yankari has focused on protecting the remaining elephants by providing support for anti-poaching patrols. In 2014, WCS signed a co-management agreement with Bauchi State government, and since then levels of protection have improved dramatically. Before 2014, an average of 10 elephant carcasses were discovered each year, most of them believed to have been killed illegally for their ivory. None have been recorded since May 2015. It is likely that elephants would be extinct in Yankari without the WCS's intervention.

Law enforcement and the ranger programme

Having recognised that it lacked the capacity and resources to manage support staff effectively, and having signed a co-management agreement with WCS in 2014, the Bauchi State government

Above

Rangers on patrol in Yankari.

Right

A key focus of Yankari's work is the elephants. Here WCS Landscape Director (and chapter author) Nachamada Geoffrey uses a radio receiver to help locate a collared Yankari elephant.

now employs about 100 rangers at Yankari. The WCS provides training, equipment, field rations, camping allowances and arrest bonuses. Four long-distance foot patrols are sent out each week from the central camp at Wikki, with close vehicle support. All rangers are armed with shotguns.

A key factor in the success of the ranger programme has been ensuring that rangers are paid a small camping allowance per night spent on patrol, and are provided with sufficient field rations to last for the duration of the patrol. These allowances supplement meagre salaries and have increased motivation and loyalty while reducing levels of corruption.

WCS developed an informer network which, since 2014, has provided valuable information leading to a number of arrests. In 2015 and 2017, with assistance from Conservation Outcomes, they also provided rangers with essential anti-poacher training, which has radically improved their effectiveness. Before training, rangers became demoralised and some were involved in corrupt practices, leasing out areas of the reserve for livestock grazing. Ranger morale is now much better and, thanks to improved training, the number of arrests has increased: 71 poachers arrested in 2014, 141 in 2015, and 97 in 2016. More importantly, the number

of elephant carcasses has declined dramatically: five in 2014, two in 2015, zero in 2016 and 2017.

WCS provides new uniforms, boots, rucksacks and basic field supplies, such as tents and sleeping mats. They lead by example and undisciplined rangers get transferred. Everyone on patrol has a functional firearm and sufficient ammunition. As a result, they are no longer afraid of poachers and none have been killed since 2012/2013. A reward system, where rangers receive a cash bonus for each successful arrest (roughly $25 for each hunter and $100 for elephant poachers), had proved to be a cost-effective means of improving protection programmes at other protected areas in the region, and was introduced in Yankari in 2010. The emphasis is on the arrest of elephant poachers and seizure of ivory.

The use of joint patrols with the army has also proved to be an effective strategy, helping to maintain the park's territorial integrity. Sambisa Game Reserve, only 300 kilometres from Yankari, is used as a safe haven for the militant Islamist group Boko Haram. Local security and early detection of armed gangs and cattle rustlers have kept them out of Yankari.

The importance of technology

The CyberTracker monitoring system was introduced in 2009 with support from the North Carolina Zoo. It dramatically improved the situation by enabling management to monitor and map coverage of anti-poaching patrols, as well as chart the frequency of illegal human activity in the reserve. The information is used to plan more focused patrols and to guide the reviews and revision of the reserve's patrol strategy. The system also provided improved knowledge of the geographical distribution and seasonal movements of elephants by tracking ranger encounters with them, or signs of their death.

In 2016, with ongoing technical support from North Carolina Zoo, the CyberTracker technology was replaced with SMART (Spatial Monitoring and Reporting Tool), a more powerful system, but still using the same hardware devices.[2] The technology generates a monthly patrol coverage map and is used for law-enforcement monitoring. It has had a major impact on the WCS's work in the reserve.

Livestock grazing

With rapidly expanding human populations across northern Nigeria and loss of traditional livestock-grazing refuges, illegal grazing of livestock inside the reserve is a growing problem, displacing elephants and other wildlife. To make matters worse, as livestock populations have increased, traditional grazing reserves have been converted to agriculture, further depleting available grazing. Each dry season, nomadic Fulani arrive with their cattle on the edge of Yankari, often sending cattle into the reserve. With reduced availability of wild prey for lions, these predators are now more likely to follow livestock herds, leading to human-lion conflict. The outcome is predictable, with lions being killed, often through poisoning, and probably in retaliation for livestock losses.

Relieving human-elephant conflict

Though the reserve boundary is generally respected by local farmers, groups of elephants from Yankari regularly cause damage to crops on surrounding farms and some individuals have been shot by irate farmers. With support from the Tusk Trust, WCS initiated an Elephant Guardian Programme in 2015 to provide training and support to locals approved by village chiefs. WCS also assisted local communities to mitigate elephant crop damage. Twelve elephant guardians are presently actively working in six villages generally considered to be those most vulnerable to crop raids. Guardians are provided with a monthly phone credit to alert the WCS office or Yankari rangers as soon as elephants are seen near their village. The guardians act as the first line of defence, deflecting elephants from crops before any damage is done.

WCS holds monthly meetings with all the elephant guardians to share information and discuss possible solutions. Further responsibilities of the guardians are to help to explain the importance of elephant conservation to the local community, and also to report cases of poaching. Over a 6-month period, they lodged 147 elephant-alert calls concerning 30 cases of illegal hunting, some of which resulted in arrests. With additional funding, it is hoped to build on this successful programme by exploring additional mitigation tools such as beehives, the use of chilli pepper fences and bright lights to deter crop-raiding elephants at night, as well as to engage with more communities experiencing crop raiding.

Education and schools outreach

Funding restraints have caused community outreach and support for conservation to be neglected. Some human-elephant conflict issues remain unresolved, and there's a general lack of environmental awareness. This has encouraged local communities to support elephant poachers from outside the region. In 2014, an elephant with a satellite collar was shot, probably in retaliation for crop raiding and as a result of local anger over the non-payment of compensation.

In 2017, WCS started a programme to enable groups of local schoolchildren to visit the reserve and to raise awareness about elephant conservation. The programme is popular and is empowering students with knowledge about elephant conservation.

Tourism potential

Yankari, which employs people from all surrounding communities, has great tourism income potential. It could indirectly benefit the state's population of over 4 million people through revenue generation. Local people may benefit from hotel employment and from selling crafts and farm produce to tourists. Strengthened law enforcement in Yankari has also improved security in surrounding areas, ensuring the park's integrity and keeping Boko Haram away.

Conclusion

Yankari's elephants are of inestimable value to both Nigeria and Bauchi State. Without them, the tourism

potential would be greatly diminished. A period of neglect saw the population plummet but they are now recovering. However, the reserve is still drastically under-funded, and rangers remain poorly paid for their efforts. In the long term, Yankari needs significantly more international funding and must develop stronger relationships with the surrounding communities. At the same time, Nigeria's domestic ivory trade must be banned in order to safeguard the country's remaining elephants.

Land is a key issue. Violent clashes between herders and farmers over disputed land are a regular occurrence, causing some states to propose drastic new anti-grazing legislation. The shortage of grazing land has meant that areas such as Yankari are subject to intense pressure, which is expected to grow. A long-term solution to this problem is required, perhaps through the creation of grazing reserves.

Climate is another problem. Insecurity caused by prolonged drought in the Lake Chad Basin is thought to have contributed to the rise of Boko Haram in the area, and increasingly extreme and unpredictable climate-change patterns are a risk for everyone. Both people and elephants will need to adapt to the changing conditions.

Opposite

Yankari elephants have very small tusks, probably because those with large tusks have already been poached.

Below

Drinking deep at a waterhole in Yankari.

© Natalie Ingle

Lake Chad's elephants

Kate Brooks

In 2014, African Parks searched by land and air, trying to account for all of the remaining elephants in Chad. Between the country's far-off borders with the Central African Republic, Cameroon and Nigeria, there were only small groups and herds identified outside of Zakouma National Park, where poachers killed 90% of the elephant population between 2005 and 2010.

The elephants living around Lake Chad near the country's eastern border with Nigeria are unique to the Sahel: they inhabit the desert, moving between water sources. From the time I first heard of their existence, I dreamed of seeing these elephants with my own eyes. I know they are special to this geographic location, and am also aware that the lake's surface is shrinking due to climate change, and that Boko Haram operate but a short distance away. After 2 days of flying over the desert and around the lake with Rian Labuschagne, I spotted the herd from the plane window and managed to capture this image.

Elephants Without Borders informed me 2 years later that the Great Elephant Survey concluded there were fewer than 650 elephants remaining in Chad, nearly 500 of which live in Zakouma. I wept, remembering the elephants that I had documented being treated for bullet wounds on that gruelling trip around the country, and the little groups we referred to as refugee populations on the borders. Chad's elephants are fighting for their survival, while becoming genetically isolated in different corners of the country.

A patroller and his hardy Kanem Bornu horse in Zakouma National Park.

@ Kyle de Nobrega

40

Zakouma: an elephant success story

Each elephant area in Africa is different, but in Zakouma, after a long battle in a dangerous region, the rescue recipe has worked.

Lorna & Rian Labuschagne

In the decade before we moved from Tanzania to Zakouma National Park in south-eastern Chad in January 2011, poaching of elephants there had been horrific. The park, covering 3 054 square kilometres, is one of the last-remaining intact wildlife strongholds in the Sudano-Sahelian ecoregion, a band across North Africa between tropical forests in the south and the dry Sahel in the north.

Armed poachers were coming from as far afield as the Darfur region of the Sudan, the *modus operandi* of the horseback hunters not having changed much in 100 years. In the past, these hunters would isolate an elephant and stab it in the groin with long spears, harassing and lancing the animal until it slowed down from internal damage. These days they use automatic rifles and shoot indiscriminately into a tightly packed, panicked herd, causing massacres of up to 60 animals of all ages at a time.

Based on historical survey information gleaned from Zakouma's archives, in the year 2000, there was an influx of elephants into the park, boosting the numbers by over 2 000 animals. Where did they come from and why? There was a possibility that they had fled the Gounda/Saint Floris and Bamingui-Bangoran national parks in the Central African

Republic (on the border with Chad) because of heavy ivory poaching there. At the time, Zakouma had offered more security. In 2002, Zakouma's elephant population peaked at an estimated 4 350 individuals. Over the following decade, it was in continuous decline, with the highest losses between 2006 and 2008 when around 2 000 elephants were killed.

Although the African savanna elephant is the mega-herbivore that receives much of the conservation focus, the park and surrounding reserves are critical for the protection of many other rare or threatened species in northern Central Africa, many of which are extremely rare, or extinct, elsewhere in their range. These include African wild dogs, cheetahs, lions, red-fronted gazelles, Kordofan giraffes and red-necked ostriches.

We arrived at Zakouma two months after African Parks Network had taken over park management, in partnership with the Chadian Ministry of Environment. A year prior there had been yet another massacre of elephants in the west of the park with at least 27 individuals of all ages killed in a single incident. With a known 39 elephants killed by poachers that year, the situation was dire. At that rate, Zakouma's elephants could be extinct in a few years.

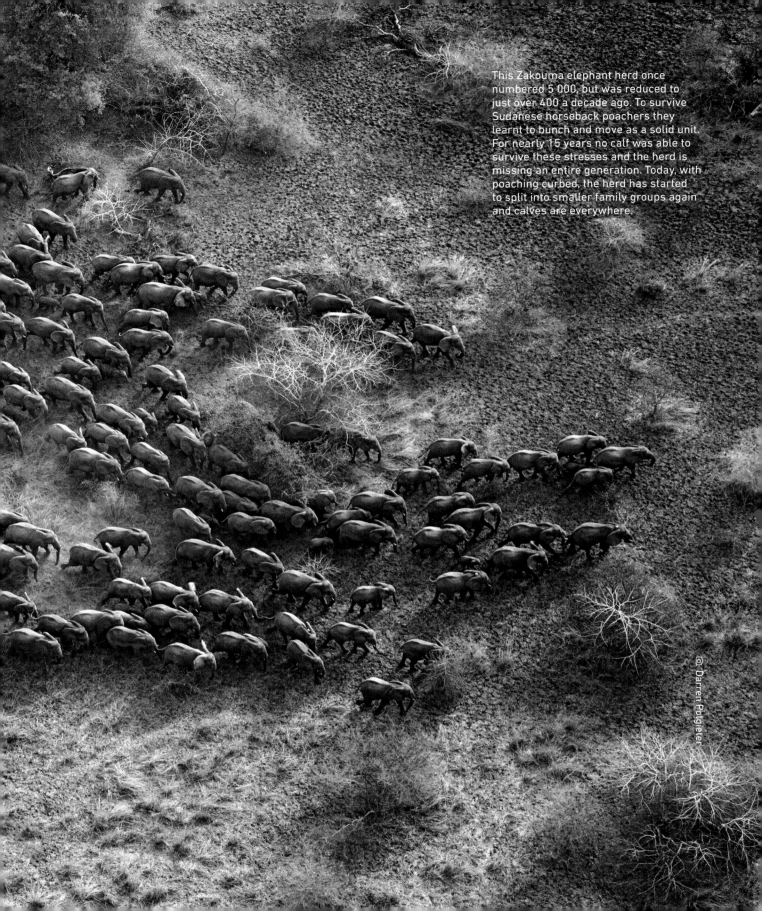

This Zakouma elephant herd once numbered 5 000, but was reduced to just over 400 a decade ago. To survive Sudanese horseback poachers they learnt to bunch and move as a solid unit. For nearly 15 years no calf was able to survive these stresses and the herd is missing an entire generation. Today, with poaching curbed, the herd has started to split into smaller family groups again and calves are everywhere.

© Darren Potgieter

Rian Labuschagne at his command and communication centre from where the park's anti-poaching activities are co-ordinated.

The Public-Private Partnership signed between the Chadian government and African Parks was the first positive step in a new direction. It gave us the mandate to fully manage the park, with the European Union providing the critical core funding, as they had done since 1989. But there were many more steps to go on the long road towards the target of zero ivory poaching.

Rudimentary information on elephant movements was that herds migrated out of the park to the north and west at the beginning of each wet season and returned for the dry season. We were also told it was impossible to work in the park in the wet season because of flooding, and that park staff generally moved out before the end of May and returned only in early November.

Having read the survey reports, in particular Mike Fay's dry-season surveys of the mid-2000s – where he counted 865 fewer elephants in 2006 than in 2005 and saw many carcasses – we determined that the wet season was the time when the elephants were being targeted. This was when they were being afforded little or no protection and were dispersed over a massive area. The tragic movements of Annie, an elephant cow collared by Mike Fay and his team in 2006, made us guess that the herds were being followed and poached incessantly.[1] A female elephant with a young calf would not move such great distances unless she was being relentlessly hounded. We realised we had to operate all year round, despite what we were being told.

We arrived only 4 months before the rains started and the herds began their outward journey. The need to prioritise kicked in hard and fast. First, we needed to collar some elephants so we could follow their movements. The second was to make park headquarters accessible throughout the year.

In April 2011, we fitted 10 satellite GPS collars to individuals in several different herds. The Zakouma elephants are known for massing into one big herd in the dry season, so the choice of individuals was largely potluck, with the hope of there being

© Colin Bell

Left

African Parks' armoury leaves no doubt that poaching means war.

© African Parks archives

Left, bottom

In February 2014, Chad burnt its ivory stockpile of just over a tonne at Goz Djarat, the small town near the entrance to Zakouma National Park. The burn was witnessed by President Déby Itno and a delegation of Chadian cabinet ministers. The burn was a symbolic message to the international markets that ivory had no value.

at least one collar in each of the herds once they split and dispersed. Work began on upgrading the headquarters airstrip to an all-weather facility, and an additional four airstrips were made: two in the western migration zone and two in the northern migration zone.

The VHF radio system was upgraded to cover a larger area and equipment was procured to enable the park rangers to operate in the wet season. An additional aircraft was brought in and a radio-control room built, with operators trained to work this 24/7. The elephant collars were monitored from there and, once the animals started migrating, we deployed patrols to be near them.

The elephants dispersed widely that year, moving up to 170 kilometres north and to the west of their dry-season range, covering an area of some 9 000 square kilometres. Working with the regional law enforcement agencies and using the outlying airstrips and the newly improved all-weather strip at HQ, we were able to monitor and protect the elephants that wet season and picked up no signs of poaching. Sadly, poachers followed some of the elephants as they returned to the park in October and killed seven, removing the ivory of five (which was later retrieved).

This pattern of wet-season migration continued the following year, with the elephants again migrating long distances north and west. The third year, however, saw a change. They started heading north, then turned around and headed back into the park and have not migrated since. We can only speculate on the reason for this. Certainly, the smaller population meant density was no longer a driver for migration. They had learnt that safety and security existed inside the park, so perhaps the migration was driven by poaching anxiety rather than lowered environmental pressure. Through a land-use plan, we have also now secured the western and northern migration corridors in case elephants use these areas again as their numbers increase.

Earlier photographs of the Zakouma herds showed a balanced age class composition. But when we started in 2011 there was an age class missing: there were no calves under 4 or 5 years of age. The birth of a single calf in the second half of that year caused great excitement. But because of this near-zero growth rate, the population continued slowly to decline, even though poaching had been drastically reduced. It took another 2 years before we started seeing an uptick in newborn calves. By the end of 2016, at least 125 calves of under 3½ years were counted.

© Kyle de Nobrega

© Kyle de Nobrega

© Kyle de Nobrega

Camp Nomade – the spectacular safari camp within Zakouma.

What caused this change in behaviour, where several hundred elephants simultaneously ceased reproducing? It was almost certainly the ongoing poaching. The trauma of massacres for such intelligent animals is unimaginable, and we can only speculate that this is what made them stop reproducing. At the height of the elephant poaching in Zakouma, it was found baby elephants constituted an incredible 24% of the diet of lions, the result of calves either being orphaned or separated from the herd during a stampede and subsequently killed by lions.

This is something we witnessed in the three elephant-poaching incidents since calving recommenced in 2013. In the aftermath of each incident, some small calves were found wandering alone. Our attempt to raise one of these orphans failed, so we then 'walked' calves back to the main herd where they stood a better chance of survival.

The change in the behaviour of the elephants over the past 6 years since African Parks took over management can only be described as astounding. Apart from the turnaround described here, a change was also observed in the way they spend their time each day. In 2011, herds would be compact and tight, displaying definite unease, rarely even going down to drink during daylight hours. Even the bachelor bulls were in one tight herd, rarely found far from each other. Today the breeding herds are often still together in a single herd of about 500, but now they are spread out over several kilometres, with the calves playing, and much interaction between the adults. The bachelor bulls are now divided into numerous small groups, joining and separating as they please.

The light at the end of the tunnel is that today Zakouma's elephant herd is again on the increase. The combination of highly motivated staff, a good park management team and the goodwill of government – both regional and central – together with the all-important support from local villagers, will ensure their survival. There's no room for complacency, however – the threat is always there. But interventions are now in place to detect a potential incident and, it's hoped, to prevent it from happening, or to react quickly if one does take place.

Even though there have been a few poaching incidents in the past 6 years, very little ivory has left the park, which we consider a success. Each elephant area in Africa is different, but Zakouma's recipe has worked. We hope that lessons can be learnt and replicated in other parts of Africa that are experiencing the sort of decline Zakouma did just a few years ago.

Zakouma turns full circle

Zakouma is a truly remarkable success story. Both African Parks and the Chad government must be commended for turning around one of Africa's greatest wildlife sanctuaries from the depth of despair and the ravages of uncontrolled poaching in the late 1990s and early 2000s to the thriving sanctuary it is today. As a reminder of the desecration, the local elephant herds are missing an entire generation from those poaching decades, when not a single calf survived the stresses that engulfed the park and the region.

Today, matters have stabilised and poaching is firmly under control, so much so that elephant numbers are rising again and rhinos have been successfully re-introduced. These photos show the arrival of six black rhinos in May 2018 after an absence of nearly 50 years – and the dust of the first one to be released into its holding boma, where they will be allowed to settle before being introduced into the wilds and freedom of Zakouma.

@ Kyle de Nobrega

Education is an essential part of conservation, but can be challenging with nomadic communities. African Parks addressed this by creating 'Secko' schools – low-cost establishments, with a single teacher assigned to each community.

© Marcus Westberg

'Tusks of tears, tusks of trust: this image symbolises the story of Zakouma's success. As I stamd before this massive wild bull elephant, providing him with fresh water directly from the hose I am holding, I am aware of the horrific scenes of slaughter he has witnessed during his life. There is total trust in this moment and yet there must be instances of great sadness and terror as he thinks back to those massacres. But I draw a deep sense of inspiration from this animal that is prepared to put all that behind him and stand calmly, just inches away from me now that Zakouma is again a sanctuary.'

Kyle de Nobrega

41

The desert elephants of Mali

Against all odds and with innovative protection, these elephants negotiate life in a place nobody imagined they'd survive.

Vance G Martin & Dr Susan Canney

People are usually astonished when they first hear about Mali elephants near the ancient desert towns of Timbuktu and Djenne. Even among experienced conservationists, this unique herd remains virtually unknown. When people learn of their existence, not only are they surprised that this bone-dry country has any elephants, but rightly regard their survival as an enigma.

News of these elephants began to spread a little when Mali suddenly became one of the front lines in the war on terrorism. As we write, in early 2018, the range of Mali's desert elephants is at the geographic centre of continuing political instability and armed conflict instigated by jihadists, bandits and criminal syndicates.

But, we get ahead of ourselves. Context is needed for the story of how these elephants have endured and the potentially unique gene pool they offer. It's about the intersection and interconnections of human culture and wildlife.

The elephants of the Gourma region in central Mali are the most northerly African elephant population, one of just two populations of desert-adapted elephants. Probably because of the region's isolation, the elephants' small, low-quality tusks, and their tolerance of local people, the population largely escaped the intense poaching that took place across this region in the 1980s.

The population is nomadic. They have a unique migration circuit – the longest recorded among the species – to cope with the region's widely dispersed and somewhat variable food and water resources. The Gourma is a dry area, receiving around 450mm of rain a year in the south, down to only 150mm in the extreme north. There's a single rainy season – when it rains at all – mostly between late June and late August, followed by a dry season lasting around 8 months. As the dry season advances, elephants are forced into one small area of permanent water, Lake Banzena, increasing both their vulnerability and the potential for conflict with local villagers and transhumant herders. How have they survived?

Elephants once ranged widely across the Sahel. Humans hunted and harassed them as they modified, cleared and fragmented the habitat, and the remaining elephants were forced constantly to adapt to finding resources while avoiding humans. This is not all in the distant past: we know that in the 1970s elephants were still using the lakes of the inner delta of the Niger River to the west of the Gourma region. Elephant 'scouts' can still be seen occasionally investigating this area. The families have long memories.

The Gourma elephants are all that are left of this once extensive population. While the human cultures with which they share central Mali can be regarded as a source of potential conflict, they

Chengeta Wildlife was asked to set up an anti-poaching unit in Mali. With the resurgence of extremism in the region since 2014, it is not just bandits and traffickers with whom they have to contend, but ISIS and Al-Qaeda as well. Their combined army/ranger/anti-poaching brigade was trained in 2016 and activated in 2017, and is responsible for the anti-poaching and counter-trafficking successes in south-east Mali throughout 2017 and into 2018.

From up here you can see villages, the oases that the elephants frequent and the sandy beginnings of the Sahara Desert. In these cliffs are nests of Rüppel's vultures, which were long thought to be extinct from this area in the Sahel.

© Nigel Kuhn

also represent a model of understanding based on local ecological knowledge and cultural tolerance. Within the Gourma area are very diverse types of land-use practised by different ethnic and cultural communities, including:

- The **pastoral** (nomadic) system of the Tuareg.
- The **agropastoral** system of the Peulh, Sonrhai, Bellah and Dogon. Animals are kept around the villages during the dry season and, in the wet season, are moved away from the villages and/or northwards to the grazing in the dunes of the non-cultivated regions. They return to their villages at the end of the harvest and the beginning of the dry season. In addition to managing their cattle, these people also cultivate crops in small parcels around villages and in areas of cleared bush, often situated in bottomlands.
- With **agriculture** as their principal activity, the Dogons, Sonrhai and the Peulh's rimaïbes (a sub-ethnicity) cultivate large fields of grain. They cache their harvest in grain storage bins in the fields or close to villages and sell their cereals in the markets of Boni and Hombori in Mali, and Djibo in Burkina Faso.
- **Gardening** is practised by sedentary populations, chiefly Bellah and Sonrhai, around perennial waterholes such as Gossi, Dimamou, Adiora and Inadiatafane. Millet, sorghum, maize and watermelons are cultivated in the wet season, and vegetables and spices for the rest of the year.[1]

In addition to these *in situ* practices, transhumant (largely Peulh) herders move through the Gourma, following traditional routes, but increasingly with large herds belonging to urban populations in the inland (Niger) delta. Herders from neighbouring countries (Niger and Burkina Faso) also use the Gourma as wet-season pasture. As well as causing pressure on the elephants, these 'outside' herders are often in conflict with agriculturalists when their herds sometimes enter their fields. Within these diverse cultures interwoven with each other and

Opposite, top and bottom

In Mali, desert-adapted elephants must deal with many challenges and learn to negotiate the region's diverse cultures, including nomads, agriculturalists and pastoralists. In this water-scarce region, waterholes have to be shared. Adapt or die is the order of the day.

with the wildlife, the region itself poses its own very specific challenges for the elephants. 'Adapt or die' has never been more relevant.

The Gourma elephants range over an area slightly smaller than Switzerland, contained within the bend of the Niger River south of Timbuktu and extending south to the border region with Burkina Faso. The north of this parched region is open sandy steppe and savanna with sparse trees, sparsely vegetated dunes and shrubby thicket-forest in the bottomlands.

The south is dominated by bands of low and relatively thick 'tiger bush'. This is a patterned vegetation community and landscape consisting of alternating bands of trees, shrubs or grass separated by bare ground or low herb cover. It runs roughly parallel to contour lines of equal elevation, alternating with open steppe and grassy dunes. Throughout the region, trees are small, with their density and height increasing from north to south. Isolated thicket-forest stands, usually surrounding waterholes and following drainage lines, provide the main elephant habitat, especially as the dry season progresses.

The population size is assessed to be 250 to 350 individuals, and we estimate it has ranged from 300 to 800 animals over the years, depending on external mortality factors such as drought, food and recent poaching.[2] A photographic mark-recapture study in 2004/2005 logged the population at around 550 animals. Compared to other regions in Africa, it's an old population with over 50% of elephants being mature adults. The harsh environment and the long migration to find food and water leads to a high mortality in newborns and young animals. Migration is a strategy the elephants use to find sufficient food and water. In this harsh and variable environment, the ability to migrate and adapt is critical to elephant survival.

The elephant migration has long been known and remarked upon anecdotally. However, despite their ecological and conservation importance, these elephants remained unstudied and poorly known to science. Estimates of the total population have been derived from interviews with local people, incomplete aerial reconnaissance, extrapolation from a short-term dung count and a photographic mark-recapture study.[3] Bruno La Marche, a French schoolteacher, studied the elephants throughout the 1970s and recorded their movements, although he never published his results.

Two traditions meet as a Tamashek nobleman warrior representing an ancient lineage meets a modern warrior fighting to protect their shared way of life and Mali's last elephants.

In the early phase of elephant conservation in the Gourma, satellite-collar data gathered by Save the Elephants was analysed to understand their migration.[4] The route was striking. The elephants spent around 95% of their time in just a few, well-defined concentration areas, where they congregated for periods of time before moving rapidly to the next area along 'corridors'. This provided clues about how they survive in this environment.

Concentration areas clearly possess features of interest to elephants, while corridors are areas where elephants don't want to linger, either because there's nothing of interest or because they feel harassed or threatened. By examining and comparing concentration and corridor areas, we learnt the type and location of these resources and at what time of year they were important. We began to understand the elephant's world and how they have adapted to this harsh region.

Different areas offer different benefits. The south has more food of better quality, but there's no surface water unless it's raining, so available water soon dries at the end of the rainy season. The north, however, contains a series of small lakes fed by surface water drainage from local rainfall that collects in depressions or 'bottomlands'. The elephants move from lake to lake. As one dries, they move on to the next, until eventually they converge on Lake Banzena, which generally holds permanent water and is the only source accessible at the end of the dry season. It is also, therefore, a target of choice for local cattle and – with far larger numbers and considerable impact – the large herds owned by distant wealthy urbanites and managed by transhumant herders.

From an observation point, rangers scan the area for suspicious movements, ready to call in scope and ground security teams for a surprise interdiction.

While the elephants have access to water, adequate food is also required to survive the long dry periods. They rely heavily on the woody vegetation of the thicketed drainage ways and bottomlands. They spend most of their time in these thickets, which provide water, food, shade and refuge, making them vital for elephant survival. But they are also key sites for agriculture and the manufacture of charcoal to be sold in urban areas – conservation issues that would have to be tackled with the local populace.

Incredibly, the elephants smell when it has rained in the south. When this happens, they begin to move south towards the rich pastures, following well-used passes and crossing stretches of grassy dunes, avoiding villages and other human habitation. At some places they linger, taking advantage of available food. In one location, the local chief (a respected pastoralist and powerful man) has prevented an area from being cleared for agriculture, offering a peaceful haven for the elephants at a time they need to replenish their fat reserves. This is also important for the females, and many give birth at this time and therefore need additional nourishment.

Eventually the elephants cross the border into Burkina Faso where they stay for a while, meandering their way eastwards, parallel to the international border. We found that this was not a coincidence. In the 1980s, Mali and Burkina were at war over this stretch of border and, after the war, Burkina Faso designated protected areas for wildlife along the border, a sort of buffer zone. It is here that the elephants find refuge.

© Nigel Kuhn

Demonstrating to local communities that they are important is crucial to winning their trust and assistance in collecting vital information. The US-based Chengeta Wildlife team stopped in this Tamashek village and towards the end of the day this old man was brought to the brigade medic on a litter pulled by donkeys. He had his wife and family gathered round as the medic worked his modern magic.

In this area, they move between the small water bodies until the rains diminish. By this time, it's normally September and they're in the far south-east of their range. As the surface pools dry, they turn north-east towards areas of permanent water once again, eventually undertaking a roughly circular migration.

This entire, complex and fascinating picture provided a basis for the Mali Elephant Project's (MEP) collaborative, community-based conservation initiative, which has been on the ground in Mali since 2006. Working in the Gourma, in the capital, Bamako, and in key conservation networks around the world, MEP has created, adapted and persisted with a unique conservation model in which humans and elephants can live together mostly peacefully and for mutual benefit. This model has also proved to have significant, positive aspects for security in remote regions.

Working in co-operation with the Mali government, MEP started with the scientific work by Save the Elephants (STE) in 2003/2006 led by Iain Douglas-Hamilton CBE, which had mapped the migration route. STE, the WILD Foundation and the US Department of State funded this early work, with WILD assuming responsibility as MEP transitioned from collaring and science to applied, elephant-centred, community-based natural resource management involving as many as possible of the multiple ethnicities in the Gourma.

Studies of the migration route showed it needed to be protected in its entirety.[5] Working with minimal funding, the MEP's first task in 2006 was, therefore, to engage with local people to understand their perceptions. It was clear that, if elephants were to survive, these people would have to play a large role. The main task was to create a shared vision throughout Mali so that elephant conservation became part of the day-to-day decision-making of people with an impact in the range.

A social survey revealed valuable information about local attitudes and traditional ecological knowledge. It showed that 78% of the local population did not want the elephants to disappear. When asked why, the most common sentiment was: if elephants disappear, it means the environment is no longer good for humans. In addition, many expressed the sentiment that every species has a right to exist and that it contributes something to the uniqueness of the ecosystem, a notion encapsulated in the word *baraka*, or blessing. Each species has its own *baraka* and if a species is lost, the system is irretrievably diminished and poorer in its ability to sustain life. Other benefits were that elephants knock down leaves, fruits and seed pods from the high branches, enabling goats to feed around their feet and the women to gather them up to sell in the market. Elephant dung was also used to cure conjunctivitis.[6]

A dialogue developed between MEP, local communities and the local government. This enabled an outreach campaign, schools programme, ecotourism guidelines, plus support for regional planning in preparation for an Elephant Management Plan.

The MEP community-engagement approach was developed initially at Lake Banzena. As human and livestock pressure on the lake increased, so did the risk of its premature desiccation. In 2009, the lake dried before the rains came, a potential catastrophe for the elephants. Fortunately, a light rain further south helped the elephants survive until the full rains began, but it was clear that something urgently needed to be done.

Information from socio-economic studies produced a common perception of the issues by communities around the lake, and this helped them determine the solutions. These studies showed us that over 96% of the cattle using the lake belonged to distant wealthy urbanites, fouling the water and causing over 50% of the local population to suffer from water-borne disease. We also found that the local people were ready to relocate outside the elephant range if the MEP could meet their needs by improving grazing for their cattle, providing fresh water and enhancing their general health and wellbeing.[7]

The Banzena process disclosed additional issues relevant across the entire elephant range. While the different ethnicities had their own systems of natural resource management, there were no single, specific systems recognised by all. The result was that access to resources of water, grass, browse, salt, fuel-wood and non-timber forest products resulted in a free-for-all. Over the last few decades, this unregulated exploitation, coupled with increasing impact from urban centres and in-migration, had led to progressive degradation.

To prevent this, MEP drew on the provisions of Mali's decentralisation legislation to help the local community create a unified convention that allowed the resident population to devise community systems to regulate resource use. This enabled communities to work closely with government foresters to patrol and enforce the regulations in the relocation area around Banzena, and empowered them to charge a fee to outsiders for the use of resources such as water, hay or grazing access rights.

The structure embraced key community groups elected by the people. These included a management committee of elders as well as surveillance brigades (reporting to the management committee) of young men for monitoring infringements of both local resource-management rules and national laws. Members of the brigades were trained in all aspects of their operation, including the law, record-keeping, how to deal with offenders and how to communicate the new systems to others. This new approach was supported by a wider outreach to other communities, including such activities as regular radio broadcasts, a schools programme and an awareness campaign alerting herd owners to the new rules.

In 2010, MEP was joined by a significant new partner, the International Conservation Fund of Canada, and was able to greatly increase the empowerment and betterment of the local communities, while still using the elephants as a central symbol of their collective efforts.

The management committee of local elders created an initial 400 000-hectare pastoral reserve, which was immediately increased by adjacent communes to form an area of 923 800 hectares. Protected by firebreaks made by teams of integrated ethnicities, this was one of the few areas of the northern Gourma that did not lose its pasture to fire in 2010. News travels fast and, as the benefits began to flow to communities relocated from Banzena, many other communities across the elephant range asked for help to replicate the situation. Additionally, while studying the sociological impact, we were told repeatedly that the price of charcoal in Timbuktu had increased several times because wood was no longer being collected from the Banzena area and this had re-ignited interest in urban woodlots. The MEP model was working. Then the conflicts began.

In 2011, after the fall of Libya's Gadaffi, Tuareg mercenaries returned to Mali from the anarchy of that country, heavily laden with tons of armaments, and fuelled the resurgence of the Tuareg secessionists. Meanwhile, jihadist presence in northern Mali (AQIM, Al-Qaeda in the Islamic Mahgreb) had been building over several years. In March 2012, a coup by the Malian Army resulted in government abandoning the northern and central areas of the country. The perfect storm had occurred.

Almost overnight, the elephant range became lawless, occupied and controlled by armed groups of different allegiances. The first poaching incident occurred in January 2012. It seemed impossible to protect the elephants and our only option was to use our greatest asset – the relationship with local communities – to build a local response. The MEP field team (entirely Malian) worked with the elders to convene a 4-day community meeting, enabling local people to discuss their problems, and us to raise the plight of the elephants. People were primarily worried about not being able to obtain grain as all transport was being hijacked by armed groups. They were also concerned about their young men joining these groups.

The MEP pledged to help them procure grain by bringing it in and distributing it by donkey cart. Local leaders and elders, in turn, pledged to communicate an edict whereby anyone killing an elephant would be branded a thief, something that carries great dishonour. At the same time, the MEP would recruit young men to form a vigilance

network to watch over elephants, watching for strangers in the area, registering incidents of elephant poaching and discovering the identities of poachers. They would also conduct resource protection activities for the communities. Although the 520 appointed young men were paid only the equivalent of food for them and their families, not one joined the armed groups. The reason was that their role carried more local status and was less risky, despite the jihadists paying local recruits the astronomical sum of $30–50/day. [8]

The jihadists moved relentlessly towards the Malian capital, Bamako. When it appeared that they were within reach, French airstrikes in January 2013 brought the advance to an end. But the Mali government never fully returned to the Gourma, the lawlessness and banditry continued, and the jihadists regrouped. The social empowerment developed by MEP controlled poaching for 3 years. Then, in 2015, insecurity increased yet again as extremist groups re-activated. This, plus the targeting of the local elephant population by international trafficking networks, resulted in a sudden escalation in poaching. In the first 6 months of 2015, 64 elephants were killed, more than three times the 20 elephants killed from January 2012–2015.

Mali has never had an armed anti-poaching unit (APU) and one was urgently needed. MEP worked closely with the government and other partners to train and deploy an APU, but these efforts were complicated by lack of in-country capacity and continuing deterioration of security. After many challenges, a mixed forester-army unit of anti-poaching rangers was deployed in February 2017. Since then, no elephants have been poached.

This was made possible by the support of a variety of partners, most notably MINUSMA – the UN peace-keeping force in Mali – together with the unwavering, highly professional commitment of the US-based Chengeta Wildlife organisation, whose dedicated personnel insisted on conducting training under a high risk of kidnap and attack. By recognising the role of community work in anti-poaching, Chengeta's doctrine combined principles of community conservation with intelligence-driven arrest and deterrence operations to maximise the use of existing skills and resources while minimising negative impact and cost.

© Rory Young

As they migrate across an area the size of Switzerland, Mali's desert-adapted elephants have to share scarce water sources with large herds of domestic livestock. Water sources like Lake Banzena and this one are vital for their survival, but as competition for scarce resources intensifies, so does human-elephant conflict.

The APU uses a doctrine adapted to the conditions of insecurity specific to the Gourma, thwarting daily threats from jihadist groups, armed bandits and poachers. The doctrine is particularly effective, being balanced with the operating philosophy successfully adapted to Malian culture. Engaging and supporting the local population ensures they are directly involved in, and receive benefits from, elephant conservation.[9] These successes were recognised when MEP was named as one of the 15 winners of the 2017 UNDP Equator Prize.[10]

The two-pronged approach to anti-poaching also has relevance for security and stabilisation efforts. It demonstrates the importance of balancing investment in appropriate security-focused responses with investment in responses to core socio-economic drivers of conflict. It's about peacebuilding and development approaches at the local level.

The challenges continue. Sabotage of the few new boreholes by fleeing jihadists and bandits meant that people had to return to Lake Banzena. Meanwhile, social divisions created by the conflict, plus the absence of the rule of law create additional work. However, thanks to the 10 years of successful community engagement, there are solutions. MEP has developed viable plans to overcome all these challenges, deliver elephant conservation, and improve local livelihoods. Our work continues. While often precarious, it has demonstrated persistence and produced solutions. Of course, as always, solutions just need financing.

The Mali Elephant Project is a project of the WILD Foundation (**www.wild.org**) and International Conservation Fund of Canada (**http://icfcanada.org**), a long-term, community-based, elephant conservation initiative working in partnership with the Malian Directorate of Eaux et Forêts (and other Mali agencies), and many other partners (see **www.wild.org/mali-elephants/partners/**).

42

The right time to die

Colin Bell

Circling vultures gave away the location of this carcass in a remote dry grassland along the fringes of Botswana's Okavango Delta. We approached on foot and found that hyenas and vultures were busy doing their work. More importantly, the tusks were intact and there were no bullet holes. The elephant had died of natural causes. Flying past a month later, after the Okavango's annual floodwaters had arrived, we came across this tranquil scene of bones and tusks beneath clear waters. This is how an elephant ought to die: of natural causes, in its own time and in the place where it was born.

We hope the stories in this book will help persuade parties that make up CITES to heed the call of governments and NGOs to uplist all elephants to full trade protection under Appendix I.

May our grandchildren never again see a butchered carcass with its tusks hacked out to be sold to people far from Africa, who don't understand the implications of their desire for pretty trinkets.

Get involved

It does matter.

Ian Michler

Current population levels and modern lifestyles are significantly affecting our planet. Nearly all of us are now sedentary, thrown together in massive metropolises, disconnected from the natural world. And these societies we live in are characterised by high levels of consumption and waste production, activities that have severely scarred the planet and left many of our fellow species in danger of extinction.

We know this from decades of solid research and science, as well as the very visible impacts of this behaviour, which relates to every aspect of the way we live, including our travel choices. These are facts that demand we reflect on the role we are playing. Some of us are alarmed, others merely concerned. Whichever it is, getting involved is essential. It does matter. In today's world, doing nothing is inexcusable.

As a traveller to Africa – and being someone who wants to make a contribution – I have found that species protection and taking care of their habitats is a great sphere of involvement. And elephants, one of the most charismatic and loved of all animals, are the torch-bearers. They are also indicators of the health of all other species: their status is a barometer of the condition of biodiversity and ecosystems in general.

By getting involved, you will make a difference.

Become involved with Africa's elephants

- By travelling in Africa and going on a wildlife safari, you will already be making a significant contribution through park and other fees you pay, as well as the jobs that you will support. Be sure to select a reliable and ethical operator and include in your itinerary places that have elephants. A well-constructed safari will provide both a wonderful, uplifting experience and make available much-needed funds for conservation. That way you become simultaneously a tourist and a conservationist.

- Become aware of the issues regarding the conservation of elephants and stay informed. This includes finding out about the poaching crisis and why trading in ivory is intimately linked to the alarming decline in elephant populations. Trade is not a solution to the long-term survival of the species. Never buy ivory.

- Consider all the different stakeholders involved: government and other regional decision-makers, the scientific community, conservation agencies, the role of ecotourism. Think carefully about the choices you make, bearing in mind the possible outcomes of such choices, and the impact beyond your own particular involvement.

- Become an outspoken advocate for elephants. Make your voice heard. Informed and inspired by your safari experiences to Africa, use your new-found knowledge to stand up for elephants. Thus empowered, go out and spread the news of their plight among family, friends, colleagues and governments on behalf of the animals that are otherwise voiceless. This includes writing and petitioning.
- Nearly all the contributors or organisations represented or listed in *The Last Elephants* work with elephants and their conservation. If you want to offer support with a donation, start with contributing to any one of them.
- Join campaigns advocating the right to survival of elephants through securing and expanding their habitat, including corridor and transfrontier initiatives. Without the space they require, all efforts to rescue these animals are diminished. This can be done through donations, as a volunteer in range states or as an educator and advocate. Whichever way you choose, be aware that these efforts are long-term and be prepared to dig in on behalf of elephants.
- If you wish to be involved on the ground, then join an internship or volunteer programme. Typically, these offer a variety of activities from administrative work to being a research assistant and educator. But be careful: there are many volunteer organisations that do great work – and there are others that are merely business-oriented, no matter what they say on their websites.
- Certain products such as coffee, wood and various fruits may have been procured and produced to the detriment of elephants. Before buying such products, determine their status.
- Become a citizen scientist. There are a number of initiatives, often using apps and tracking, downloading or photographic techniques and methods that enable you, while on safari, to provide valuable information on elephants, and so contribute to conservation efforts.

What you should not do

- Don't support operations that offer elephant-back riding or other forms of animal exploitation using elephants in captivity – they are not regarded as conservation initiatives. By supporting them through visits or donations, you will not be contributing to the survival of the species. Instead, you will be entrenching a cycle of use and abuse.
- Don't visit zoos or so-called sanctuaries that keep elephants under poor conditions.
- When in Africa, or anywhere else, do not buy any products – jewellery, crafts or antiques, for example – that use ivory, elephant hair or any other body parts as this promotes the killing of these animals.
- And finally, a plea to hunters: while trophy hunting of elephants still remains in place in some countries, we ask you to avoid shooting large tusked elephants as they are the gene pool for future generations.

Suggested NGOs to support

A list of some of the better NGOs in alphabetical order (and not in order of importance). Apologies for any worthy organisations we may have omitted in error.

African Parks	www.african-parks.org
African Wildlife Foundation	www.awf.org
Bhejane Trust	www.bhejanetrust.org
Born Free	www.bornfree.org.uk
Big Life	www.biglife.org
BirdLife	www.birdlifebotswana.org.bw &
	www.birdlife.org.za
Chengeta	www.chengetawildlife.org
COMACO Trust	www.itswild.org
Conservation Action Trust	www.conservationaction.co.za
Conservation Lower Zambezi	www.slczambia.org
David Sheldrick Wildlife Trust	www.sheldrickwildlifetrust.org
EAGLE Enforcement	www.eagle-enforcment.org
Eden to Addo	www.edentoaddo.co.za
Elephant Trust	www.elephanttrust.org
Elephant Voices	www.elephantvoices.org
Elephants Alive	www.elephantsalive.org
Elephants Without Borders	www.elephantswithoutborders.org
Endangered Wildlife Trust	www.EWT.org.za
e'Pap	www.epap.co.za
Frankfurt Zoological Society	www.fzs.org
Game Rangers' Association of Africa	www.gameranger.org
LAGA Wildlife Law Enforcement	www.laga-enforcement.org
Lion Guardians	www.lionguardians.org
Maasai Wilderness Conservation Trust	www.maasai.com/conservation/the-trust
Mara Elephant Project	www.maraelephantproject.org
Niassa Wilderness Trust	www.niassawilderness.com
Northern Rangelands Trust	www.nrt-kenya.org
PAMS Foundation	www.pamsfoundation.org
Peace Parks Foundation	www.peaceparks.org
Rhino Conservation Botswana	www.rhinoconservationbotswana.com
Save The Elephants	www.savetheelephants.org
Saving The Survivors	www.savingthesurvivors.org
Southern Tanzania Elephant Program	www.stzelephants.org
Space for Elephants Foundation	www.spaceforelephants.com
Tourism Conservation Fund	www.tourismconservationfund.org
Tsavo Trust	www.tsavotrust.org
Tusk	www.tusk.org
Victoria Falls Anti-poaching Unit	www.vfapu.com
Wild Aid	www.wildaid.org
WILD Foundation	www.wild.org
Wilderness Foundation Africa	www.wildernessfoundation.co.za
Wildlife Conservation Society	www.wcs.org
WildlifeDirect	www.wildlifedirect.org
WWF South Africa	www.wwf.org.za
Zambezi Society	www.zamsoc.org

© Chris Fallows

Wayne Lotter

A remarkable man

Assassinated in Dar es Salaam, Tanzania, on 16 August 2017

From Will Travers OBE

If they, whoever they are, think that Wayne Lotter's passing will silence Wayne's voice, or bring his mission to a halt, or will intimidate those who follow his example – they could not have got it more wrong. It will have exactly the opposite effect. But they should know this also: that their evil will be met with true justice and the rule of law. Criminality will never prosper while there are good women and good men, like Wayne, who are prepared to do what's right. We will follow his example and, ultimately, we will win.

From Chris Bakkes

Africa has lost one of its greatest and most dedicated conservationists. Wayne Lotter was killed by cowards who are plundering our magnificent continent. He went to another country to make a difference and made great personal sacrifices. He and his colleagues faced incredible odds against an implacable enemy and succeeded. He disrupted their designs and filled their hearts with fear, so they sent an errand runner to collect the butcher bill.

He was one of a rare breed of men. His iron resolve, principles and beliefs stood first and foremost above all else. He was never scared to challenge authority where he felt it was wrong. He was a free, unconventional thinker who had

no regard for cheap popularity. He wanted to get the job done.

I know of no other conservationist who got that job done more effectively than Wayne. The high-profile arrests of top-level ivory traffickers in Tanzania were testimony to that. It made him feared and loathed by wildlife crime syndicates. And it cost him his life.

Wayne and I were trail rangers on the Sweni Wilderness Trail in the Kruger Park for 3 years. Our shared passion for the wilderness and its wildlife bonded us. Days spent tracking elephants, facing lions and revelling in the wild nature around us made us friends. His eccentricities and unique wit would often crack me up.

We parted ways and lost contact for years. The present continental poaching crisis brought us together again. I was proud of his success and wanted to be like him. We met and discussed the issues. I realised that the old carefree Wayne had crystallised into an unwavering, no-nonsense wildlife crime-fighter. There were no grey areas in his approach. It was either right or wrong.

Our last encounter was in 2017 on a wooden boat on Lake Malawi, sailing along the shore, watching fish eagles swooping. Our conversation varied from the present crisis to the memories of our youthful days in the veld. It was good. I'm glad that I had the opportunity to tell him, before he left, how proud I was of him and that our friendship was sound and unbreakable.

Wayne Lotter's death must not be in vain. He must serve as the example that we must all live by. All of us who dedicate our lives to wildlife conservation.

Wayne, you were the conservationist we all should aspire to be. If there were more of you, our troubles here would be fewer. I love you as a friend and mourn with your family and friends. Rest well, my brother. I salute you.

From Ian Michler

Wayne Lotter was a remarkable man. And in a relatively short time within the world of conservation, he achieved remarkable results. Passionate, committed, courageous and above all else, a man of immense integrity, Wayne has left an exemplary legacy that will always serve as an inspiration to conservationists and lovers of wilderness from around the world.

His assassination serves as a stark reminder: vested interests and criminal forces, working in cohorts, are looting the natural world. They remain a scourge and are as ruthless as they are crooked and cowardly, a situation that makes protecting habitat and biodiversity so challenging.

Wayne's life and work as a dedicated naturalist and conservationist started with his studies. Armed with an MA in Nature Conservation, he worked his way through various South African government and private sector organisations. He immersed himself in almost every aspect of environmental management, from finance and community development to field ecology and wildlife crime. His professional ethic was apparent from the outset.

It was these experiences and an urge to play a more meaningful role that equipped him so thoroughly for his move to Tanzania where he co-founded the Protected Area Management Solutions Foundation (PAMS) with his friend Krissie Clark and Ally Namangaya in 2006.

With the organisation based primarily on gathering intelligence on the ground and raising awareness amongst communities, Wayne set about introducing his unique style of operating to stem and then reverse what could only be described as a monumental poaching crisis. Tanzania and the Selous Game Reserve, in particular, were losing around 20 elephants daily, a massacre that accounted for over 60% of the country's entire elephant herd within a 10-year period.

His approach and attitude having been variously described by colleagues and supporters as strategic, lean, innovative and even ground-breaking, there are many lessons to be taken. Of these, the one that stands out most, is the care and concern he bestowed upon people. Protecting elephants and other wild species propelled Wayne's efforts, but his deep and genuine consideration for Tanzania's rural communities and those that he worked with became central to the success of his endeavours. Unassuming, but charming and fiercely determined, he instilled both a sense of urgency and pride amongst Tanzanians about caring for their national heritage.

Under Wayne's tenure, PAMS achieved what few other conservation agencies manage to do in a lifetime of work. By the time of his death it had been instrumental in the confiscation of over 1 100 illegal firearms, and ensured the arrest of over 2 300 poachers and traffickers. Even more impressive was a conviction rate exceeding 80%, which included some of the region's most sought-after wildlife crime bosses. In the process, PAMS also trained hundreds of Tanzanian scouts and brought environmental awareness to thousands upon thousands of young school children.

The ultimate measure of PAMS's success was that, by early 2014, Tanzania's elephant poaching crisis had been slowed and by 2015, census data was indicating a slight recovery in populations. Wayne never doubted this was going to happen.

Most writers involved in this book knew Wayne as a warm and valued friend or colleague, but all knew about the magnitude of his work. Having set about securing the survival of one of Africa's most iconic species, fearless and utterly focused, he was a leader and the unsung hero amongst us. Wayne, you and your work will never be forgotten.

Opposite

A young Okavango elephant touches, sniffs and pays tribute to a relative – as we pay tribute to Wayne Lotter and the many rangers and anti-poaching staff around Africa who have lost their lives protecting the continent's wildlife.

Elephants are extremely sensitive and alert to their surroundings. This old bull is trying not to disturb the dove perched comfortably, but precariously on his foot.

© Peter & Beverly Pickford

Understanding elephant behaviour

Audrey Delsink

Elephants are intelligent and emotional and they want to be left in peace. Like humans, they have a personal space that they do not like invaded. Remember, you are in their territory. They always have right of way. Elephants appreciate silence, patience and slow, consistent movements. Respect for the animals and common sense must always prevail. For your own safety, it's important to read their body language.

Classic elephant behaviours

© Tim Driman

Standing tall (threat behaviour)
Elephants normally stand or move about with their eyes cast down. A direct gaze with eyes open is a component of many displays. A typical posture used mainly by females in challenging non-elephant threats, such as predators and people, would be standing or moving with the head held well above the shoulders, the chin raised (as opposed to tucked in) and looking down at her adversary over her tusks with an **eyes-open stare** and the ears maximally forward. The animal appears to increase in height and sometimes will deliberately stand upon an object such as a log or anthill in order to increase its height. This elephant means: I've got you in my sights, so watch it!

© Bobby-Jo Vial

Head-shake (threat behaviour)
An abrupt shaking of the head, which causes the ears to flap sharply and dust to fly, is a sign of an individual's annoyance with or disapproval of an individual or circumstance. The **head-shake** usually starts with the head twisted to one side and then rapidly rotated from side to side. The ears slap against the side of the face or neck, making a loud smacking sound. This can also be used in play to feign annoyance. Head-jerking (a single upward movement followed by a slower return) and head-tossing (the head is lowered and then lifted sharply so that the tusks make an arc) are also mild threat displays.

© Colin Bell

Ear-spreading (threat behaviour)
Facing an opponent or predator head-on, with ears fully spread (at 90 degrees to body), presumably for the purpose of appearing larger, is another threat display. Elephants may also spread their ears when they're excited, surprised or alarmed.

© Colin Bell

Forward-trunk-swing (threat behaviour)
Swinging or tossing the trunk in the direction of an adversary, typically while blowing out forcefully. Elephants swing their trunk at other smaller animals (e.g. egrets, ground hornbills, warthogs and people) to frighten them away – or simply for amusement. A high-intensity version of the **forward-trunk-swing** is the **aggressive-whoosh** made by musth males, who toss or swing their trunk in an exaggerated manner in the direction of an adversary while blowing loudly through it with a loud whooshing sound.

© Michelle Henley

Throw-debris (threat behaviour)
Lifting or uprooting objects and using the trunk to throw them in the direction of an opponent or predator. This display may also be observed in play. An elephant's aim can be very accurate, even at some distance.

© Michelle Henley

Bush-bash (threat behaviour)
Tossing the head and tusks back and forth through bushes or other vegetation, creating noise and commotion and demonstrating strength. It's probably an expression of 'look what I can do with you', but it is also used in play.

© Michelle Henley

Tusk-ground (threat behaviour)
Bending or kneeling down to tusk the ground, and sometimes lifting and tossing vegetation, can be be directed towards human observers, possibly as a demonstration of 'look what I will do with you'. It's usually seen when two males – and especially musth males – are manoeuvring during an escalated contest, but may also be directed towards people. In some cases, tusking the ground may be a form of redirected aggression but is also seen during play. In females, a vigorous scraping/trampling or tusking of the ground may be observed in cows after they have given birth.

@ Guts Swanepoel

Trunk-twisting (showing apprehension)
Twisting the tip of the trunk back and forth in situations where an elephant is apprehensive or unsure of what action to take.

@ Sabine Stols

Distant-frontal-attitude (in play or showing submission)
In expectant or playful situations, elephants may pause with the trunk up in a periscope or S-shape, waiting for an adversary, in preparation for duelling or a play partner's next move. As two individuals approach one another with intent to duel or spar, one or both may raise their trunk above the head and curl the tip towards the other. Apart from the context, this display appears very similar in form to periscope-sniff.

@ Colin Bell

Touch-face (showing apprehension)
Self-directed touching of the face, apparently for reassurance, very often in the context of interaction with another elephant, may also be seen in any situation where an elephant feels uneasy. Touch-face includes self-touching of mouth, face, ear, trunk, tusk or temporal gland.

@ Garth Thompson

Foot-swinging (showing apprehension)
Raising and holding or tentatively swinging the foreleg intermittently when unsure of what action to take. Swinging of the hind foot may also be observed, although this is less common than the forefoot.

Displacement-feeding (showing apprehension)

Plucking at vegetation, as if foraging, but without actually eating any of the material, is behaviour for the purpose of monitoring a situation. When in this mode, if the animal actually eats, it does so in a desultory or distracted fashion. The elephant may also slap vegetation against a foot or other part of its body. This is performed in conflict situations, such as during fighting or sparring, or when an individual feels indecisive, such as weighing up fleeing versus fighting. It can also be a sign of defensiveness or despondency. This behaviour is often displayed by young males near an oestrous female: by pretending to do something else, they hope not to provoke aggression by the guarding male.

Mock-charge (threat behaviour)

Rushing towards an adversary or predator, full stature (standing-tall) and with ear-spreading, but stopping short of its target. An elephant may forward-trunk-swing or aggressively kick dust as it abruptly stops. A mock-charge is often associated with a shrill trumpet blast.

Real-charge (aggressive behaviour)

Rushing towards a predator or other adversary while ear-spreading, head raised or lowered with the apparent intention of following through. The trunk may be slightly curved under so that tusks can make contact first. A real-charge is usually silent.
(This photo was taken from the safety of an underground photo hide.)

Signs of musth in reproductively active bulls

Musth walk and a strong smell
Head is carried high with chin tucked in and an overall swaggering gait. Musth has a distinct, strong smell that is easily discernible to humans.

Swollen temporal gland
Glands are at least the size of an orange at the peak of musth. Musth males often drape their trunk over their tusks to relieve pressure on the glands.

Temporal gland secretion
Gland secretes an oily fluid that runs down the cheek to the chin and eventually leaves stained streaks. Musth males often rub the glands against trees. (Note that temporal gland secretions alone are not a reliable indicator of musth – they can also indicate stress or excitement in non-musth elephants. Look for co-occurrence with other signs.)

Frequent urine dribbling
Occurs during full musth with penis kept inside sheath so urine sprays the hind legs. The sheath is eventually stained yellow-green and legs have dark streaks running down them.

How to view elephants from a vehicle

- Slow down as soon as you see elephants. Do not rush into the sighting.
- Assess the situation regarding escape routes, terrain and the apparent animal behaviour before settling into a sighting.
- Elephants should not be approached by closer than 50m. Switch the engine off. If the elephants are comfortable, they will gradually approach you. If they choose to do so, do not switch the engine on, but sit quietly and enjoy the sighting.
- If, however, the elephants approach to within 20m of your vehicle, switch on the engine, wait a few seconds and slowly back away. NEVER allow an elephant to touch the vehicle.
- In order to prevent the elephants feeling boxed or caged in, a maximum of three vehicles (although two vehicles is preferable) is advisable per sighting unless the driver in charge feels there should be fewer.
- As elephants are diurnal animals (active during the day), never shine spotlights on them at night.
- When viewing elephants, do not stand up abruptly or make sudden movements in the vehicle. This may frighten the animals and cause a threatening or aggressive response.
- Never take fruit on a game drive.

What to do when elephants display threat behaviour

- Switch on the engine, wait a few seconds and slowly back away to give the elephant space.
- If switching the engine on appears to aggravate the situation, switch the engine off immediately; wait a few minutes and then try switching on again when the situation is calmer, before retreating.
- In the case of bulls in musth, keep well clear as their behaviour can be unpredictable, erratic and often aggressive.

About the authors

We asked scientists, game guards, poets, activists, academics, journalists, lodge owners and leaders of NGOs involved in elephant work throughout Africa if they would be prepared to write chapters for this book. The response was extraordinary and heartening. We are indebted to all these people who took time to pen the thousands of words that make up this book. They have done it free of charge and for the creatures they love and respect: elephants.
We salute you and thank you.

Karl Ammann is an independent filmmaker, photographer and author who has lived in Kenya for 40 years, documenting conservation issues and, more recently, concentrating on the illegal wildlife trade. **www.karlammann.com**

Kathi Lynn Austin is the founder and executive director of the Conflict Awareness Project (CAP). She leads CAP's investigations into global trafficking networks fuelling conflict, organised crime and the exploitation of natural resources and wildlife. She previously served as an arms trafficking expert at the United Nations. Links to her documentary exposé on poaching kingpins Follow the Guns can be found on CAP's website. **www.conflictawareness.org**

Carina Bruwer is a PhD student at the Centre of Criminology at the University of Cape Town. Her research explores responses to organised crime around eastern Africa, one of which is the illicit ivory trade in Kenya, Tanzania and Mozambique. She holds an LLM in Public International Law from the University of Stellenbosch, South Africa, and is an admitted attorney.

Colin Bell has been fortunate to work and guide throughout much of Africa from the early 1970s right through to the present day. He has been at the receiving end of angry, stressed, harassed elephants, and witnessed calm, relaxed elephants during their golden glory years when poaching levels were minimal. **www.africasfinest.co.za**

Luca Belpietro is a conservationist who has been living with the Maasai in southern Kenya for more than 2 decades. He founded a unique boutique eco-lodge, Campi ya Kanzi, and a pioneer community-rooted conservation organisation, Maasai Wilderness Conservation Trust. He studied economics and did his thesis on *Wildlife as a renewable resource in Kenya: sustainable development and environment conservation.* **www.maasai.com**

Dr Susan Canney is director of the Mali Elephant Project, with extensive experience in West Africa, in numerous nature conservation projects elsewhere in Africa, Asia and Europe and as a research officer at the Green College Centre for Environmental Policy & Understanding. She is a research associate of the Department of Zoology at the University of Oxford, a trustee of Tusk Trust, a member of the Sahara Conservation Fund's Conservation and Science Committee and a member of the African Elephant Specialist Group. Susan also co-authored the book *Conservation* for Cambridge University Press, which offers a global perspective, placing conservation at the centre of sustainability and environmental policy. **www.zoo.ox.ac.uk**

Dr Michael Chase is the founder and director of Elephants Without Borders, based in Botswana. He is the country's first ecologist to receive a doctorate dedicated to elephant ecology, and received the prestigious Presidential Order of Meritorious Service Award in 2015. He was the visionary behind, and principal investigator of the Great Elephant Census, the first systematic continental count of African savanna elephants. He has published numerous scientific publications, supervised dozens of PhD and Masters students and appears in many wildlife documentaries. **www.elephantswithoutborders.org**

Romy Chevallier is a senior researcher with the Governance of Africa's Resources Programme (GARP) at the South African Institute of International Affairs. She leads GARP's work on climate change and resilience, and holds an MA in International Relations from the University of the Witwatersrand, South Africa. **www.saiia.org.za**

Ian Craig is director of conservation for the Northern Rangelands Trust in Kenya. He was a founder and CEO of Lewa Wildlife Conservancy, Kenya's leading private-sector land-management conservation organisation. **www.nrt-kenya.org**

Adam Cruise is an international journalist, conservation commentator and author specialising in wildlife. He is a PhD candidate in Environmental Philosophy at Stellenbosch University in South Africa, specifically concentrating on the ethical considerations of wildlife conservation and management. www.travel-hack.com

James Currie is a wildlife TV host, producer, author and conservationist. He has a passion for the last remaining big tusker elephants. In addition to working for Wilderness Safaris, James is an expert in sustainable conservation and holds an MSc in Sustainable Environmental Management. www.bigtuskers.com

Audrey Delsink is executive director of Humane Society International – Africa and a member of the Elephant Specialist Advisory Group in South Africa. She is a registered ecologist (SACNASP Professional Natural Scientist) and field director of the Elephant Immunocontraception programmes in South Africa. She is currently working on elephant spatial demography and sociality with the Amarula Elephant Research Programme. www.hsi.org

Andrew Dunn is the country director for the Wildlife Conservation Society (WCS) in Nigeria and leads efforts to save Cross River gorillas, lions and elephants in the country through science, conservation action and education and by inspiring people to value nature. He has an MSc from Edinburgh University, UK. www.wcs.org

Dr Richard Fynn is a senior research scholar in Rangeland Ecology at the Okavango Research Institute in Maun, Botswana. His research is on grazing ecosystem ecology and he holds a PhD from the University of KwaZulu-Natal, South Africa. www.ori.ub.bw

Dr Marion E Garaï is chairperson of the Elephant Specialist Advisory Group, South Africa, with a PhD in Ethology, and has specialised in orphaned-elephant behaviour. She is a trustee of the Space for Elephants Foundation and the Elephant Reintroduction Trust, is on the Scientific Advisory Board of the European Elephant Group, and is a long-standing member of the IUCN/African Elephant Specialist Group. www.esag.co.za

Nachamada Geoffrey is the landscape director for the Wildlife Conservation Society (WCS) in Yankari Game Reserve, working to save some of Nigeria's last remaining elephant and lion populations. Nacha has an MSc from Oxford Brookes University, UK. **www.wcs.org**

Ross Harvey is a senior researcher with the Governance of Water and River Basins in Africa (a GARP project) at the South African Institute of International Affairs. He leads GARP's work on the extractive industries and illegal wildlife trade, and holds an MPhil in Public Policy from the University of Cape Town, South Africa. He is currently pursuing a PhD in Economics. **www.saiia.org.za**

Dr Michelle Henley is a co-founder, director and principal researcher at Elephants Alive. She has been studying elephants for more than 20 years and has a PhD from the University of the Witwatersrand, South Africa. She has been monitoring elephant movements and their social interactions within the Great Limpopo Transfrontier Park, straddling South Africa, Mozambique and Zimbabwe. Michelle won the Wildlife and Environmental Society of South Africa (WESSA)'s National Award for an Individual (2013) and was elected one of the 10 most inspiring women in South Africa by Culture Trip. She is a member of the IUCN's African Elephant Specialist Group. **www.elephantsalive.org**

Naftali Honig works for African Parks Network in Garamba National Park in the Democratic Republic of the Congo. Across the African continent, he has been working for a decade combating the poaching and illegal trafficking of wildlife. He holds a degree from Cornell University, USA. **www.garamba.org**

Dr Paula Kahumbu is CEO of WildlifeDirect and has an enduring passion for preserving threatened wildlife and habitats in Kenya and beyond. Since 2008 she has also spearheaded the successful Hands Off Our Elephants Campaign with Margaret Kenyatta, Kenya's First Lady, to mobilise the justice sector in combatting elephant poaching and trafficking of ivory. She has a PhD from Princeton University, USA, in Ecology and Evolutionary Biology and has conducted field research on elephants in Kenya. Kahumbu has run the CITES office of the Kenya Wildlife Service as well as Lafarge Eco Systems, a company responsible for the restoration of mined-out lands in East Africa. She is a documentary film producer and has co-authored globally marketed books that inspire children about conservation. Awards for her conservation work include the Princeton Medal (2017), the Tribeca Disruptive Innovations Award, the Whitley Award (2014), the Presidential Award, Order of the Grand Warrior of Kenya (2014) and the National Geographic/Buffet Award for Leadership in Conservation in Africa (2011). She was named Brand Kenya Ambassador (2013) and National Geographic Emerging Explorer (2011). **www.wildlifedirect.org**

Lorna and **Rian Labuschagne.** Rian and Lorna are the project leaders of the Serengeti Conservation Project, based in the park. Before this, they worked for African Parks Network as managers of Zakouma National Park in south-eastern Chad. Between 2002 and 2008, Rian worked as the MD of Grumeti Reserves in the Serengeti District of Tanzania, with Lorna in charge of GIS. Previously they were Frankfurt Zoological Society technical advisers for the rhino conservation projects in the Ngorongoro Crater and Serengeti National Park. Rian has worked in Liwonde National Park and for South African National Parks as a section ranger. They have worked together as a team in conservation for around 35 years.

Kelly Landen moved to Botswana in 2002 after a long career on the oceans, as her true passion is wildlife. Working with Mike Chase, she co-founded Elephants Without Borders in Botswana, with a supportive office in the USA. Together, they have built up the organisation, extending the elephant-monitoring research to include research on many large herbivores and taking the lead in the Great Elephant Census. They are also involved in community education, human-wildlife co-existence initiatives and wildlife rescue operations. Kelly kindly contributed many of the photographs in this book. **www.elephantswithoutborders.org**

Dr Keith Leggett is a founding member of the Namibian Elephant and Giraffe Trust and has conducted research on elephants in Zimbabwe, Botswana, Ethiopia, Malaysia and Namibia. The desert-dwelling elephants of the Kunene Region have always been his main interest and he has published many scientific and popular articles on these unique animals. He has also been involved in elephant and conservation politics, but now prefers the relative sanity of supervising PhD students studying large mammals in Africa. **www.fowlersgap.unsw.edu.au**

Vance Martin has worked globally on conservation issues since 1975, and in 1983 became President of the WILD Foundation, an NGO born in Africa and established formally in the USA. Committed to collaboration, WILD works across cultures, governments and national boundaries to effect positive change for wilderness, wildlife and people. Working in conservation field projects, policy, communications and culture, Martin has helped initiate many conservation groups, served on their boards, and is convenor of the World Wilderness Congress. Among many responsibilities, he is currently a director of Wilderness Foundation Global (which he co-founded), and chairman of the Wilderness Specialist Group (IUCN/WCPA). **www.wild.org**

Dr Ian McCallum is a psychiatrist, analyst, author, passionate amateur naturalist and specialist wilderness guide. He is a founder member of the International League of Conservation Writers and his book *Ecological Intelligence – Rediscovering Ourselves in Nature* won the Wild Literary Award in 2009. Ian has published two anthologies of wilderness poems: *Wild Gifts* (1999) and *Untamed* (2012). He was the writer/poet for the Dylan Lewis UNTAMED exhibition at the Kirstenbosch National Botanical Garden (2010–2012). He is a trustee of the Cape Leopard Trust and a long-time associate of the Wilderness Leadership School. In 2016, Ian was a recipient of the Wildlife and Environmental Association of South Africa Lifetime Conservation Achievement Awards. **www.ian-mccallum.co.za**

Ian Michler has spent the last 28 years working as a safari operator, specialist guide, environmental photojournalist and ecotourism consultant across Africa, including 13 years living in the Okavango Delta, Botswana. His writing has appeared in a broad range of local and international publications. For 15 years he was a features writer and columnist for *Africa Geographic*, covering conservation, wildlife management and ecotourism issues of the day. He is also author and photographer of seven natural history and guide books on Africa. Ian lives along the Garden Route in South Africa and is enrolled as a part-time student for an MSc in Sustainable Development at Stellenbosch University, South Africa. **www.inventafrica.com**

Dr Neil Midlane joined Singita in 2014 and he handles conservation aspects of all current and planned Singita reserves. He has a PhD in Conservation Biology from the University of Cape Town; and an Honours in Accounting and an MA in Environmental Management, both from Stellenbosch University, South Africa. He's a member of the African Lion Working Group, the Mozambique Carnivore Working Group and the Transboundary Conservation Specialist Group. **www.singita.com**

Dr Tim O'Connor is a scientist with the South African Environmental Observation Network and an Honorary Professor with the University of the Witwatersrand, Johannesburg, where the subject of elephant-plant relations has been his research interest for the past 20 years.

Sharon Pincott is an elephant specialist, conservationist and author of five books, including *Elephant Dawn*. For 13 years she monitored and fought for the lives of Zimbabwe's 'presidential elephants', while recording their social structure and population dynamics. She chose to use the popular press to spread awareness, rather than working towards a scientific degree. She features in an award-winning international documentary and formed one of the most remarkable relationships with wild elephants ever documented. **www.sharonpincott.com**

Dr Don Pinnock is a criminologist and investigative journalist specialising in environmental issues. He has held lectureships in journalism and criminology at Rhodes University and the University of Cape Town (both in South Africa), and has published 17 books covering political biography, natural history, travel, crime, youth gangs, urban history and a novel for adolescents. He was editor of *Getaway* magazine, during which time he explored much of Africa. He has a PhD in Political History. **www.pinnock.co.za**

Greg Reis, through his conservation entity One Africa, supports a range of key environmental initiatives, one of which is the Niassa Wilderness project in the Niassa Reserve, Northern Mozambique. He has been in professional services executive management for the past 25 years and now applies tried-and-tested business principles to conservation projects. He holds degrees in Computer Science and Electrical and Electronic Engineering from the University of Cape Town, South Africa. **www.oneafrica.net**

Patricia Schonstein is an award-winning novelist and poet whose work has been translated into seven languages. She curates various anthologies and is co-editor of a poetry quarterly. She holds an MA in Creative Writing from the University of Cape Town, South Africa. **www.patriciaschonstein.com**

Dr Jeanetta Selier is an elephant scientist with more than 20 years' experience in the field of elephant conservation and management. She leads the Central Limpopo Valley Elephant Research Project, is a member of the IUCN African Elephant Specialist Group, and is a research fellow at the School of Life Sciences, University of KwaZulu-Natal, South Africa. She holds a PhD from the same university. **www.sanbi.org**

Clive Stockil was born and raised on a large cattle ranch in the Chiredzi district in the south-east corner of Zimbabwe. He is fluent in both Shona and Xangaan. In the 1980s he assisted mediation between the Mahenye community and the Gonarezhou National Park, paving the way for the first CAMPFIRE project. He is working on the establishment of the first community wildlife conservancy in Zimbabwe. He owns Senuko Ranch and, with neighbouring ranchers, formed the Save Valley Conservancy in 1992 and was its first chairman. In 2013, he was given the inaugural Prince William Award for Conservation in Africa. **www.chilogorge.com**

Garth Thompson has been a professional guide in Africa for the past 38 years. He works in over 15 countries on the continent, guiding visitors. His main passion and focus are elephants. He has given photographic presentations on elephants globally since 1983. **www.garththompsonsafaris.com**

Will Travers OBE co-founded the Born Free Foundation (1984), which works to stop individual wild animals' suffering, protect threatened species, promote compassionate conservation and engage and empower local communities worldwide. He is president of the foundation, a board member of Born Free USA and president and board chair of the Species Survival Network. He lived in Kenya while his mother, Virginia McKenna OBE, and his late father, Bill Travers MBE, made the film *Born Free* (1966) and has since dedicated his adult life to keeping wildlife in the wild. **www.bornfree.org.uk**

Andrea K Turkalo was, until recently, an associate conservation scientist at the Wildlife Conservation Society. Her research was concentrated on the forest elephant population at the Dzanga Clearing in the Central African Republic, where she spent close to 3 decades monitoring and conducting the first long-term study on this species. She holds an MA in Science Education from Columbia University, USA. **Blog: https://blog.wcs.org/photo/author/aturkalo**

Hugo van der Westhuizen has been a project leader with the Frankfurt Zoological Society since 1997, initially in North Luangwa National Park, Zambia. In 2007, he started the Gonarezhou Conservation Project in Zimbabwe. He holds an MSc in Conservation Biology from the University of Kent in Canterbury, UK. **www.fzs.org**

Dan Wylie teaches English at Rhodes University, Grahamstown, South Africa. He has published three books on the Zulu leader Shaka, including *Myth of Iron: Shaka in History* (UKZN Press). He has also published a memoir, *Dead Leaves: Two Years in the Rhodesian War* (UKZN Press); *Elephant* and *Crocodile* in the Reaktion Books Animals Series; and several volumes of poetry. Most recently, he has concentrated on Zimbabwean literature and on ecological concerns in literature.
www.danwyliecriticaldiaries.blogspot.com

About the photographers

We are indebted to the many photographers who contributed so willingly to this book, and to the many others who offered their images, but couldn't be included. Every one of the photographers (and one artist) gave their images for free, as they know how much elephants are in crisis and want to be part of the solution. We salute and thank every one of you.

Dana Allen is a naturalist, author, guide, educator, artist and professional wildlife photographer. Originally from California, he has lived and worked in southern, Central and East Africa for the past 25 years and currently lives in Zimbabwe. He founded PhotoSafari in 1991 and has specialised in photographing wildlife, the environment and tourism activities ever since. www.PhotoSafari-Africa.net

Grant Atkinson is a wildlife photographer and guide based in Cape Town. He spent 12 years working in Botswana's Okavango Delta and the ˙inyanti and Chobe regions. He and his wife, ˙ena, currently lead photographic expeditions in ˙˙a and abroad. www.grantatkinson.com

˙˙ustinus is a renowned wildlife artist ˙˙tings are on display and whose ˙˙ad throughout the world. He grew ˙˙˙d has spent most of his 65 years ˙˙˙motest parts of the continent, ˙˙˙ation for his paintings. ˙˙us.com

˙˙˙r investigative ˙˙e Sharna are influential ˙˙nservation ˙˙r photography, they ˙˙ct the wilderness

for future generations. They guide small groups into some of the wildest and most remote areas on the planet. www.wildphotossafaris.com / www.afripics.com

Colin Bell has been photographing elephants around Africa for the past 40 years during his career as a safari guide and camp operator. One of the advantages of his chosen career is the privilege of spending quality time in the field – across the continent, on a regular basis, armed with cameras. www.africasfinest.co.za

Joerg Boethling is a freelance photographer based in Hamburg, Germany. He started taking photographs in 1985 as a seaman on voyages to Asia and Africa. Today he travels and works worldwide, focusing on social, economic and environmental issues. He has a special passion for India, but in the last years has also turned his attention to Africa. www.visualindia.de

Kate Brooks is an international photojournalist turned filmmaker, who chronicled conflict and human rights issues for nearly 2 decades before turning her lens to conservation. Her photographs have been extensively published in *TIME*, *Newsweek*, *Smithsonian* and *The New Yorker*, and exhibited around the world. Brooks' passion for filmmaking was sparked from working as a cinematographer on the documentary *The Boxing Girls of Kabul* in 2010. In 2012–2013, she was a Knight Wallace Fellow at the University of Michigan, USA. There she researched the global wildlife-trafficking crisis before embarking on directing *The Last Animals*. Brooks' drive and passion for conservation come from the fundamental belief that time is running out and that we are at a critical moment in natural history. www.thelastanimals.com / www.katebrooks.com

Will Burrard-Lucas is a British wildlife photographer known for developing innovative devices that allow him to gain unique photographic perspectives. He is the founder of Camtraptions, RAW Exposure and WildlifePhoto. com. His images can be found on his website or on Instagram. www.burrard-lucas.com / @willbl

Peter Chadwick is an award-winning photographer who specialises in photographing and writing about conservation and environmental issues in Africa. He is a Fellow of the International League of Conservation Photographers and a key supporter of the Game Rangers' Association of Africa. www.peterchadwick.co.za / www.gameranger.org / www.conservationphotography.co.za

David Chancellor is an award-winning documentary photographer. His interests are mapping the jagged and contested line where humans and wild creatures meet, and highlighting the commodification of wildlife. He has exhibited in major galleries and museums, and published worldwide. In 2012 he published the monograph *Hunters*. www.davidchancellor.com

Shem Compion is a naturalist and professional wildlife photographer. His work is published worldwide and his images have won national and international awards. He has published five books. Shem runs www.c4photosafaris.com, leading photo safaris throughout Africa.

Ross Couper's unique view of life is interpreted through his photography. As both the resident photographer and a safari guide for Singita Game Reserve, his field experience allows him to depict wildlife behaviour artistically through photographic imagery. www.rosscouper.com

Kyle de Nobrega has been a nature guide for over a decade. He takes his guests on safaris that not only epitomise true wilderness experiences and wildlife of Africa, but also highlight conservation and the sustainability of the natural world. www.naturalistphoto.com

Deon de Villiers has managed the operations for some of Africa's most sought-after and remote luxury safari camps. Before this, he ran business development for global software solution providers into South-east Asia. He uses his experience in developing sales strategies for purpose-driven wildlife adventures in the Asia-Pacific region, focusing on awareness and education for conservation and sustainability. www.safagraphics.com

Tim 'Gonondo' Driman has been involved in wildlife conservation for over 35 years. He is a wildlife photographer, Sony Shooter, Wild Coolers ambassador and FGASA field and trails guide. He travels sub-Saharan Africa with a big grin and a clutch of mirror-less cameras. www.timdrimanphotography.com

Daniel Dugmore was born in Southampton, England, and was introduced to photography at a young age while visiting family in Botswana. He later emigrated there to pursue his passion of safari guiding and conservation photography. He leads wildlife photography workshops and safaris across Africa. www.danieldugmore.com

Chris Fallows' work has been featured internationally in the form of articles, documentaries and award-winning imagery. Together with his wife, Monique, he specialises in showcasing wildlife in all its diversity, and highlighting the need for its conservation. They live in Cape Town, South Africa. www.apexpredators.com

Dr Paul Funston works for Panthera as senior director of their Lion Programme, and is based in Namibia's Zambezi Region, where herds of elephants still roam. He works to save lions and their prey in key protected areas across Africa, with a regional focus in the KAZA region. www.panthera.org

Martin Harvey has been a wildlife and travel photographer for more than 20 years and, more recently, a videographer. His images have been published in magazines and books throughout the world, and he is frequently commissioned to photograph travel destinations and environmental issues. He has recently changed from stills to video. www.wildimagesonline.com

Andrew Howard was born in Johannesburg in an era when television ran for an hour a day and there was time to explore the natural world. Over time, he developed a deep emotional attachment to Africa's wild places. He has poured this passion into design, brand building and photography. www.andrewhowardphoto.com

Natalie Ingle is an avid traveller and environmental activist. While working for the Wildlife Conservation Society, she helped document and fight some of the most intense conservation challenges of our time, and has supported some of wildlife's greatest champions – such as those in Yankari. Her writing has appeared in the *New York Times* and her photography has been used to support a variety of social and ecological causes.

Lets Kamogelo is a Botswana-based professional guide and wildlife photographer who uses his skills to make a positive contribution to the conservation of Botswana's pristine wilderness areas. For this he uses all platforms of social media and is featured in a range of magazines. www.goo.gl/a5Hyve

Elsen (Elk) Karstad is resident in Kenya and an eco-entrepreneur involved in sustainable charcoal production, seed-ball reforestation and fish farming in the Mount Kenya region. His MSc at the University of Alberta, Canada, explored the ecology of the hippos in the Mara River. He has been involved in wildlife photography for over 40 years and holds regular exhibitions in Nairobi. seedballskenya@gmail.com

Nigel Kuhn is a soldier, wildlife guide, photographer and teacher. He joined Chengeta Wildlife in 2016 as an anti-poaching trainer, based in Mali. By combining his photography, film, military experience and knowledge of the bush, he is able to capture special moments in time. www.nigelkuhn.com

Hannes Lochner has been photographing wildlife professionally since 2007. He has produced five photographic books, three of which were dedicated to the Kalahari. To achieve this, he lived in the Kalahari for 6 years. His latest book, *Planet Okavango*, is as much art as photography. He is a graduate of Stellenbosch University, South Africa. www.hanneslochner.com

Dr Johan Marais is a veterinary surgeon and CEO of the non-profit Saving the Survivors, dealing with endangered wildlife, and rhinos, in particular. He has travelled extensively in southern, East and Central Africa in search of Africa's legendary great tusker bulls, which he memorialised in two books. www.facebook.com/johan.marais.55 / www.savingthesurvivors.org

Ian Michler has spent the last 28 years working as a safari operator, specialist guide, environmental photojournalist and ecotourism consultant across Africa, including 13 years living in the Okavango Delta, Botswana. His writing has appeared in a broad range of local and international publications. For 15 years he was a feature writer and columnist for *Africa Geographic*. He is also the author and photographer of seven natural history and guide books on Africa. www.inventafrica.com

Martin Middlebrook trained as a wildlife artist before beginning a career as a photographer and photojournalist. He has worked throughout the world on subjects as varied as the human consequences of the war in Afghanistan, the tribes of Omo Valley in Ethiopia and elephant conservation in Africa. He is based in Paris. www.martinmiddlebrook.com

Anthony Njuguna is one of three 'urban nomads' who operate under the collective Routes Adventure banner, documenting the cultural and natural beauty of Kenya through visual storytelling. They met 4 years ago on a 450-kilometre wildlife conservation walk across two countries, raising awareness for African wildlife rangers. www.routes.co.ke

Vuthlare Nyathi is a Wild Shots Outreach student from Mabine Primary School in Limpopo, South Africa. The Outreach engages young local people in wildlife and conservation through photography. Vuthlare's photo is from an Outreach course run in partnership with Elephants Alive. c/o mkmikekendrick@gmail.com

Peter and **Beverly Pickford** are professional wildlife and natural history photographers and conservationists based in Africa. They began their careers as game reserve and lodge managers, moved to conservation and wilderness photojournalism and have published nine books. Their latest book, *Wild Land*, is a celebration of the last great wilderness areas on Earth. www.wildlandphoto.com

Dr Don Pinnock is a photojournalist specialising in environmental issues. He has published 17 books and held three photographic exhibitions. During his time as editor of *Getaway* magazine, he explored much of Africa. www.pinnock.co.za

Christophe Pitot grew up in Mauritius where he became a voluntary conservationist at the age of 14. After graduating as lead trails guide in 2011 in South Africa, he went on to work for Elephant-Human Relations Aid in Namibia. He is now Field Operations Manager and leads the tracking, movement and identification study of Namibia's desert-adapted elephants. www.sacredworld.co.uk

Michael Poliza is a World Wide Fund for Nature ambassador and has been portraying the wonders of Africa and beyond for the past 20 years. His huge coffee-table books are sold in over 70 countries. In 2011 he founded Michael Poliza Private Travel, taking travel and nature enthusiasts to the most secluded and untouched spots on Earth. www.michaelpoliza.com

Darren Potgieter was born just south of the Kruger National Park, where he developed his passion for wildlife. After completing studies in environmental science and ecology he started a not-for-profit enterprise that provided aerial resource surveys and services to the conservation sector, successfully completing projects in Mozambique, Namibia, Botswana, Zambia and South Africa. His career has included working as field operations manager and establishing Zakouma National Park's elephant monitoring and protection programme, and being section manager in the Niassa Reserve in Mozambique. He now lives in South Africa and continues to support conservation efforts throughout Africa.

Thierry Prieur is a teacher in Montreal, Canada, and has a Masters degree in music, psychology and education. He has published a manual for pre-school teachers, showing how to integrate arts and culture in the school curriculum. Photography and travel are his favourite hobbies.

Scott Ramsay is a photographer and writer focusing on conservation and national parks, and has travelled to around 75 African protected areas. His photographs, articles and interviews with conservationists have been published internationally. His photography seeks out the sacredness of Africa's wild places. www.LoveWildAfrica.com

Pieter Ras is a safari lodge manager who has travelled and worked in some of the most remote places on Earth. He loves wildlife and portrays it through his photographs. You can see more of his work on his website. www.raisinphoto.com

Robert J Ross is a New Yorker who spends part of each year living in Africa. For the past 20 years he has been capturing beautiful moments in Africa and around the world. Rob has published *The Selous in Africa – A Long Way From Anywhere*, in which he showcases Tanzania's Selous Game Reserve, one of Africa's last great wilderness areas. www.rjrossphoto.com

Dave Southwood grew up on a farm in Botswana and was introduced by his father, from a very early age, to nature and photography. During his guiding career he developed a passion for storytelling through his imagery. He travelled the world searching for an African equivalent, and realised that none existed. He mentors fellow guides, photographers and guests.

Keith Stannard was raised in a family of avid wildlife enthusiasts who spent most of their weekends and holidays exploring parks and reserves around southern Africa. With wildlife in his DNA, he was destined to spend his life working in East and southern Africa as a

safari guide, lodge manager and now as an eco-tourism practitioner.

Brent Stapelkamp is a wildlife photographer with a deep passion for lions. Studying these cats for the last decade or so (including Cecil) has given him an understanding of the rich complexities around human-wildlife relationships, which drives him to tell those stories in his photos. He founded The Soft Foot Alliance Trust with his wife, Laurie Simpson, and they work towards just that. www.softfootalliance.org

Sabine and **Charl Stols** are husband-and-wife photographic safari guides based in Kasane, Botswana. German-born Sabine and her South African husband Charl, along with **Guts Swanepoel**, lead photo safaris and workshops for Pangolin Photo Safaris in Chobe National Park, the Okavango Delta and other wild places around Africa. www.pangolinphoto.com

Brent Stirton is a photographer who focuses on conservation issues for 8 months of the year, and on more conventional photojournalism for the remaining 4 months. He does most of his work for *National Geographic* magazine, but also works for the *New York Times*, *Le Figaro*, *Stern* and other international media titles. He works for Human Rights Watch on a regular basis, and is primarily interested in the interrelations between humans and the environment. www.brentstirton.com

Waldo Swiegers is a professional photographer based in South Africa. He works mostly on the African continent and tells visual stories through various international agencies. His work is featured in publications all over the world. His online portfolio is available on his website. www.waldoswiegers.com.

Steven Thackston is a photographer based in Atlanta, Georgia. His work reflects a variety of interests, including the preservation of Africa's wildlife. He has worked with the rangers of the Northern Rangelands Trust in Kenya. www.steventhackston.com

Garth Thompson is regarded by many as one of the best guides in Africa. For close on 40 years he has been travelling to Africa where he guides visitors to the region, following his main passion – elephants. www.garththompsonsafaris.com

Andrea K Turkalo has worked with the Ba'Aka people in the Central African Republic for 22 years, studying and photographing the elephants that frequented the Dzanga Baï clearing. During this time, she was able to identify over 3 000 individual elephants. Her study contributed immensely to the understanding of their vocalisations and habits. Blog: https://blog.wcs.org/photo/author/aturkalo

Heinrich Van Den Berg is an award-winning wildlife and nature photographer, having published more than 30 acclaimed photographic books. He is founder of the publishing company HPH Publishing. www.heinrichvandenberg.com / www.hphpublishing.co.za

Bobby-Jo Vial has been photographing wild elephants for over a decade, inspiring many people to take an interest in wildlife and conservation. Her images have featured in many international publications, among them the *Sydney Morning Herald*, *The Times*, *The New York Post*, *Paris Match* and *Africa Geographic*. Her recent publication, *Reflections of Elephants*, was released in 2016, co-inciding with World Elephant Day. The book raises funds for the protection of some of Africa's last big tuskers through her work as part of The Askari Project. www.dumasafaris.com.au

John Vosloo is an amateur photographer with a passion for wildlife and photography. His images have been published locally and internationally.

He says that sitting quietly in the bush in the early morning and knowing an elephant is passing by at close quarters, hearing its skin chafing as it walks, its stomach rumbling, and then looking into those wise, all-seeing eyes inspires him to pick up his camera. www.johnvosloophotography.com

Tami Walker's photography is the product of her passion, love and admiration for her beautiful country, Zimbabwe. She hopes that her images engage, inspire and encourage true exploration of Africa – not only to experience its magnificence but also to understand its fragility. www.tamiwalkerphotography.com

Carlton Ward Jr is a National Geographic Explorer who has photographed the Mali Elephant Project and continues to assist with its communications. He uses photography to inspire conservation of Florida's nature and culture: by connecting the public to panthers, in particular, he is currently seeking to encourage the habitat protection needed to expand the panther population and keep Florida wild. www.carltonward.com

Marcus Westberg is a Swedish photographer, writer and guide focusing mainly on conservation issues in sub-Saharan Africa. A Wildlife Photographer of the Year finalist with a background in environmental science, Marcus spends much of his time in the field working with non-profit organisations such as African Parks. www.lifethroughalens.com / maptia.com/marcuswestberg/store

Prof Lee White CBE has worked as a scientist and a conservation and environmental policy maker throughout the rainforest countries of West and Central Africa for over 30 years. For almost 20 years Lee worked for the Wildlife Conservation Society, and then shifted to work for the Gabonese government, initially as a climate-change scientist and UNFCCC negotiator, and from 2009 as head of their national parks agency. www.wcs.org

References

Chapter 3: Imagining Africa's elephants

Dan Wylie

1: http://researchspace.csir.co.za/dspace/handle/10204/2091
2: Bleek-Lloyd archive.
3: Chapman (2012).
4: Le Vaillant (2007 edn).
5: Stigand (1913).
6: Gary (1958).
7: Douglas-Hamilton (1975).
8: Brettell (1994).

Chapter 6: A tale of two elephants

Audrey Delsink

1: Hopkinson et al. (2008).
2: Van Aarde et al. (2008).
3: A moratorium on culling and the relocation of orphans was applied in 1994. Slotow et al. (2008).
4: Fayrer-Hosken et al. (2000).
5: Delsink & Kirkpatrick (2012).
6: In captive individuals and in the KNP trials. In Bertschinger et al. (2008).
7: Cohn & Kirkpatrick (2015).
8: Delsink et al. (2006).
9: Delsink et al. (2013a).
10: The vaccine works on an individual animal's immunity and thus there is a small percentage that may not demonstrate an immune response.
11: Bertschinger et al. (2018).
12: Department of Environmental Affairs and Tourism (DEAT) (2008).
13: Slotow et al. (2008).
14: Such activity is not restricted in the Norms and Standards or TOPS (Threatened or Protected Species) regulations, even when it is used to repeatedly harass the animal.
15: There has to be a better way to manage elephants that speaks to their biological requirements and responses.
16: Poole (1994).
17: Pinnock, D (2018) The problem of an elephant that just wants to stay home. Daily Maverick https://www.dailymaverick.co.za/article/2018-04-20-the-problem-of-an-elephant-that-just-wants-to-stay-home.

Chapter 12: Constant gardeners of the wild

Garth Thompson

1: Dunham et al. (2004).

Chapter 13: Ensuring elephant survival through community benefit

Romy Chevallier & Ross Harvey

1: Leithead, A (2016).
2: Songhurst, A, McCulloch, G & Coulson, T (2015).
3: Lindsey, PA et al. (2013).
4: Chase, MJ et al. (2016).
5: Wittemyer, G et al. (2014).
6: Massay, GE (2017).
7: Ibid.: 7.
8: Harvey, R, Alden, C & Wu, Y (2017).
9: Harvey, R (2017).
10: Harvey, R (2015).
11: Bennett, EL (2014).
12: Challender, DWS & Macmillan, DC (2014).
13: Wu, Y, Rupp S & Alden, C (2016).
14: This is not a philosophical endorsement of the view that elephants should only stay 'if they pay their way'. It is simply recognition of the fact that in a world where competing land-use options is a development reality, the intrinsic value of elephants and wilderness landscapes is not necessarily fully appreciated. For further discussion on these topics, see Duffy, R (2014).
15: Orr, T (2016).
16: Van der Duim, R et al. (2015).
17: For instance, the members of one CBO whom these researchers interviewed in September 2015 expressed their frustration at the centrally imposed loss of hunting revenues in Botswana, and asserted that poaching was a definite response option.
18: It must be noted that the game-hunting debate remains polarised across the region, with countries such as Kenya and Botswana having banned hunting tourism in 1977 and 2014 respectively.
19: Crookes, DJ & Blignaut, JN (2016).
20: Chevallier, R & Harvey, R (2016b).
21: Bunney, K, Bond, WJ & Henley, M (2017).
22: Harvey, R (2015): 23.
23: Hiedanpää, J & Bromley, DW (2014).
24: Chevallier, R & Harvey, R (2016).
25: Gujadhur, T (2001).
26: Chabal, P & Daloz, JP (1999); Van de Walle, N (2001).
27: Orr, T (2016) also makes the important point that rent capture occurs not only at the local but also at the international level. Western consultants earn large rents from devising management plans and research briefings – revenue that could go straight into the pockets of local community members who could be paid to preserve wildlife. Obviously the options are not quite as simple as this, but the point remains that there is inefficient and inconsistent capture at all levels.
28: Email correspondence with Ian Craig, Director of Conservation at the Northern Rangelands Trust, Kenya, 24 August 2016.
29: www.theguardian.com/environment/2015/aug/03/delta-bans-hunting-trophies-cecil-the-lion
30: Some academics are skeptical as to whether chillies are an effective deterrent. See Hedges, S & Gunaryadi, D (2010). But it looks as though variated use may be appropriate, along with other deterrents such as capsicum oleoresin. See also Songhurst, A, McCulloch, G & Coulson, T (2015).
31: Levy, B (2014); Acemoglu, D & Robinson, JA (2013).
32: On this question of local historical institutions of democracy, see Hillbom, E (2012): 'To keep the chief accountable and to hinder corruption, he was checked by the kgotla, a semi-democratic system building on public meetings, where male members of the tribe could air their opinions regarding the chief's actions': 78.
33: On the importance of better communication, see Snyman, S (2014).
34: Orr, T (2016).
35: Email correspondence with Ian Craig, op. cit.
36: Ripple, WJ (2015)
37: For a practical set of policy recommendations, see Chevallier, R & Harvey, R (2016).

Chapter 14: Funding elephant conservation

Dr Don Pinnock

1: Huismann, W (2014).
2: Ibid.: 18.
3: Ibid.: 36.
4: https://news.mongabay.com/2016/04/big-conservation-gone-astray
5: https://news.mongabay.com/2016/05/big-donors-corporations-shape-conservation-goals
6: Huismann (2014).
7: Huismann, ibid.: 78.
8: Peck, J & Tickell, A (2002); Castree, N (2010); Foucault, M (2004).
9: https://cer.org.za/wp-content/uploads/1999/01/Draft-National-Biodiversity-Offset-Policy.pdf
10: Berry, T (1999).

org/reports/corporate-war-crimes-prosecuting-pillage-natural-resources

24: http://www.gameranger.org/news-views/media-releases/170-media-statement-the-use-of-military-and-security-personnel-and-tactics-in-the-training-of-africa-s-rangers.html; https://theconversation.com/foreign-conservation-armies-in-africa-may-be-doing-more-harm-than-good-80719?utm_source=twitter&utm_medium=twitterbutton

25: https://theconversation.com/foreign-conservation-armies-in-africa-may-be-doing-more-harm-than-good-80719?utm_source=twitter&utm_medium=twitterbutton; for a few examples of private security companies and non-profit organisations involved in anti-poaching operations, see: Rhula Intelligent Solutions (http://www.rhula.net/); Maisha Consulting (http://maishaconsulting.com/environmental-security/); VETPAW (http://vetpaw.org/)

26: Author interviews and observations during the course of field research, 2014–2017.

27: See, for example, http://www.newsweek.com/2017/08/18/trophy-hunting-poachers-rhinos-south-africa-647410.html; https://adamwelz.wordpress.com/2017/08/14/commentary-on-nina-burleighs-newsweek-article-on-race-war-centered-on-rhinos-in-south-africa; https://www.militarytimes.com/veterans/2015/05/07/reports-nonprofit-vetpaw-kicked-out-of-tanzania

28: For examples of proposed standards and criteria, see: https://www.asisonline.org/About-ASIS/Who-We-Are/Presidents-Perspective/Pages/Standards-for-Private-Security-Contractors.aspx; http://www.gameranger.org/news-views/media-releases/170-media-statement-the-use-of-military-and-security-personnel-and-tactics-in-the-training-of-africa-s-rangers.html; http://psm.du.edu/national_regulation/united_states/laws_regulations/

29: http://lowvelder.co.za/220456/intensive-protection-zone-set-become-safe-haven-rhino/; https://www.dailymaverick.co.za/article/2017-07-24-poaching-sa-heads-for-1000-rhino-killings-for-the-fifth-year-in-a-row/

30: 40mm Under Barrel Grenade Launcher (UBGL) systems: see, for example, https://www.sanparks.org/about/news/?id=56814; http://www.timeslive.co.za/sundaytimes/stnews/2016/07/22/SANParks-plans-to-lob-grenades-at-rhino-poachers

31: https://www.nytimes.com/2016/11/28/science/a-forgotten-step-in-saving-african-wildlife-protecting-the-rangers.html?mcubz=3&_r=0

32: https://www.environment.gov.za/mediarelease/molewa_worldrangerday2017

33: http://news.nationalgeographic.com/2015/11/151124-zimbabwe-elephants-cyanide-poaching-hwange-national-park-africa/; https://voices.nationalgeographic.org/2014/08/17/poisons-and-poaching-a-deadly-mix-requiring-urgent-action/; http://www.ifaw.org/united-states/news/more-poisoned-arrows-used-poach-tsavo-elephants

34: https://www.theguardian.com/environment/2017/aug/19/super-gangs-africa-poaching-crisis?CMP=Share_iOSApp_Other

Chapter 19: CITES and trade: is this the organisation to save elephants?

Adam Cruise

1: CITES.org What is CITES.
2: Reeve, R (2002): 88.
3: CITES.org How CITES Works.
4: CITES.org What is CITES?
5: CITES.org The CITES Appendices.
6: CITES.org Conference of the Parties.
7: CITES.org The CITES Appendices.
8: CITES.org The CITES Secretariat.
9: CITES.org CITES Compliance Procedures.
10: Nowak, K (2016).
11: CITES.org.
12: Lemieux, AM & Clarke, RV (2009).
13: Humane Society International (2017).
14: Currey, D & Moore, H (1994).
15: CITES.org Address by the President of the Republic of Zimbabwe, CDE RG Mugabe.
16: CITES.org Resolution Conf. 10.10 (Rev. CoP17).
17: CITES.org MIKE and ETIS.
18: Wasser, SK et al. (2007).
19: Traffic.org (2017).
20: WWF.org.
21: CITES.org CoP14 Prop. 6 Consideration of Proposals for Amendment of Appendices I and II.
22: Hsiang, S & Sekar, N (2016).
23: Chase, MJ et al. (2016).
24: IUCN (2016).
25: Cruise, A (2016).
26: CITES.org Trade in Live Elephants from Zimbabwe to China.
27: CITES.org Conf. 11.20 (Rev. CoP17) Definition of the term 'Appropriate and Acceptable Destinations'.
28: FWS.gov (2016).
29: CITES.org CoP17 Prop. 16 Consideration of Proposals for Amendment of Appendices I and II.
30: CITES.org Reservations.
31: CITES.org Gaborone Amendment to the text of the Convention.
32: CITES.org CITES Trade Database.
33: Cruise, A (2016).
34: Ibid.

Chapter 20: Translocating elephants: are welfare and conservation in conflict?

Dr Marion E Garaï

1: Biggs, HC (2003).
2: National Norms & Standards for Elephants (2008).
3: Garaï, ME et al. (2004).
4: Pretorius, Y, Garaï, ME & Bates, LA (2018).
5: Junker, J, Van Aarde, RJ & Ferreira, SM (2008).
6: Scholes, RJ & Mennell, KG (2008).
7: 13% from 1992 to 1994; Garaï unpublished survey data.
8: Pretorius, Y, Garaï, ME & Bates, LA (2018).
9: Sukumar, R (1993).
10: Garaï, ME et al. (2004).
11: Garaï, ME & Töffels, O (2011).
12: Range: 8.9–22 years; Whitehouse, AM & Hall-Martin, AJ (2000); Moss, CJ (2001).
13: Poole, JH (1987).
14: Garaï, ME & Töffels, O (2011).
15: Evans, K, Moore, R & Harris, S (2013).
16: Slotow, R & Van Dyk, G (2001).
17: Garaï, ME & Carr, RD (2001); see also Fernando, P, Leimgruber, P & Pastorini, J (2012).
18: Garaï, ME (1997).
19: Personal communication with owners and managers.
20: Garaï, ME (1997).
21: Lee, PC & Moss, CJ (2012).
22: Garaï, ME et al. In preparation.
23: Keese, N (2012).
24: Bradshaw, GA (2009).
25: Abe, EL (1994).
26: Archie, EA & Chiyo, PI (2011).
27: Kurt, F & Garaï, ME (2007).
28: Garaï, ME & Töffels, O (2011).
29: Shannon, G et al. (2013).
30: McComb, K et al. (2000).
31: Woolley, L-A et al. M (2008).
32: Poole, JH & Moss, CJ (2008).
33: Viljoen, JJ et al. (2008); Bradshaw, GA & Shore, AN (2007).
34: Garaï, ME et al. In preparation.
35: Pretorius, Y (2004).
36: Elephant Specialist Advisory Group (ESAG) (2017).
37: Bradshaw, GA (2009).
38: Tingvold, HG et al. (2013).
39: Van Aarde, RJ & Jackson, TP (2007).
40: As defined by Hanski, I & Gilpin, M (1991).
41: Loarie, SR, Van Aarde, RJ & Pimm, SL (2009).

Chapter 22: Making a safe haven

Colin Bell

- Van Aarde, R (2013).
- Conservation Ecology Research Unit Publications.
- www.fws.gov/le/pdf/CITES-and-Elephant-Conservation.pdf
- CITES: Consideration of Proposals for Amendment of Appendices I and II.
- cites.org/sites/default/files/eng/cop/17/prop/KE_Loxodonta.pdf
- Marais, J & Ainslie, A (2010).
- Junker, J (2008).
- Purdon, A *et al.* (2018).
- World Travel & Tourism Council (WTTC) (2018).

Chapter 23: Botswana's sanctuary

Kelly Landen

1: Continent-wide survey reveals massive decline in African savannah elephants: https://peerj.com/articles/2354/
2: The shared nature of Africa's elephants: https://www.sciencedirect.com/science/article/pii/S0006320717303890

Chapter 28: Desert-dwelling elephants of north-west Namibia

Dr Keith Leggett

1: Movement data: Leggett (2006) Independent research; genetic analysis: Ishida *et al.* (2011).
2: Viljoen (1987).
3: *Ibid*.
4: Owen-Smith (1970).
5: Viljoen (1987).
6: *Ibid*.
7: *Ibid*.
8: Lindeque & Lindeque (1991).
9: Killian (2017).
10: Laws, 1970; Leuthold & Sale (1973); Kerr & Fraser (1975); Western & Lindsay (1984); Ruggerio (1992); Tchamba (1993); White (1994); Thouless (1995); Dublin (1996); Babaasa (2000).
11: Babaasa (2000).
12: Viljoen (1989); Leggett, Fennessy & Schneider (2002).
13: Jacobson *et al.* (1995); Fennessy, Leggett & Schneider (2001).
14: Leggett, Fennessy & Schneider (2003).
15: Poole (1996).
16: Lindeque & Lindeque (1991); Viljoen (1987); Moss & Poole (1983); Poole (1996).
17: e.g. Poole (1996).
18: Douglas-Hamilton (1972); Moss & Poole (1983).
19: Viljoen (1987).
20: Owen-Smith (pers. com.).

21: Leggett, KEA, Ramey, RR & MacAlister-Brown, L (2011).
22: All the genetic information presented in this chapter was provided by Dr Rob Ramey of the Denver Natural History Museum.
23: Wyatt & Eltringham (1974); Guy (1976); Kabigumila (1993).
24: Leggett (2006).
25: Guy (1976).
26: Kabigumila (1993).
27: Guy (1976).
28: Poole (1982).
29: Guy (1976).
30: Barnes (1982).
31: Wyatt & Eltringham (1974); Guy (1976); Kabigumila (1993).
32: Leggett *et al.* (2001); Viljoen & Bothma (1990).
33: Leggett (2006).
34: Guy (1977).
35: Stoinski *et al.* (2000).

Chapter 29: Selous Game Reserve: paradise lost?

Colin Bell

1: 30 000 poached elephants annually = 60 000 tusks. At a conservative 7 kilograms per tusk, that equates to around 420 000 kilograms annually.

Chapter 31: Beneath Kilimanjaro: elephant conservation in Kenya

Dr Paula Kahumbu

1: http://www.traffic.org/home/2016/5/9/sophisticated-poachers-could-undercut-bold-kenyan-fight-agai.html
2: http://www.savetheelephants.org/project/mike/
3: https://cites.org/eng/prog/mike/data_and_reports
4: https://eawildlife.org/resources/reports/Report_of_the_task_force_on_WildLife_Security.pdf
5: http://spaceforgiants.org/giantsclub/
6: http://wildlifedirect.org/hands-off-our-elephants/
7: https://wildlifedirect.org/wp-content/uploads/2017/03/Rapid-Reference-Guide-2016.pdf
8: http://wildlifedirect.org/wp-content/uploads/2017/05/WL-Digest-2016.pdf
9: Kahumbu *et al.* (2014).
10: Kahumba *et al.* (2014).
11: http://www.savetheelephants.org/project/mike/
12: http://wwf.panda.org/what_we_do/endangered_species/elephants/human_elephant_conflict.cfm
13: http://elephantsandbees.com/deterring-elephants-with-the-sound-of-bees/

Chapter 33: The elephant and the kid

Luca Belpietro

1: Stephen Cobb estimated there to be 35 000 elephants in 1973 and 1974. Simon Trevor, Tsavo Warden under David Sheldrick in 1960, told me of 45 000 elephants in 1974.

Chapter 34: Urgent intervention needed to save forest elephants

Wynand Viljoen

1: Poulsen *et al.* (2017).

Chapter 35: Garamba National Park: conservation on the continental divide

Naftali Honig

1: Hillman Smith, K & Kalpers, J (n.d.): Garamba.

Chapter 39: The elephants of Yankari

Nachamada Geoffrey & Andrew Dunn

1: It was established in 1895 to save wildlife and wild places through science, conservation action and education and to inspire people to value nature.
2: To find out more, visit www.smartconservationsoftware.org

Chapter 40: Zakouma: an elephant success story

Lorna Labuschagne

1: www.youtube.com/watch?v=wn56eKvbqcs

Chapter 41: The desert elephants of Mali

Vance G Martin & Susan Canney

1: Maiga, M (1996); Ganame, N (1999); Canney, S *et al.* (2007).
2: Blake, S *et al.* (2003).
3: Sayer, JA (1977); Douglas-Hamilton, I (1979); Bouché, P *et al.* (2009); Dias, J *et al.* (2015); Jachmann, H (1991); Canney, S *et al.* (2007).
4: Canney, S *et al.* (2007); Wall, J *et al.* (2013).
5: Canney, S *et al.* (2007).
6: Canney, S (2015).
7: Ganame, N *et al.* (2009).
8: Canney, S & Ganame, N (2014).
9: Canney, S (2017).
10: Hill, M (2017).

Bibliography

Abe, EL (1994) *The behavioural ecology of Elephant Survivors in Queen Elizabeth National Park, Uganda.* PhD dissertation, Cambridge University.

Acemoglu, D & Robinson, JA (2013) The pitfalls of policy advice. *Journal of Economic Perspectives* 27(2): 173–192.

Adams, WM *et al.* (2004) Biodiversity conservation and the eradication of poverty. *Science* 306: 1146–1149.

Alden Wily, L (2011) *Rights to resources in crisis: Reviewing the fate of customary land tenure in Africa.* Rights and Resources Group, Washington DC.

Alexander, RD (1974) The evolution of social behavior. *Annual Review of Ecology and Systematics* 5: 325–383. Available at: http://www.jstor.org/stable/2096892

All Africa (2013) *Tanzania: Five Tanzania Port Authority officials fired for corruption.* [Online] Available at: http://allafrica.com/stories/201301230166.html [Accessed 29-08-2017].

Archie, EA & Chiyo, PI (2011) Elephant behaviour and conservation: Social relationships, the effects of poaching, and genetic tools for management. *Molecular Biology.* Available at: doi: 10.1111/j.1365-294X.2011.05237

Asner, GP & Levick, SR (2012) Landscape-scale effects of herbivores on treefall in African savannas. *Ecology Letters* 15(11): 1211–1217.

Babaasa, D (2000) Habitat selection by elephants in Bwindi Impenetrable National Park, south-western Uganda. *African Journal of Ecology* 38: 116–122.

Baillie, J, Hilton-Taylor, C & Stuart, SN (2004) *IUCN Red List of Threatened Species. A Global Species Assessment.* IUCN, Gland, Switzerland and Cambridge, United Kingdom.

Baldus, R (2009) *Wild Heart of Africa.* Rowland Ward, Johannesburg.

Barnard, PJ *et al.* (2007) Differentiation in cognitive and emotional meanings: An evolutionary analysis. *Cognition and Emotion* 21: 1155–1183.

Barnes, RFW (1982) Elephant feeding behaviour in Ruaha National Park, Tanzania. *African Journal of Ecology* 20: 123–126.

Bates, LA, Poole, JH & Byrne, RW (2008) Elephant cognition. *Current Biology* 18(13): 544–546.

Bechky, A (1990) *Adventuring in East Africa.* Sierra Club Books, San Francisco.

Bennett, EL (2014) Legal ivory trade in a corrupt world and its impact on African elephant populations. *Conservation Biology* 29(1): 54–60.

Berry, T (1999) *The Great Work.* Three Rivers Press, New York.

Bertschinger, HJ *et al.* (2008) Reproductive control of elephants. In RJ Scholes & KG Mennell (eds.), *Elephant Management: A Scientific Assessment for South Africa*: 257–328. Wits University Press, Johannesburg.

Bertschinger, HJ *et al.* (2018) Porcine Zona Pellucida vaccine immunocontraception of African elephant (*Loxodonta africana*) cows: A review of 22 years of research. *Bothalia: African Biodiversity & Conservation Biology* 48(2): 8.

Biggs, HC (2003) *The Kruger experience: Ecology and management of savanna heterogeneity.* Island Press, Washington DC.

Blake, S *et al.* (2003) *The Last Sahelian Elephants: Ranging behaviour, population status and recent history of the desert elephants of Mali.* Save the Elephants. Unpublished report.

Bleek-Lloyd Archive. Available at: http://www.aluka.org/stable/10.5555/al.ch.document.lydblkp30072

Blignaut, J, De Wit, M & Barnes, J (2008) The economic value of elephants. In RJ Scholes & KG Mennell (eds.), *Elephant Management: A Scientific Assessment for South Africa*: 446–476. Wits University Press, Johannesburg.

Bookbinder, MP *et al.* (1998) Ecotourism's support of biodiversity conservation. *Conservation Biology* 12: 1399–1404.

Bouché, P *et al.* (2009) Les éléphants du Gourma, Mali: Statut et menaces pour leur conservation. *Pachyderm* 45: 47–56.

Bradshaw, GA (2009) *Elephants on the edge.* Yale University Press, Sheridan Books.

Bradshaw, GA & Shore, AN (2007) How elephants are opening doors: Development neurology, attachment and social context. *Ethology* 113: 426–436.

Brashares, JS, Arcese, P & Sam, MK (2001) Human demography and reserve size predict wildlife extinction in West Africa. *Proceedings of the Royal Society B: Biological Sciences* 268: 2473–2478.

Brashares, JS *et al.* (2004) Bushmeat hunting, wildlife declines, and fish supply in West Africa. *Science* 306: 1180–1183.

Brennan, AJ & Kalsi, JK (2015) Elephant poaching & ivory trafficking problems in Sub-Saharan Africa: An application of O'Hara's principles of political economy. *Ecological Economics* 120: 326.

Brettell, NH (1994) *Selected Poems.* Snailpress, Cape Town.

Bunnefeld, N *et al.* (2013) Incentivizing monitoring and compliance in trophy hunting. *Conservation Biology* 27: 1344–1354.

Bunney, K, Bond, WJ & Henley, M (2017) Seed dispersal kernel of the largest surviving megaherbivore – the African savanna elephant. *Biotropica* 49(3): 395–401.

Burn, RW, Underwood, FM & Blanc, J (2011) Global trends and factors associated with the illegal killing of elephants: A hierarchical Bayesian analysis of carcass encounter data. *PLOS ONE* 6(9): e24165.

Campbell, J (2015) *Tackling the illicit African wildlife trade.* Council on Foreign Relations. Available at: http://www.cfr.org/africa-sub-saharan/tackling-illicit-african-wildlife-trade/p37031

Canney, S (2015). Locals benefit from elephant protection. WILD Mali Elephant Project blog. Available at: http://www.wild.org/blog/locals-benefit-elephant-protection/

Canney, S (2017). Ground-breaking initial success in protecting Mali's elephants – but it must be sustained. *National Geographic* blog. Available at: http://voices.nationalgeographic.org/2017/04/07/ground-breaking-initial-success-in-protecting-malis-elephants-but-it-must-be-sustained/

Canney, S & Ganame, N (2014) Engaging youth and communities: Protecting the Mali elephants from war. *Nature and Faune* 28(1): 51–55.

Canney, S *et al.* (2007) *The Mali elephant initiative: A synthesis of knowledge, research and recommendations concerning the population, its range and the threats to the elephants of the Gourma.* WILD Foundation, Save the Elephants, Environment & Development Group, USA.

Carlson, K, Wright, J & Donges, H (2015) In the line of fire: Elephant and rhino poaching in Africa. In *Small Arms Survey 2015*: 6–35. Cambridge University Press, United Kingdom.

Caro, T (2011) On the merits and feasibility of wildlife monitoring for conservation: A case study from Katavi National Park, Tanzania. *African Journal of Ecology* 49: 320–331.

Castree, N (2010) Neoliberalism and the biophysical environment 1. *Geography Compass* 4: 121.

Chabal, P & Daloz, JP (1999) *Africa Works: Disorder as Political Instrument.* The International African Institute in association with James Currey, Oxford, & Indiana University Press, Bloomington, Illinois.

Chaiklin, M (2010) Ivory in world history – early modern trade in context. *History Compass* 8(6): 535.

Challender, DWS & Macmillan, DC (2014) Poaching is more than an enforcement problem. *Conservation Letters* 7(5): 1–11.

Chamaillé-Jammes, S et al. (2008) Resource variability, aggregation and direct density dependence in an open context: The local regulation of an African elephant population. *Journal of Animal Ecology* 77(1): 135–144.

Chamaillé-Jammes, S, Fritz, H & Madzikanda, H (2009) Piosphere contribution to landscape heterogeneity: A case study of remote-sensed woody cover in a high elephant density landscape. *Ecography* 32(5): 871–880.

Chamaillé-Jammes, S, Valeix, M & Fritz, H (2007) Managing heterogeneity in elephant distribution: Interactions between elephant population density and surface-water availability. *Journal of Applied Ecology* 44(3): 625–633.

Chapman, M (ed.) (2002) *The New Century of South African Poetry*. Ad Donker, Johannesburg.

Chapron, G et al. (2014) Recovery of large carnivores in Europe's modern human-dominated landscapes. *Science* 346: 1517–1519.

Chase, MJ & Griffin, CR (2009) Elephants caught in the middle: Impacts of war, fences and people on elephant distribution and abundance in the Caprivi Strip, Namibia. *African Journal of Ecology* 47: 223–233.

Chase, MJ et al. (2016) *The Great Elephant Census*.

Chase, MJ et al. (2016) Continent-wide survey reveals massive decline in African savannah elephants: *Peerj* 4: e2354. Available at: https://peerj.com/articles/2354/#p-3 [Accessed 05-02-2017].

Chen, F (2015) Poachers and snobs: Demand for rarity and the effects of antipoaching policies. *Conservation Letters* 9(1): 65.

Chevallier, R & Harvey, R (2016) Is community-based natural resource management in Botswana viable? *SAIIA Policy Insights* 31. SAIIA, Johannesburg.

Child, BA et al. (2012) The economics and institutional economics of wildlife on private land in Africa. *Pastoralism: Research, Policy and Practice* 2: 1–32.

CITES *Address by the President of the Republic of Zimbabwe, CDE RG Mugabe*. Available at: https://cites.org/sites/default/files/eng/cop/10/E10-open.pdf [Accessed 04-02-2017].

CITES Available at: /sites/default/files/eng/cop/17/prop/KE_Loxodonta.pdf

CITES *Conference of the Parties*. Available at: https://cites.org/eng/disc/cop.php [Accessed 05-02-2017].

CITES CoP14 Prop. 6 *Consideration of Proposals for Amendment of Appendices I and II*. Available at: https://www.cites.org/eng/cop/14/prop/E14-P06.pdf [Accessed 05-02-2017].

CITES CoP17 Prop. 16 *Consideration of Proposals for Amendment of Appendices I and II*. Available at: https://cites.org/sites/default/files/eng/cop/17/prop/060216/E-CoP17-Prop-16.pdf [Accessed 05-02-2017].

CITES *Gaborone Amendment to the Text of the Convention*. Available at: https://cites.org/eng/disc/gaborone.php [Accessed 05-02-2017].

CITES *MIKE and ETIS*. Available at: https://cites.org/eng/prog/mike_etis.php [Accessed 05-02-2017].

CITES *Reservations*. Available at: https://www.cites.org/eng/app/reserve_intro.php [Accessed 05-02-2017].

CITES *Resolution Conf. 10.10 (Rev. CoP17)*. Available at: https://cites.org/sites/default/files/document/E-Res-10-10-R17.pdf [Accessed 04-02-2017].

CITES *Trade Database*. Available at: https://trade.cites.org/ [Accessed 05-02-2017].

CITES (2014) *Elephant conservation, illegal killing and ivory trade*:19. CITES Standing Committee, Geneva. Available at: http://www.cites.org/sites/default/files/eng/com/sc/65/E-SC65-42-01_2.pdf [Accessed 01-03-2017].

CITES (2016) *Consideration of Proposals for Amendment of Appendices I and II*. Available at: www.cites.org/eng/cop/17/prop/index.php

CITES (2016) *Criteria for Amendment of Appendices I and II*. Annex 5, Resolution Conf, 9.24 (Rev. CoP 16).

CITES (2016) *Report on the Elephant Trade Information System (ETIS)*. Proceedings of the Seventeenth Meeting of the Conference of the Parties, 24 September–5 October 2016: 26. Available at: doi: 10.1016/S0378-777X(84)80087-6

CITES (2017) Annex 5, Resolution Conf, 9.24 (Rev. CoP 17).

CITES (2017) *The CITES Appendices*. Available at: https://cites.org/eng/app/index.php

CITES (n.d.) *What is CITES?* Available at: https://cites.org/eng/disc/what.php [Accessed 04-02-2017].

CITES, IUCN, TRAFFIC (2013) *Status of African elephant populations and levels of illegal killing and the illegal trade in ivory*. Available at: http://goo.gl/Z3uuZE [Accessed 28-02-2017].

Ciuti, S et al. (2012) Effects of humans on behaviour of wildlife exceed those of natural predators in a landscape of fear. *PLOS ONE* 7: e50611.

Clegg, BW & O'Connor, TG (2016) Harvesting and chewing as constraints to forage consumption by the African savanna elephant (*Loxodonta africana*). *PeerJ* 4: e2469.

Cohn, P & Kirkpatrick, JF (2015) History of the science of wildlife fertility control: Reflections of a 25-year international conference series. *Applied Ecological Environmental Science* 3: 22–29.

Crookes, DJ & Blignaut, JN (2015) Debunking the myth that a legal trade will solve the rhino horn crisis: A system dynamics model for market demand. *Journal for Nature Conservation* 28: 11–18. Available at: http://www.econrsa.org/system/files/publications/working_papers/working_paper_520.pdf [Accessed 30-08-2016].

Crosta, A, Beckner, M & Sutherland, K (2015) *Blending Ivory: China's Old Loopholes, New Hopes*. Elephant Action League, Los Angeles.

Cruise, A (2016) Fighting illegal ivory trade, EU lags behind. *National Geographic* blog. Available at: http://news.nationalgeographic.com/2016/06/ivory-trafficking-european-union-china-hong-kong-elephants-poaching/ [Accessed 05-02-2017].

Ciuti, S et al. (2012) Effects of humans on behaviour of wildlife exceed those of natural predators in a landscape of fear. *PLOS ONE*.

Cumming, GS, Cumming, DHM & Redman, CL (2006) Scale mismatches in social-ecological systems: Causes, consequences, and solutions. *Ecology & Society* 11: 14.

Curran, LM et al. (2004) Lowland forest loss in protected areas of Indonesian Borneo. *Science* 303: 1000–1003.

Currey, D & Moore, H (1994) *Living Proof: African Elephants; The Success of the CITES Appendix 1 Ban*. Environmental Investigation Agency (EIA), London.

De Boer, WF et al. (2013) Understanding spatial differences in African elephant densities and occurrence, a continent-wide analysis. *Biological Conservation* 159: 468–476.

Defenders of Wildlife (n.d.) *Basic facts about elephants*. Available at: http://www.defenders.org/elephant/basic-facts [Accessed 03-02-2017].

Delsink, AK & Kirkpatrick, JF (2012) *Free-Ranging African Elephant Immunocontraception: A New Paradigm for Elephant Management*. 1st edn. Humane Society International, Cape Town.

Delsink, AK et al. (2002) Field applications of immunocontraception in African elephants (*Loxodonta africana*). *Reproduction* 60: 117–124.

Delsink, AK et al. (2006) Regulation of a small, discrete African elephant population through immunocontraception in the Makalali Conservancy, Limpopo, South Africa. *South African Journal of Science* 102: 403–405.

Delsink, AK *et al.* (2013a) Lack of spatial and behavioural responses to immunocontraception application in African elephants (*Loxodonta africana*). *Journal of Zoo and Wildlife Medicine* 44(4S): S52–74.

Delsink, AK *et al.* (2013b) Biologically relevant scales in large mammal management policies. *Biological Conservation* 167: 116–126.

Department of Environmental Affairs and Tourism (DEAT) (2008) National Environmental Management: Biodiversity Act, 2004 (Act 10 of 2004): National Norms and Standards for the Management of Elephants in South Africa. Government Notice No. 251. *Government Gazette* No. 30833, South Africa.

Dias, J *et al.* (2015) *Gourma Elephants Survey, Mali, 2015.* Wildlife Conservation Society, New York.

Di Minin, E & Toivonen, T (2015) Global protected area expansion: Creating more than paper parks. *BioScience* 65(7): 637–638.

Di Minin, E *et al.* (2013a) Understanding heterogeneous preference of tourists for big game species: Implications for conservation and management. *Animal Conservation* 16: 249–258.

Di Minin, E *et al.* (2013b) Creating larger protected areas enhancing the persistence of big game species in the Maputaland-Pondoland-Albany biodiversity hotspot. *PLOS ONE* 8: e71788.

Di Minin, E, Leader-Williams, N & Bradshaw, CAJ (2016) Banning trophy hunting will exacerbate biodiversity loss. *Trends in Ecology and Evolution* 31(2): 99–102.

Douglas-Hamilton, I (1972) *On the ecology and behaviour of the African elephant.* PhD thesis, University of Oxford.

Douglas-Hamilton, I (1979) *The African elephant action plan.* Final report to USFWS. IUCN, Nairobi.

Douglas-Hamilton, I & Douglas-Hamilton, O (1975) *Among the Elephants.* Book Club Associates, London.

Du Toit, JG (2001) *Veterinary care of African elephants.* South African Veterinary Foundation, Pretoria.

Dublin, H & Niskanen, LS (2003) *Guidelines for the in situ translocation of the African elephant for conservation purposes.* IUCN/SSC AfESG.

Dublin, HT (1996) Elephants of Masi Mara, Kenya: Seasonal habitat selection and group size patterns. *Pachyderm* 22: 25–35.

Duffy, R (2014a) Interactive elephants: Nature, tourism and neoliberalism. *Annals of Tourism Research* 44: 88–101.

Duffy, R (2014b) Waging a war to save biodiversity: The rise of militarized conservation. *International Affairs* 90(4): 833.

Duffy, R *et al.* (2015) Toward a new understanding of the links between poverty and illegal wildlife hunting. *Conservation Biology* 30(1): 19.

Dunham, KM *et al.* (2014) *Aerial survey of elephants and other large herbivores in the Sebungwe (Zimbabwe)* WWF-SARPO Occasional Paper, 12. World Wide Fund for Nature, Harare.

Elephant Action League (EAL) (2015) *Pushing ivory out of Africa: A criminal intelligence analysis of elephant poaching and ivory trafficking in East Africa*: 2. Available at: https://eia-international.org/wp-content/uploads/EIA-Vanishing-Point-lo-res1.pdf [Accessed 28-02-2017].

Elephant Action League (EAL) & Wildleaks (2015) *Flash Mission Report: Port of Mombasa, Kenya.*

Elephant Specialist Advisory Group (ESAG) (2017) *Understanding Elephants. Guidelines for Safe and Enjoyable Viewing.* Struik Nature, Cape Town.

Emslie, RH (2013) African rhinoceroses – Latest trends in rhino numbers and poaching. *African Indaba Newsletter* 1(2).

Environmental Investigation Agency (EIA) (2014) *Vanishing point: Criminality, corruption and the devastation of Tanzania's elephants.* EIA, London.

Equator Prize (UNDP) (2017) *Meet the winners.* Available at: http://www.equatorinitiative.org/2017/06/27/ep-2017-meet-the-winners/

Evans, DN (2010) *An eco-tourism perspective of the Limpopo River Basin with particular reference to the Greater Mapungubwe Transfrontier Conservation Area given the impact thereon by the proposed Vele Colliery*: 18. Tourism Working Group of the Greater Mapungubwe Transfrontier Conservation Area (GMTFCA).

Evans, K, Moore, R & Harris, S (2013) The social and ecological integration of captive-raised adolescent male African elephants (*Loxodonta africana*) into a wild population. *PLOS ONE* 8(2): e55933. Available at: doi: 10.1371/journal.pone.0055933

Eyewitness News (EWN) (2016) *Kenya seizes nearly 2 tonnes of ivory from shipment bound for Cambodia.* 23 December (Online). Available at: http://ewn.co.za/2016/12/23/kenya-seizes-nearly-2-tonnes-of-ivory-from-shipment-bound-for-cambodia [Accessed 30-03-2017].

Fattebert, J *et al.* (2013) Long-distance natal dispersal in leopard reveals potential for a three-country metapopulation. *South African Journal of Wildlife Research* 43: 61–67.

Fayrer-Hosken, RA *et al.* (2000) Immunocontraception of African elephants. *Nature* 407: 149.

Fennessy, JT, Leggett, KEA & Schneider, S (2001) Faidherbia albida, *distribution, density and impacts of wildlife in the Hoanib River catchment, Northwestern Namibia.* Desert Research Foundation of Namibia. Occasional Paper, 17: 1–47.

Fernando, P, Leimgruber, P & Pastorini, J (2012) Problem-Elephant Translocation: Translocating the Problem and the Elephant? *PLOS ONE* 7(12) e50917. Available at: doi: 10.1371/journal.pone.0050917

Festa-Bianchet, M (2003) Exploitative wildlife management as a selective pressure for life-history evolution of large mammals. In M Festa-Bianchet & M Apollonio (eds.), *Animal Behavior and Wildlife Conservation*: 191–210. Island Press, Washington DC.

Foley, JA *et al.* (2005) Global consequences of land use. *Science* 309: 570–574.

Forest Resources Management and Conservation Act 10 of 1996; Department of Environmental Affairs (2014): Environmental Impact Assessment Regulations. *Government Gazette* No. 10328, South Africa.

Foucault, M (2004) *The Birth of Biopolitics.* Picador, New York.

Ganame, N (1999) *Conservation et valorisation ecotouristique des elephants.* Association Française des Volontaires du Progrès, Bamako. Unpublished report.

Ganame, N *et al.* (2009) *Study on the Liberation from Human and Livestock Pressure of Lake Banzena in the Gourma of Mali.* WILD Foundation, Boulder, Colorado.

Gandiwa, E *et al.* (2013) Illegal hunting and law enforcement during a period of economic decline in Zimbabwe: A case study of northern Gonarezhou National Park and adjacent areas. *Journal for Nature Conservation* 21: 133–142.

Garaï, ME (1992) Special relationships between female Asian elephants (*Elephas maximus*) in Zoological Gardens. *Ethology* 90: 197–205.

Garaï, ME (1997) *The development of social behaviour in translocated juvenile African elephants* (Loxodonta africana). PhD thesis, University of Pretoria.

Garaï, ME & Carr, RD (2001) Unsuccessful introductions of adult elephant bulls to confined areas in South Africa. *Pachyderm* 31: 51–57.

Garaï, ME & Töffels, O (2011) Afrikansiche Elefanten im Zoo und im Freiland: Ein Vergleich. *Das Elefanten-Magazin* 19: 60–66.

Garaï, ME *et al.* (2004) Elephant reintroductions to small fenced reserves in South Africa. *Pachyderm* 37: 28–36.

Gary, R (1958) *The Roots of Heaven.* Penguin, Harmondsworth, United Kingdom.

Graham, MD *et al.* (2009) The movement of African elephants in a human-dominated land-use mosaic. *Animal Conservation* 12: 445–455.

Gujadhur, T (2001) *Joint venture options for communities and safari operators in Botswana*. IUCN/SNV CBNRM Support Programme. Occasional Paper, 6. IUCN/SNV, Gaborone.

Guldemond, R & Van Aarde, R (2008) A meta-analysis of the impact of African elephants on savanna vegetation. *Journal of Wildlife Management* 72: 892–899.

Guy, PR (1976) Diurnal activity patterns of elephant in the Sengwa Region, Rhodesia. *East African Wildlife Journal* 14: 285–295.

Guy, PR (1977) Coprophagy in the African elephant (*Loxodonta africana*, Blumenbach). *East African Wildlife Journal* 15(2): 174.

Hall-Martin, A & Bosman, P (1986) *Elephants of Africa*. Struik Publishers, South Africa.

Hanks, J (2003) Transfrontier Conservation Areas (TFCAs) in southern Africa: Their role in conserving biodiversity, socioeconomic development and promoting a culture of peace. *Journal of Sustainable Forestry* 17: 127–148.

Hanski, I & Gilpin, M (1991) Metapopulation dynamics: Brief history and conceptual domain. *Biological Journal of the Linnean Society* 42: 3–16.

Harvey, R (2015) *Preserving the African elephant for future generations*. SAIIA Occasional Paper, 219. South African Institute of International Affairs, Johannesburg.

Harvey, R (2017) China's ban on domestic ivory trade is huge, but the battle isn't won. *The Conversation* (Africa). Available at: https://theconversation.com/chinas-ban-on-domestic-ivory-trade-is-huge-but-the-battle-isnt-won-71090 [Accessed 09-08-2017].

Harvey, R, Alden, C & Wu, Y (2017) Speculating a fire sale: Options for Chinese authorities in implementing a domestic ivory trade ban. *Ecological Economics* 141: 22–31.

Hayman, G & Brack, D (2002) *International environmental crime: The nature and control of international black markets*: 7. Workshop report, Royal Institute of International Affairs (RIIA), London.

Hedges, S & Gunaryadi, D (2010) Reducing human-elephant conflict: Do chillies help deter elephants from entering crop fields? *Oryx* 44(1): 139–146.

Helm, CV & Witkowski, ETF (2012) Continuing decline of a keystone tree species in the Kruger National Park, South Africa. *African Journal of Ecology* 51: 270–279.

Hiedanpää, J & Bromley, DW (2014) Payments for ecosystem services: Durable habits, dubious nudges, and doubtful efficacy. *Journal of Institutional Economics* 10(2): 175–195.

Hill, M (2017) *Mali elephants win big*. Available at: http://www.wild.org/blog/mali-elephants-win-big/

Hillbom, E (2012) Botswana: A development-oriented gate-keeping state. *African Affairs* 111: 67–89.

Hillman Smith, K & Kalpers, J (n.d.) *Garamba: Conservation in Peace and War*. Published by the authors.

Hoare, RE (1999) Determinants of human-elephant conflict in a land-use mosaic. *Journal of Applied Ecology* 36: 689–700.

Hoare, RE (2000) African elephants and humans in conflict: The outlook for co-existence. *Oryx* 34: 34–38.

Hopkinson, L, Van Staden, M & Ridl, J (2008) National and International Law. In RJ Scholes & KG Mennel (eds.), *Elephant Management: A Scientific Assessment for South Africa*: 477–536. Wits University Press, Johannesburg.

Hsiang, S & Sekar, N (2016) *Does legalization reduce black market activity? Evidence from a global ivory experiment and elephant poaching data*. Available at: http://www.nber.org/papers/w22314.pdf

Huismann, W (2014) *PandaLeaks: The Dark Side of the WWF*. Nordbook, Germany.

Humane Society International (2017) *Elephant ivory trade-related timeline with relevance to the United States*. Available at: http://www.hsi.org/assets/pdfs/Elephant_Related_Trade_Timeline.pdf

Ihwagi, FW *et al.* (2015) Using poaching levels and elephant distribution to assess the conservation efficacy of private, communal and government land in northern Kenya. *PLOS ONE* 10: e0139079.

International Fund for Animal Welfare (IFAW) (2013) *Criminal nature: The global security implications of the illegal wildlife trade*. Yarmouth Port, Massachusetts.

International Union for Conservation of Nature (IUCN) (2008) *Loxodonta africana*. [Online] Available at: http://www.iucnredlist.org/details/12392/0 [Accessed 03-02-2017].

Ishida, Y *et al.* (2011) Distinguishing forest and savanna African elephants using short nuclear DNA sequences. *Journal of Heredity*. Available at: doi: 10.1093/jhered/esr073

IUCN (2016) *Poaching behind worst African elephant losses in 25 years*. Available at: https://www.iucn.org/news/poaching-behind-worst-african-elephant-losses-25-years-%E2%80%93-iucn-report [Accessed 05-02-2017].

Jachmann, H (1991). *Current status of the Gourma elephants in Mali: A proposal for an integrated resource management project*. IUCN, Gland, Switzerland.

Jacobson, PJ, Jacobson, KM & Seely, MK (1995) *Ephemeral Rivers and Their Catchments: Sustaining People and Development in Western Namibia*. Desert Research Foundation of Namibia, Windhoek.

Junker, J (2008) *An Analysis of Numerical Trends in African Elephant Populations*. University of Pretoria.

Junker, J, Van Aarde, RJ & Ferreira, SM (2008) Temporal trends in elephant *Loxodonta africana* numbers and densities in northern Botswana: Is the population really increasing? *Oryx* 42(1): 58–65.

Kabigumila, J (1993) Feeding habits of elephants in Ngorongoro Crater, Tanzania. *African Journal of Ecology* 31: 156–164.

Kahumbu, P *et al.* (2014). *Scoping study on the prosecution of wildlife related crimes in Kenyan courts – January 2008 to June 2013*. WildlifeDirect.

Kahumbu, P (n.d.) *Beneath Kilimanjaro: Elephant conservation in Kenya*. Occasional Paper.

Keese, N (2012) Anketten, Freilauf, Gruppengeburt? *Das Elefanten-Magazin* 21: 36–38.

Kerley, GIH & Landman, M (2006) The impacts of elephants on biodiversity in the eastern Cape subtropical thickets. *South African Journal of Science* 102: 395–402.

Kerr, MA & Fraser, JA (1975) Distribution of elephant in a part of the Zambezi Valley, Rhodesia. *Arnoldia* 7: 1–14.

Killian, W (2017) Elephant numbers up, conflict down. *The Sun Newspaper*, Namibia. Available at: https://www.namibiansun.com/news/elephant-numbers-up-conflicts-down

Knappert, J (1981) *Namibia: Land and Peoples, Myths and Fables*. EJ Brill, London.

Kodandapani, N, Cochrane, MA & Sukumar, R (2004) Conservation threat of increasing fire frequencies in the Western Ghats, India. *Conservation Biology* 18: 1553–1561.

Kurt, F & Garaï, ME (2007) *The Asian Elephant in Captivity: A Field Study*. Cambridge University Press India Pvt. Ltd. Under the Foundation Books imprint, New Delhi.

Lavorgna, A (2014) Wildlife trafficking in the internet age. *Crime Science* 3(5): 2.

Laws, RM (1970) Elephants as agents of habitat and landscape change in East Africa. *Oikos* 21: 1–15.

Laws, RM, Parker, ISC & Johnstone, CB (1970) Elephants and habitats in North Bunyoro, Uganda. *East African Wildlife Journal* 8: 163–180.

Le Vaillant, F (2007) *Travels into the Interior of Africa via the Cape of Good Hope*. I Glenn (ed.). Van Riebeeck Society, South Africa.

Lee, PC & Moss, CJ (2012) Wild female elephants (*Loxodonta africana*) exhibit personality traits of leadership and social integration. *Journal of Comparative Psychology* 126(3): 224–232.

Leggett, KEA (2006) Home range and seasonal movement of elephants in the Kunene Region, northwestern Namibia. *African Zoology* 41:17–36.

Leggett, KEA, Fennessy, JT & Schneider, S (2001) *A preliminary study of elephants in the Hoanib River catchment, northwestern Namibia.* DRFN Occasional Paper, 16: 1–52.

Leggett, KEA, Fennessy, JT & Schneider, S (2002) Does land use matter in an arid environment? A case study from the Hoanib River Catchment, northwestern Namibia. *Journal of Arid Environments* 53: 529–543.

Leggett, KEA, Fennessy, JT & Scheider, S (2003) Seasonal distributions and social dynamics of elephants in the Hoanib River catchment, northwestern Namibia. *African Zoology* 38: 305–316.

Leggett, KEA, Ramey, RR & MacAlister-Brown, L (2011) Matriarchal associations and reproduction in a remnant subpopulation of desert-dwelling elephants in Namibia. *Pachyderm* 49: 20–32.

Leithead, A (2016) *Why elephants are seeking refuge in Botswana.* BBC News. Available at: http://www.bbc.com/news/world-africa-37230700 [Accessed 05-09-2017].

Lemieux, AM & Clarke, RV (2009) The international ban on ivory sales and its effects on elephant poaching in Africa. *British Journal of Criminology* 49: 451.

Leuthold, W & Sale, JB (1973) Movements and patterns of habitat utilization of elephants in Tsavo National Park, Kenya. *East African Wildlife Journal* 11: 369–384.

Levy, B (2014) *Working with the Grain: Integrating Governance and Growth in Development Strategies.* Oxford University Press, New York.

Lindeque, M & Lindeque, PM (1991) Satellite tracking of elephants in northwest Namibia. *African Journal of Ecology* 29: 196–206.

Lindsay, K et al. (2017) The shared nature of Africa's elephants. *Biological Conservation* 215: 260–267.

Lindsey, PA et al. (2013) The bushmeat trade in African savannas: Impacts, drivers, and possible solutions. *Biological Conservation* 160: 80–96. Available at: doi: 10.1016/j.biocon.2012.12.020

Linnell, J, Salvatori, V & Boitani, L (2008) *Guidelines for population level management plans for large carnivores in Europe.* A Large Carnivore Initiative for Europe report prepared for the European Commission.

Loarie, SR, Van Aarde, RJ & Pimm, SL (2009) Elephant seasonal vegetation preferences across dry and wet savannas. *Biological Conservation* 142: 3099–3107.

Lochner, H (2018) *Planet Okavango.* 1st edn. HPH Publishing, Johannesburg.

Lunstrum, E (2014) Green militarization: Anti-poaching efforts and the spatial contours of Kruger National Park. *Annals of the Association of American Geographers.* Available at: doi: 10.1080/00045608.2014.912545

Maiga, M (1996) *Enquête socio-économique sur les interactions homme-elephants dans le Gourma malien.* Institut Supérieur de Formation et de Recherche Appliquée, Bamako.

Maisels, F et al. (2013) Devastating decline in forest elephants in Central Africa. *PLOS ONE* 8(3): e59469. Available at: doi: 10.1371/journal.pone.0059469

Majani, F (2013) Corrupt officials ensure the battle against poaching remains futile. *Mail & Guardian.* [Online] Available at: https://mg.co.za/article/2013-08-08-00-corrupt-officials-ensure-the-battle-against-poaching-remains-futile/ [Accessed 29-08-2017].

Makhabu, SW, Skarpe, C & Hytteborn, H (2006) Elephant impact on shoot distribution on trees and on rebrowsing by smaller browsers. *Acta Oecologica* 30(2): 136–146.

Marais, J & Ainslie, A (2010). *In Search of Africa's Great Tuskers.* 1st edn. Penguin Books, South Africa.

Massay, GE (2017) *In search of the solution to farmer-pastoralist conflicts in Tanzania.* SAIIA Occasional Paper, 257. South African Institute of International Affairs, Johannesburg. Available at: https://www.saiia.org.za/occasional-papers/1209-in-search-of-the-solution-to-farmer-pastoralist-conflicts-in-tanzania/file

Matthiessen, P (1982) *Sand Rivers.* Bantam Dell Publishers, New York.

McComb, K et al. (2000) Unusually extensive networks of vocal recognition in African elephants. *Animal Behaviour* 29: 1103–1109.

McLellan, E et al. (2014) Illicit wildlife trafficking: An environmental, economic and social issue. *Perspectives* No. 14. United Nations Environment Programme (UNEP), Nairobi.

Montesino Pouzols, F et al. (2014) Global protected area expansion is compromised by projected land-use and parochialism. *Nature* 516: 383–386.

Moss, CJ (2001) The demography of an African elephant (*Loxodonta africana*) population in Amboseli, Kenya. *Journal of Zoology* 255: 145–156.

Moss, CJ & Poole, JH (1983) Relationships and social structure in African elephants. In RA Hinde (ed.), *Primate Social Relationships: An Integrated Approach.* Blackwell Scientific Publications, Oxford.

Mwambingu, R (2017) Mombasa Port fails to tame corruption cartels. *Mediamax.* Available at: http://www.mediamaxnetwork.co.ke/business/286491/286491/

Naidoo, R et al. (2015) Complementary benefits of tourism and hunting to communal conservancies in Namibia. *Conservation Biology* 30(3): 628–638.

National Norms & Standards for Elephants (2008) *Government Gazette* No. 30833, South Africa.

Naughton, L, Rose, R & Treves, A (1999) *The social dimensions of human-elephant conflict in Africa: A literature review and case studies from Uganda and Cameroon.* African Elephant Specialist Group, Human-Elephant Conflict Task Force of IUCN, Gland, Switzerland.

Naylor, RT (2005) The underworld of ivory. *Crime, Law and Social Change* 42: 261–295.

Nicholson, B (n.d.) *The Last of Old Africa.* Safari Press, California.

Norton-Griffiths, M (2007) How many wildebeest do you need? *World Economics* 8(2): 41–64.

Nowak, K (2016) *CITES Alone cannot combat Illegal Wildlife Trade.* South African Institute of International Affairs, Cape Town.

Nyhus, P & Tilson, R (2004) Agroforestry, elephants, and tigers: Balancing conservation theory and practice in human-dominated landscapes of Southeast Asia. *Agriculture, Ecosystems and Environment* 104: 87–97.

O'Connor, TG (2017) Demography of woody species in a semi-arid African savanna reserve following the re-introduction of elephants. *Acta Oecologica* 78: 61–70.

Orr, T (2016) *Re-thinking the application of sustainable use policies for African elephants in a changed world.* SAIIA Occasional Paper, 241. South African Institute of International Affairs, Johannesburg.

Owen-Smith, GL (1970) *The Kaokoveld: An ecological base for future development planning.* Unpublished report.

Owen-Smith, RN (2002) *Adaptive herbivore ecology: From resources to populations in variable environments.* Cambridge University Press, United Kingdom.

Packer, C et al. (2013) Conserving large carnivores: Dollars and fence. *Ecology Letters* 16: 635–641.

Peck, J & Tickell, A (2002) Neoliberalizing space. *Antipode* 34(3): 380–404.

Pickford, P & Pickford, B (1998) *The Miracle Rivers, the Okavango and Chobe of Botswana.* Southern Books, South Africa.

Pinnock, D (2018) The problem of an elephant that just wants to stay home. *Daily Maverick.* Available at: https://www.dailymaverick.co.za/article/2018-04-20-the-problem-of-an-elephant-that-just-wants-to-stay-home/ [Accessed 25-04-2018].

Plumptre, AJ et al. (2007) Transboundary conservation in the greater Virunga landscape: Its importance for landscape species. *Biological Conservation* 134: 279–287.

Poole, JH (1982) *Musth and male-male competition in African elephants*. PhD thesis, University of Cambridge, United Kingdom.

Poole, JH (1987) Rutting behaviour of African elephants. *Behaviour* 102: 283–316.

Poole, JH (1996) The African elephant. In K Kangwana (ed.), *Studying Elephants, Nairobi, Kenya: AWF Technical Handbook Series* No. 7. African Wildlife Foundation.

Poole, JH (2004) Sex differences in the behaviour of African elephants. In RV Short & E Balaban (eds.), *The Differences between the Sexes*: 331–346. Cambridge University Press, 1994.

Poole, JH & Moss, CJ (2008) Elephant sociality and complexity. In C Wemmer & CA Christen (eds.), *Elephants and Ethics: Toward a morality of coexistence*: 69–98. The John Hopkins University Press, Baltimore.

Poulsen, JR et al. (2017) Poaching empties critical Central African wilderness of forest elephants. *Current Biology* 27(4): R134–R135.

Pretorius, Y (2004) *Stress in the African elephant on Mabula Game Reserve, South Africa*. Masters dissertation, University of KwaZulu-Natal, Durban.

Pretorius, Y, Garaï, ME & Bates, LA (2018) The status of African elephant *Loxodonta Africana* populations in South Africa. *Oryx* 1–7. Available at: https://www.cambridge.org/core and at doi: 10.1017/S0030605317001454

Purdon, A et al. (2018) *Partial Migration in Savanna Elephant Populations Distributed across Southern Africa*. National Center for Biotechnology Information, Bethesda, Maryland.

Reeve, R (2002) *Policing International Trade in Endangered Species: The CITES Treaty and Compliance*. The Royal Institute of International Affairs. Earthscan Publications Ltd, London.

Ripple, WJ et al. (2015) Collapse of the world's largest herbivores. *Science Advances* 1(4): 1–12.

Roberts, AM (2014) Detailed look at the ivory trade and the poaching of elephants. *Quinnipiac Law Review* 33.

Ross, K (2010) *Okavango, Jewel of the Kalahari*. Struik Nature, Cape Town.

Ross, RJ (2015) *The Selous in Africa, a Long Way from Anywhere*. Officina Libraria, Milan.

Ruggerio, RG (1992) Seasonal forage utilization by elephants in central Africa. *African Journal of Ecology* 30: 137–148.

Rutina, LP, Moe, SR & Swenson, JE (2005) Elephant *Loxodonta africana* driven woodland conversion to shrubland improves dry-season browse availability for impalas *Aepyceros melampus*. *Wildlife Biology* 11(3): 207–213.

Sayer, JA (1977) Conservation of large mammals in the Republic of Mali. *Biological Conservation* 12(6): 245–263.

Scholes, RJ & Mennell, KG (2008) *Elephant Management. A Scientific Assessment for South Africa*. Wits University Press, Johannesburg.

Scovronick, NC & Turpie, JK (2009) Is enhanced tourism a reasonable expectation for transboundary conservation? An evaluation of the Kgalagadi Transfrontier Park. *Environmental Conservation* 36: 149–156.

Selier, SAJ & Di Minin, E (2015) Monitoring required for effective sustainable use of wildlife. *Animal Conservation* 18: 131–132.

Selier, SAJ et al. (2014) Sustainability of elephant hunting across international borders in southern Africa: A case study of the Greater Mapungubwe Transfrontier Conservation Area. *Journal of Wildlife Management* 78: 122–132.

Selier, SAJ et al. (2016) The legal challenges of transboundary wildlife management at the population level: The case of a trilateral elephant population in southern Africa. *Journal of International Wildlife Law and Policy* 19: 101–135.

Selier, SAJ, Slotow, R & Di Minin, E (2015) Elephant distribution in a transfrontier landscape: trade-offs between resource availability and human disturbance. *Biotropica* 47: 389–397.

Selier, SAJ, Slotow, R & Di Minin, E (2016) The influence of socioeconomic factors on the densities of high-value cross-border species. *Peerj* 4: e2581.

Shannon, G et al. (2008) The utilization of large savanna trees by elephant in southern Kruger National Park. *Journal of Tropical Ecology* 24(3): 281–289.

Shannon, G et al. (2011) Relative impacts of elephant and fire on large trees in a savanna ecosystem. *Ecosystems* 14: 1372–1381.

Shannon, G et al. (2013) Effects of social disruption in elephants persist decades after culling. *Frontiers in Zoology*. Available at: http://www.frontiersinzoology.com/content/10/1/62 and at doi: 10.1186/1742-9994-10-62

Short, RV & Balaban, E (1994) *The Differences between the Sexes*: 331–346. Cambridge University Press, United Kingdom.

Sianga, K et al. (2017) Spatial refuges buffer landscapes against homogenisation and degradation by large herbivore populations and facilitate vegetation heterogeneity. *Koedoe* 59(2): 1–13.

Sitati, NW et al. (2003) Predicting spatial aspects of human-elephant conflict. *Journal of Applied Ecology* 40: 667–677.

Skarpe, C et al. (2004) The return of the giants: Ecological effects of an increasing elephant population. *Ambio* 33: 276–282.

Slotow, R & Van Dyk, G (2001) Role of delinquent young 'orphan' male elephants in high mortality of white rhinoceros in Pilanesberg National Park, South Africa. *Koedoe* 44: 85–94.

Slotow, R et al. (2008) Lethal management of elephants. In RJ Scholes & KG Mennell (eds.), *Elephant Management: A Scientific Assessment for South Africa*: 370–405. WITS University Press, Johannesburg.

Smith, RJ et al. (2003) Governance and the loss of biodiversity. *Nature* 426: 67–70.

Snyman, S (2014) Partnership between a private sector ecotourism operator and a local community in the Okavango Delta, Botswana: The case of the Okavango Community Trust and Wilderness Safaris. *Journal of Ecotourism* 13(2–3): 1–20.

Somerville, K (2017) *The toxic legacy of Tanzania's ivory*. The Marjan Centre. [Online] Available at: https://themarjancentre. wordpress.com/ [Accessed 02-03-2017].

Songhurst, A, McCulloch, G & Coulson, T (2015) Finding pathways to human-elephant coexistence: A risky business. *Oryx* 50(4): 713–720.

Stigand, CH (1913) *Hunting the Elephant in Africa and Other Recollections of Thirteen Years' Wanderings*. Macmillan, New York.

Stoinski, TS et al. (2000) A preliminary study of the behavioral effects of feeding enrichment on African elephants. *Zoo Biology* 19: 485–493.

Stokes, EJ et al. (2010) Monitoring great ape and elephant abundance at large spatial scales: Measuring effectiveness of a conservation landscape. *PLOS ONE* 5: e10294.

Sukumar, R (1993) Minimum viable populations for elephant conservation. *Gajah* 11: 48–52.

Tchamba, MN (1993) Number and migration patterns of savanna elephants (*Loxodonta africana africana*) in northern Cameroon. *Pachyderm* 16: 66–71.

Thouless, CR (1995) Long distance movements of elephants in northern Kenya. *African Journal of Ecology* 33: 321–334.

Thouless, CR et al. (2016) *African Elephant Status Report 2016: An update from the African Elephant Database*. Occasional Paper, series of the IUCN Species Survival Commission, No. 60. IUCN/SSC/African Elephant Specialist Group, Gland, Switzerland.

Tingvold, HG et al. (2013) Determining adrenocortical activity as a measure of stress in African elephants (*Loxodonta africana*) in relation to human activities in Serengeti ecosystem. *African Journal of Ecology.* Available at: doi: 10.1111/aje.12069

Tinley, K (1973) *An ecological reconnaissance of the Moremi Wildlife Reserve.* Okavango Wildlife Society.

TRAFFIC (2017) *First Southern African Auction takes Place.* Available at: http://www.traffic.org/home/2008/10/28/first-ivory-auction-from-southern-africa-takes-place.html [Accessed 04-02-2017].

Tremblay, S (2017) Leading elephant conservationist shot dead in Tanzania. *The Guardian.* Published 17 August 2017.

Trouwborst, A (2015) Global large carnivore conservation and international law. *Biodiversity and Conservation* 24(7): 1567–1588.

Turkalo, AK, Wrege, PH & Wittemyer, G (2013) Long-term monitoring of Dzanga Bai forest elephants: Forest clearing use patterns. *PLOS ONE* 8(12): e85154. Available at: doi: 10.1371/journal.pone.0085154

Turkalo, AK, Wrege, PH & Wittemyer, G (2016) Slow intrinsic growth rate in forest elephants indicates recovery from poaching will require decades. *Journal of Applied Ecology.* Available at: doi: 10.1111/1365-2664.12764

United Nations Office on Drugs and Crime (UNODC) (2003) *Convention against Corruption:* iii. UNODC, Vienna.

United Nations Office on Drugs and Crime (UNODC) (2013) *Transnational Organized Crime in Eastern Africa: A Threat Assessment:* 30. UNODC, Vienna.

United Nations Office on Drugs and Crime (UNODC) (2016) *World Wildlife Crime Report.* UNDOC, Geneva.

United Nations Office on Drugs and Crime (UNODC) and Elephant Action League (EAL) (2015) *Pushing ivory out of Africa: A criminal intelligence analysis of elephant poaching and ivory trafficking in East Africa:* 2. Available at: https://eia-international.org/wp-content/uploads/EIA-Vanishing-Point-lo-res1.pdf [Accessed 28-02-2017].

Valeix, M *et al.* (2011) Elephant-induced structural changes in the vegetation and habitat selection by large herbivores in an African savanna. *Biological Conservation* 144(2): 902–912.

Van Aarde, RJ (2009) *Elephants: Facts and Fables.* 1st edn. Conservation Ecology Research Unit, University of Pretoria & IFAW.

Van Aarde, RJ (2013) *Elephants: A Way Forward.* 1st edn. Conservation Ecology Research Unit, University of Pretoria.

Van Aarde, RJ (2017) *Elephants: A Way Forward.* CERU/IFAW, University of Pretoria.

Van Aarde, RJ & Jackson, TP (2007) Megaparks for metapopulations: Addressing the causes of locally high elephant numbers in southern Africa. *Biological Conservation* 134: 289–297.

Van Aarde, RJ *et al.* (2008) Elephant population biology and ecology. In RJ Scholes & KG Mennell (eds.), *Elephant Management: A Scientific Assessment for South Africa:* 84–115. Wits University Press, Johannesburg.

Van de Walle, N (2001) *African Economies and the Politics of Permanent Crisis, 1979–1999.* Cambridge University Press, United Kingdom.

Van der Duim, R, Lamers, M & Van Wijk, J (2015) Novel institutional arrangements for tourism, conservation and development in eastern and southern Africa. In R van der Duim, M Lamers & J van Wijk (eds.), *Institutional Arrangements for Conservation, Development and Tourism in Eastern and Southern Africa:* 1–16. Springer Publishing Company, New York.

Vasilijevi, M *et al.* (2015) Transboundary conservation: A systematic and integrated approach. *Best Practice Protected Area Guidelines Series* No. 23: 107. IUCN, Gland, Switzerland. Available at: doi: 10.2305/IUCN.CH.2015.PAG.23.en

Viljoen, JJ *et al.* (2008) Translocation stress and faecal glucocorticoid metabolite levels in free-ranging African savanna elephants. *South African Journal of Wildlife Research* 38(2): 146–152.

Viljoen, PJ (1987) Status and past and present distribution of elephants in Kaokoveld, South West Africa/Namibia. *South African Journal of Zoology* 22: 247–257.

Viljoen, PJ (1989) Habitat selection and preferred food plants of desert-dwelling elephant population in the northern Namib Desert, South West Africa/Namibia. *African Journal of Ecology* 27: 227–240.

Viljoen, PJ & Bothma, J du P (1990) Daily movements of desert-dwelling elephants in the northern Namib Desert. *South African Journal of Wildlife Research* 20: 69–72.

Vogt, H (2015) East Africa's biggest port emerges as major transit point for smuggling, threatening Africa's elephants. *The Wall Street Journal.* Available at: https://www.wsj.com/articles/kenyan-port-is-hub-for-illicit-ivory-trade-1447720944 [Accessed 28-02-2017].

Von Gerhardt-Weber, KEM (2011) *Elephant movements and human-elephant conflict in a transfrontier conservation area.* MSc thesis, University of Stellenbosch.

Wall, J *et al.* (2013) Characterizing properties and drivers of long-distance movements by elephants (*Loxodonta africana*) in the Gourma, Mali. *Biological Conservation* 157: 60–68.

Wasser, SK *et al.* (2007) Using DNA to track the origin of the largest ivory seizure since the 1989 trade ban. *Proceedings of the National Academy of Sciences* 104(10): 4228–4233.

Wasser, SK *et al.* (2015) Genetic assignment of large seizures of elephant ivory reveals Africa's major poaching hotspots. *Science* 349(6243): 84–87.

Western, D & Lindsay, WK (1984) Seasonal herd dynamics of a savanna elephant population. *African Journal of Ecology* 22: 229–244.

Western, D, Russell, S & Cuthill, I (2009) The status of wildlife in protected areas compared to non-protected areas of Kenya. *PLOS ONE* 4: e6140.

White, LJT (1994) *Sacoglottis gabonensis* fruiting and the seasonal movement of elephants in the Lopé Reserve, Gabon. *Journal of Tropical Ecology* 10: 121–125.

Whitehouse, AM & Hall-Martin, AJ (2000) Elephants in Addo Elephant National Park, South Africa: Reconstruction of the population's history. *Oryx* 34(1): 46–55.

Wittemyer, G *et al.* (2008) Accelerated human population growth at protected area edges. *Science* 321: 123–126.

Wittemyer, G *et al.* (2014) Illegal killing for ivory drives global decline in African elephants. *Proceedings of the National Academy of Sciences* 111(36): 13117–13121.

Woodroffe, R & Ginsberg, JR (1998) Edge effects and the extinction of populations inside protected areas. *Science* 280: 2126–2128.

Woolley, L-A *et al.* (2008) Population and individual elephant response to a catastrophic fire in Pilanesberg National Park. *PLOS ONE* 3(9): e3233. Available at: doi:10.1371/journal.pone.0003233

World Travel and Tourism Council (2018) *Travel and Tourism Economic Impact: Botswana.*

World Wildlife Fund (2007) *WWF Positions CITES CoP14.* Available at: file:///Users/adamcrui/Downloads/wwf_cites_positions_enfinal%20(1).pdf [Accessed 05-02-2017].

Wu, Y, Rupp, S & Alden, C (2016) *Values, culture and the ivory trade ban.* SAIIA Occasional Paper, 244. South African Institute of International Affairs, Johannesburg.

Wyatt, JR & Eltringham, SK (1974) The daily activity of the elephant in the Rwenzori National Park, Uganda. *East African Wildlife Journal* 12: 273–289.

Young, KD & Van Aarde, RJ (2010) Density as an explanatory variable of movements and calf survival in savanna elephants across southern Africa. *Journal of Animal Ecology* 79(3): 662–673.

Yufang, G & Clark, SG (2014) Elephant ivory trade in China: Trends and drivers. *Biological Conservation* 180: 24.

1'30 TOURISM KENYA UHURU 1963

KENYA 35/- The Big Five - Elephant

REPUBLIQUE GABONAISE Éléphant de Forêt (Loxodonta africana cyclotis) Postes 1988 50f

RARE ANIMALS OF EAST AFRICA KENYA UGANDA TANZANIA Shs 2'

1 FRANC ETAT INDEPENDANT DU CONGO

RÉPUBLIQUE DU BENIN 300 FAUNE AFRICAINE

GAMBIA 10d TEN SHILLINGS

Botswana Elephants Cooling Down in the Okavango Delta 2016 Pt.80

RÉPUBLIQUE FÉDÉRALE DU CAMEROUN ELEPHANT - CHUTES DU NTEM 1F

COMPANHIA DE MOÇAMBIQUE 1 ESCUDO 1

RÉPUBLIQUE DU TCHAD 1

SOUTH AFRICA STANDARD POSTAGE Elephant (Loxodonta Africana)

BOTSWANA ¼c POSTAGE DUE
BOTSWANA 6c POSTAGE DUE
BOTSWANA 2c POSTAGE DUE
BOTSWANA 1c POSTAGE DUE

SAFARI RALLY APRIL 7 1977 TANZANIA Sh. 5/=

REPUBLIC OF NIGERIA ELEPHANTS 1d MAURICE FIEVET

U.S. 6c AFRICAN ELEPHANT HERD

POSTES 2016 300 République CENTRAFRICAINE

SWAZILAND 3½c ELEPHANT

REPUBLIQUE GABONAISE Éléphant de Forêt (Loxodonta africana cyclotis) Postes 1988 40f

Lake Manyara National Park VIEW FROM THE LODGE TOURIST ATTRACTIONS OF TANZANIA Tanzania 400/-

6d POSTAGE & REVENUE 1910 1935 SOUTHERN RHODESIA

1d POSTAGE & REVENUE SOUTHERN RHODESIA

RÉPUBLIQUE CENTRAFRIC...

FOLKLORE ET TOURISME POSTES 20f RÉPUBLIQUE DU CONGO

KENYA UGANDA 15c TANGANYIKA

Tanzania ANIMAL GIANTS Elephant 500/=

KENYA UGANDA 5/- TANGANYIKA

3,50Z ZAIRE PARC DES VIRUNGA Éléphants

SILVER JUBILEE 1977 KENYA 5

TANZANIA 1991 Loxodonta africana 200/.

COMPANHIA DE MOÇAMBIQUE MARFIM ½ CENTAVO CENTAVO

POSTAGE REVENUE GAMBIA 1s ONE SHILLING 1s

RÉPUBLIQUE DU ZAIRE 4K

KENYA 1'30

RÉPUBLIQUE CENTRAFRIC... Les Éléphants 65

NAMIBIA N$ 2.60 Helge Denker 2002

Botswana
Elephant Herd Seeking Water · Ditshupo Mogapi
2016
P10.00

TANZANIA
Loxodonta africana 35/.

WILD ANIMALS OF TANZANIA
ELEPHANTS
TANZANIA 700/=

ELEPHANT
KENYA
1'30

Suid-Afrika
4d Posgeld Postage
South Africa

750F
REPUBLIQUE CENTRAFRICAINE

TANZANIA NORTHERN CIRCUIT
Tanzania 500

SOUTH AFRICA

ELÉPHANT AFRIQUE
aérienne
100 F REPUBLIQUE DU TCHAD

REPUBLIQUE GABONAISE
100f
Postes 1988

5/ SIERRA LEONE 5/
1883 1983
FIVE SHILLINGS

GAMBIA
£1

POSTE 750F
REPUBLIQUE CENTRAFRICAINE

REPUBLIQUE CENTRAFRICAINE
Les Éléphants
Loxodonta cyclotis
650F
POSTES 2011

750F
REPUBLIQUE CENTRAFRICAINE

GAMBIA
2/
TWO SHILLINGS

SOUTH AFRICA
R3.35

900F Postes 900F Postes
Loxodonta africana Loxodonta africana
REPUBLIQUE CENTRAFRICAINE REPUBLIQUE CENTRAFRICAINE
900F Postes 900F Postes
Elephas maximus Elephas maximus indicus
REPUBLIQUE CENTRAFRICAINE REPUBLIQUE CENTRAFRICAINE

South Africa
B5

BOTSWANA
P5.40
OKAVANGO DELTA
1000th UNESCO World Heritage Site

175.00Vt
MOÇAMBIQUE
Elefantes

REPUBLIQUE POPULAIRE DU CONGO
2F
M. MONVOISIN
LOXODONTA AFRICANA
GUILLAME
POSTES 1971

TANZANIA TANZANIA
75/ 75/
Loxodonta africana AFRICAN ELEPHANT

750F
REPUBLIQUE CENTRAFRICAINE

CENTENARY OF ETOSHA
NAMIBIA
Inland Registered Mail Paid

ZAIRE 50 NK
PARC NATIONAL

REPUBLIQUE DU NIGER
Elephas maximus 850F

REPUBLIQUE DU NIGER
Loxodonta cyclotis 850F

République du Mali
200f
TAUROTRAGUS DERBIANUS

REPUBLIQUE CENTRAFRICAINE
Les Éléphants
Loxodonta africana
650F
POSTES 2011

REPUBLIQUE CENTRAFRICAINE
Les Éléphants
Loxodonta cyclotis
650F
POSTES 2011

TANZANIA
Loxodonta africana 30/

REPUBLIQUE DU NIGER
Loxodonta africana 850F
Elephas Maximus sumatranus 850F

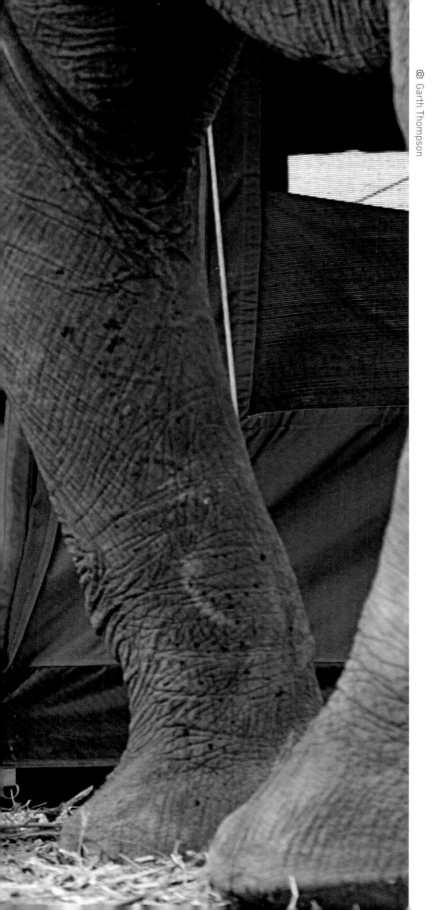

'I defy anyone to look upon
elephants without a sense
of wonder. Their enormous
size, their gigantic stature,
represents a mass of liberty
that sets you dreaming.'

Romain Gary

'It's war now. We are losing
our national heritage;
we are losing our elephants.
We have to act now.'

Dr Paula Kahumbu
WildlifeDirect

Text © 2019: Individual authors of chapters
Photographs © 2019: Individual photographers as
    indicated alongside images
Maps copyright © 2019: Elephants Without Borders /
    https://peerj.com/articles/2354/
Published edition © 2019: Don Pinnock & Colin Bell

Published in North America by Smithsonian Books
First published in 2019 by Struik Nature, an imprint
of Penguin Random House South Africa (Pty) Ltd,
reg. no. 1953/000441/07, PO Box 1144, Cape Town, 8000
South Africa, all rights reserved

Compiled by Don Pinnock & Colin Bell
Designed by Gaelen Pinnock (www.scarletstudio.net)
Publisher: Pippa Parker
Managing editor: Helen de Villiers
In-house design: Dominic Robson
Proofreader: Emsie du Plessis
Reproduction: Hirt and Carter Cape (Pty) Ltd
Printing and binding: C&C Offset Printing Co., Ltd

MIX
Paper from
responsible sources
FSC® C018179
www.fsc.org

ISBN 978-1-58834-663-6
Manufactured in China, not at government expense

23 22 21 20 19   1 2 3 4 5